AUTOMOTIVE ELECTRICAL AND ELECTRONIC SYSTEMS

AUTOMOTIVE ELECTRICAL AND ELECTRONIC SYSTEMS

James D. Halderman

Professor

Sinclair Community College

REGENTS/PRENTICE HALL
Englewood Cliffs, New Jersey 07632

Library of Congress Cataloging-in-Publication Data

HALDERMAN, JAMES D., (*date*)
  Automotive electrical and electronic systems.

  Includes index.
    1. Automobiles—Electric equipment.   2. Automobiles—
Electronic equipment.   3. Automobiles—Electronic
equipment—Maintenance and repair.   4. Automobiles—
Electric equipment—Maintenance and repair.   I. Title.
TL272.H22 1988   629.2′54   87–1264
ISBN   0-13-054362-4

Editorial/production supervision and interior design: **Kathryn Pavelec**
Cover design: **Diane Saxe**
Manufacturing buyer: **Rhett Conklin/Harry P. Baisley**
Page layout: **Meg Van Arsdale**

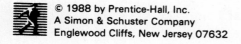
Printed in the United States of America.

10 9 8 7 6

ISBN   0-13-054362-4   025

PRENTICE-HALL INTERNATIONAL (UK) LIMITED, *London*
PRENTICE-HALL OF AUSTRALIA PTY. LIMITED, *Sydney*
PRENTICE-HALL CANADA INC., *Toronto*
PRENTICE-HALL HISPANOAMERICANA, S.A., *Mexico*
PRENTICE-HALL OF INDIA PRIVATE LIMITED, *New Delhi*
PRENTICE-HALL OF JAPAN, INC., *Tokyo*
PRENTICE-HALL OF SOUTHEAST ASIA PTE. LTD., *Singapore*
EDITORA PRENTICE-HALL DO BRASIL, LTDA., *Rio de Janeiro*

CONTENTS

3   Ohm's Law and Circuit Analysis   19

4   Capacitance and Magnetism   30

5   Semiconductors   39

6   Wiring and Circuit Diagrams   47

7   Lighting and Signaling Circuits   59

8   Analog and Digital Dash Instruments   78

9   Electrical Accessories   92

13 Battery Testing and Service Procedures   157

14 Starter Motors and the Cranking Circuit   170

15 Starting System Testing and Service   183

16   Alternators and the Charging Circuit   *202*

17   Alternator and Charging Circuit Testing and Service   *219*

18   Basic Ignition Operation   *235*

19   Electronic Ignition Operation   248

20   Ignition Timing and Timing Advance   264

21   Spark Plug Wires, Distributor Caps, and Rotors   282

22   Spark Plugs   291

23   Ignition System Testing   *306*

24   Oscilloscope Testing   *316*

25   Computerized Engine Control System Operation   *335*

26   Computerized Engine Control System Diagnosis and Testing   349

27   Body Computers   365

Glossary   375

Answers to Multiple-Choice Questions   381

Index   383

Table of Sidebars   389

PREFACE

This book is intended for anyone interested in learning about automotive electrical systems. The information will lead the reader logically through the entire field of automotive electricity and electronics. The book is organized so that selected chapters or topics can be studied individually, with only a minimal amount of material presented in earlier chapters needed before understanding the succeeding chapters.

The *reader* is kept in mind at all times through an easy-to-read style of writing. All the material is explained with photos, illustrations, and examples of troubleshooting malfunctions.

This book does the impossible! It presents automotive electrical system concepts and operation in a simple, concise format with more detail than is found in most electrical systems books. The author also explains the effects that one component or circuit can have on another, with the goal of improving diagnostic and troubleshooting skills.

The material presented in this book has been used in actual classroom instruction. As a result of this "field testing," many suggestions from students and instructors have been incorporated into the final book, including:

1. Clear, concise definitions of new terms in the text *and* in the glossary
2. Symptoms of defective components, and explanations
3. Examples of real-life diagnosis and troubleshooting
4. Photographs of actual components wherever possible
5. Examples of troubleshooting using actual problems experienced by professionals in the field
6. Electrical replacement-part purchasing considerations
7. Explanations of operations, with emphasis on "why" they work
8. Examples and practical applications in every chapter
9. Up-to-date information, including such topics as direct-fire ignition systems, permanent-magnet starters, computerized engine controls, and body computers

To assure that the needs of the reader are met, the author has presented every topic in the following format:

1. Basic operation
2. Parts involved
3. Testing methods and results using both low-cost test equipment and electronic test equipment
4. Symptoms of defective operation (characteristics)
5. Diagnosis and service procedures
6. Troubleshooting examples with solutions
7. Summary of chapter material
8. Study questions at the end of each chapter
9. Multiple-choice questions, with answers

JAMES D. HALDERMAN

ACKNOWLEDGMENTS

The author gratefully acknowledges the help of the following companies in allowing use of their illustrations:

Allen Testproducts
ALLTEST INC.
ASE
Autolite
Battery Council International
Buick Motor Division, General Motors Corporation
Cadillac Motor Car Division, General Motors Corporation
Champion Spark Plug Company
Chevrolet Motor Division, General Motors Corporation
Chrysler Corporation
Ford Motor Company
General Electric Corporation
General Motors Corporation
Lester Catalog Company
NGK Spark Plug Company
Nippodenso Spark Plug Company
Oldsmobile Division, General Motors Corporation
Pontiac Motor Division, General Motors Corporation
Robert Bosch Corporation
Sun Electric Corporation
Texas Instruments Incorporated
Toyota Motor Sales, USA, Inc.

I also wish to thank my colleagues and students at Sinclair Community College in Dayton, Ohio, for their comments and suggestions.

Most of all, I wish to thank my wife, Michelle, for her assistance in all phases of manuscript preparation.

AUTOMOTIVE ELECTRICAL AND ELECTRONIC SYSTEMS

1

Basic
Electricity

THIS CHAPTER INTRODUCES AND DEFINES BASIC ELECTRICITY WITH AN EX-
planation of terms, units of measurement, and uses in automotive applications. The topics covered in this chapter include:

1. What electricity is
2. Static electricity
3. Structure of the atom
4. Definition of "amperes"
5. Definition of "volts"
6. Definition of "ohms"
7. Conductors and insulators
8. Conventional versus electron theory
9. Sources of electrical current

ELECTRICITY

The word *electricity* comes from the Greek word *elektron,* meaning "amber" (a fossil resin). The ancients produced electric charges by rubbing amber with wool. This produced *static* electricity, which was the first known type of electricity. It is called "static" (motionless) because the charge is at rest and not moving through a wire.

It was detected that there are actually two types of electrical charges. When a rubber rod was rubbed with flannel or fur, a *negative (−) charge* was generated in the rod. When a glass rod was rubbed with silk, the glass rod had a *positive (+) charge.* See Figure 1-1.

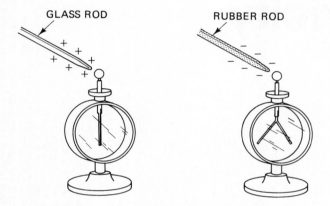

FIGURE 1-1   *A test instrument called an electroscope illustrates that positive (+) and negative (−) charges are opposite.*

It was also discovered that "like" charged objects (both positive or both negative) repelled or moved away from each other. "Unlike" charged objects (one positive and one negative) attracted or moved toward each other. The negative (−) charges were determined to be an atom with an extra number of negative (−) charged electrons. See Figure 1-2. Electricity is actually the movement of electrons from one atom to another.

THE ATOM AND ELECTRONS

To help explain electricity, the composition of an atom must be understood. An atom is the smallest unit of all matter in the universe. Our universe is composed of matter, which is *anything* that has mass and occupies space. Matter is, therefore, anything *except* the nothingness of space. Matter can be in a solid form such as a table, or in a liquid form such as water or gasoline. It can also be in a gaseous state such as water vapor (steam) or gasoline fumes. All matter is made from slightly over 100 individual components called *elements.*

Each element can be cut down or reduced in size until the smallest part of the element remaining can still be identified as that particular element. This smallest particle is called an *atom.* See Figure 1-3. There are *electrons* traveling around a dense center of each atom, called the nucleus. The nucleus contains *protons,* which have a positive (+) charge, and *neutrons,* which are electrically neutral (have no charge). In orbits surrounding the nucleus are

FIGURE 1-2   *(Courtesy of General Motors Corporation.)*

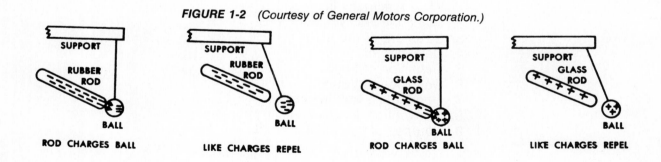

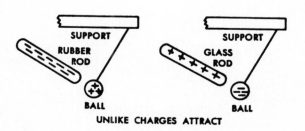

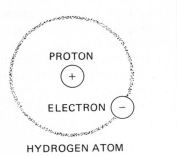

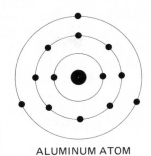

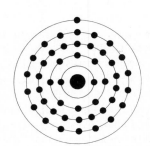

FIGURE 1-3 *The hydrogen atom is the simplest atom, with only one proton, one neutron, and one electron. More complex materials contain higher numbers of protons, neutrons, and electrons.*

electrons, which have a negative (−) charge and weigh only about 1/1800 of the weight of a proton. Each atom contains an equal number of electrons and protons. Because each negative-charged electron is balanced with the same number of positive-charged protons, an atom has a *neutral* charge (no charge).

How Big Is an Atom?

As an example of the relative sizes of the parts of an atom, if an atom were magnified so that the nucleus were the size of the period at the end of this sentence, the whole atom would be bigger than a house.

FREE ELECTRONS

Each element in the universe has its own individual characteristic atom, each having its own number of protons, neutrons, and electrons. An atom has the same number of electrons in orbits around the nucleus as there are number of protons. These orbiting electrons are in orbits of various distances from the center of the nucleus, depending on the number of electrons. These different orbits (called shells) are identified by letters: *K, L, M, N, O,* and so on.

The orbit closest to the nucleus is the *K* orbit and it has a limit of two electrons. If an atom has more than two electrons, the additional electrons have to move farther away from the nucleus. See Figure 1-4. The second orbit from the nucleus is called the *L* shell and it has a maximum capacity of eight electrons. There are many different shells and each shell has its own limit for the number of electrons that can occupy each shell. The five closest shells are listed below together with the maximum of electrons needed to fill each shell:

shell *K*: 2 electrons (filled shell)
shell *L*: 8 electrons (filled shell)
shell *M*: 18 electrons (filled shell)
shell *N*: 32 electrons (filled shell)
shell *O*: 32 electrons (filled shell)
and so on.

The element aluminum contains 13 electrons. These electrons will fill the *K* shell (2) and the *L* shell (8) and start to fill the the *M* shell with its remaining 3 electrons. Since the *M* shell has a capacity for 18 electrons, this leaves the 3 electrons almost alone in the outer orbit.

If an element has fewer than four electrons in its outer orbit, the electrons are called *free electrons* because they can be easily "bumped" out of their shell and into the shell of an identical atom next to it by an electrical force.

FIGURE 1-4 *Electron shells. As a shell close to the nucleus becomes filled, additional electrons for a particular atom must start to fill electron orbits farther from the nucleus.*

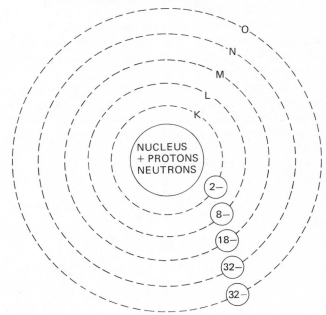

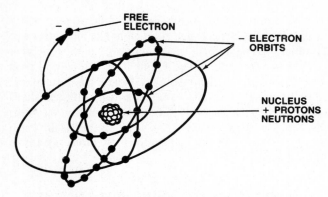

FIGURE 1-5   *Copper atom. (Courtesy of Chrysler Corporation.)*

A COPPER ATOM HAS ONE FREE ELECTRON

Copper has 29 protons and 29 electrons. Copper's electrons are placed in each shell up to their maximum number until the first three shells are completely filled: 2 + 8 + 18 = 28 electrons, and the lone remaining twenty-ninth electron is in the fourth shell by itself. This lone electron in the outer orbit is a *free electron*. Because it is all by itself in the fourth shell, it is easy for copper to lose this electron or have it bumped from one atom to another. Therefore, copper is a good electrical conductor. See Figure 1-5.

The farther the free electrons are from the nucleus, the weaker the "pull" of the proton's positive force in the nucleus on the electron's negative pull. See Figure 1-6 for an example of an atom with one free electron. The resistance of the free electrons moving from one atom to another is *lower* with elements that have the fewest electrons in their outer orbit.

This movement of free electrons explains how static electrical charges are produced. A rubber rod rubbed with flannel or fur actually transfers free electrons from the flannel or fur into the outer orbit of the rubber rod, creating

FIGURE 1-6   *Silver atom.*

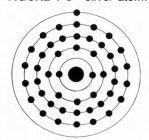

rubber atoms, which have more negative (−) electrons than positive (+) protons. The resulting rubber atoms have a negative (−) charge. If a glass rod is rubbed with silk, the silk *removes* electrons from the glass rod and gives the rod a net positive (+) charge.

CONDUCTORS

Conductors are materials with fewer than four electrons in their atom's outer orbit. Copper is an example of an excellent conductor because it has only one electron in its outer orbit. This orbit is far enough away from the nucleus of the copper atom that the pull or force holding the outer most electron in orbit is relatively weak. The price of copper is reasonable compared to the relative cost of other conductors with similar properties.

All good conductors of electricity are also good conductors of heat and cold. Conductors are also classified as *metals*. Iron, steel, copper, aluminum, silver, and gold are examples of metals (conductors). Metals can be further defined as containing iron (ferrous metals), such as cast iron or steel, and those metals not containing iron (nonferrous metals). Copper, silver, mercury, gold, and aluminum are examples of nonferrous metals.

INSULATORS

Insulators are materials with more than four electrons in their atom's outer orbit. Having more than four electrons in their outer orbit, it becomes easier for these materials to acquire (gain) electrons than to release electrons. Examples of insulators include plastics, wood, glass, rubber, ceramics (spark plugs), and varnish for covering (insulating) copper wires in alternators and starters.

SEMICONDUCTORS

Materials with exactly four electrons in their outer orbit are neither conductors nor insulators and are called *semiconductor* materials. See Chapter 5 for a further explanation and applications.

CURRENT ELECTRICITY

Movement of electrons through a conductor is called *current* (moving) electricity, in contrast to *static* electricity, where there is no electron movement. In fact, once static electricity is discharged, it becomes current electricity because the electrical charges are then in motion and no longer static.

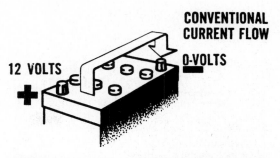

FIGURE 1-7 Current electricity is the movement of electrons through a conductor. (Courtesy of General Motors Corporation.)

FIGURE 1-9 The conventional theory says that electrical current flows from positive (+) to negative (−).

HOW ELECTRONS MOVE THROUGH A CONDUCTOR

A conductor contains neutral atoms whose electrons are constantly moving at random in the material. The electrons are normally being knocked in all directions by the atoms, which are vibrating millions of times per second. If an outside source of power, such as a battery, is connected to the ends of a conductor, a positive (+) charge (lack of electrons) is placed on one end of the conductor and a negative (−) charge is placed on the opposite end of the conductor. The negative (−) charge will repel the free electrons from the atoms of the conductor, while the positive (+) charge on the opposite end of the conductor will attract electrons. As a result of this attraction of opposite charges and repulsion of like charges, electrons will flow through the conductor. See Figure 1-7. These electrons actually travel a zigzag course from one atom to another. Because each electron bumps other electrons like a chain reaction, the overall effect is the flow of electrons through the conductor traveling near the speed of light. This electron action is similar to the movement of a row of dominoes as they are knocked down or hitting a row of pool balls. See Figure 1-8.

THAT'S A LOT OF ELECTRONS!

To help visualize the great number of atoms (the more atoms, the more free electrons), a copper wire 1 m long with a cross-sectional area of 1 mm² contains about *85 sextillion* copper atoms. This number is 85 followed by 21 zeros, or 85,000,000,000,000,000,000,000 free electrons.

CONVENTIONAL THEORY VERSUS ELECTRON THEORY

It was once thought that electricity had only one charge and moved from positive (+) to negative (−). This flow of electricity through a conductor is called the *conventional theory* of current flow. See Figure 1-9. After the discovery of the electron and its negative (−) charge, the *electron theory* indicates electron flow from negative (−) to positive (+). Most automotive applications use the conventional theory. This book will use the conventional theory unless stated otherwise.

AMPERES

The *ampere* is the unit used throughout the world as a measure of the *amount* of current flow. When 6.28 billion billion electrons (the name for this large number of electrons is a *coulomb*) move past a certain point in 1 second, this represents 1 ampere of current. See Figure 1-10. The ampere is the electrical unit for the amount of electron flow, just as "gallons per minute" is the unit that can be used to measure the quantity of water flow. It was named for a French electrician, André Marie Ampère (1775–1836). The conventional abbreviations and measurement for *ampere* are summarized as follows:

1. The ampere is the unit of measurement of the amount of current flow.
2. "A" and "amps" are acceptable abbreviations for "ampere."

FIGURE 1-8 The movement of electrons through a conductor is similar to the action of dominoes as they are knocked down.

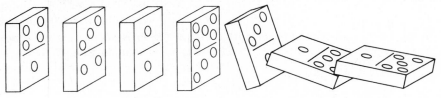

FIGURE 1-10   *One ampere is the movement of 1 coulomb (6.28 billion billion electrons) past a point in 1 second. (Courtesy of General Motors Corporation.)*

3. The capital letter *I,* for "intensity," is used in mathematical calculations to represent amperes.

4. Amperes are measured by an *ammeter* (not amp-meter).

WHAT IS A COULOMB?

A coulomb is 6.28×10^{18} electrons. The $\times 10^{18}$ indicates "scientific notation" or the method used in science to express very large or very small numbers. Instead of writing all the zeros after a large number, the scientific notation method multiplies the number by how many tens there are in the number (how many decimal places there are to the right of the decimal point). Therefore, a coulomb is 6,280,000,000,000,000,000 (6.28 with the decimal point moved to the right 18 places) electrons at *rest* (not moving). If a coulomb of electrons passes a given point in 1 second, then by definition, this is 1 ampere.

VOLTS

A *volt* is the unit of measurement of electrical *pressure*. It is named for Alessandro Volta (1745–1827), an Italian physicist. The comparable unit using water as an example would be pounds per square inch (psi). It is possible to have very high pressures (volts) and low water flow (amperes). It is also possible to have high water flow (amperes) and low pressures (volts). Voltage is also called electrical *potential* because if there is voltage present in a conductor, there is a potential (possibility) for current flow. Voltage does not flow *through* conductors, but voltage does cause current (in amperes) to flow through conductors. The conventional abbreviations and measurements for voltage are as follows:

1. The volt is the unit of measurement of the amount of electrical *pressure.*

2. *Electromotive force,* abbreviated EMF, is another way of indicating voltage.

3. "V" is the generally accepted abbreviation for "volts."

4. The symbol used in calculations is *E,* for "electromotive force."

5. Volts are measured by a *voltmeter.*

OHMS

Resistance to the flow of current through a conductor is measured in units called *ohms,* named after a German physicist, Georg Simon Ohm (1787–1854). The resistance to the flow of free electrons through a conductor results from the countless collisions the electrons make within the atoms of the conductor. The conventional abbreviations and measurements for resistance are as follows:

1. The ohm is the unit of measurement of electrical *resistance.*

2. The symbol for ohms is Ω (Greek capital letter omega), the last letter of the Greek alphabet.

3. The symbol used in calculations is *R,* for "resistance."

4. Ohms are measured by an *ohmmeter.*

WHY DOES THE RESISTANCE INCREASE WITH TEMPERATURE?

The resistance of a conductor to current flow is caused by the countless collisions of electrons with vibrating conductor atoms. If the temperature of the conductor increases, the vibrations of the atoms become stronger and the electrons have a harder time moving through the conductor.

SOURCES OF DIRECT-CURRENT ELECTRICITY

1. *Chemical.* A battery is a chemical device that produces a voltage potential between two different metal plates submerged in an acid. Lead peroxide and lead plates in a sulfuric acid electrolyte are commonly used for automotive applications.

2. *Photoelectric.* This source was discovered by Heinrich Rudolf Hertz (1857–1894), a German

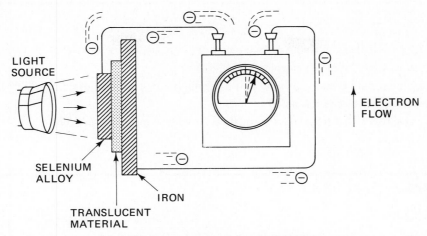

FIGURE 1-11 *Electron flow is produced by light striking a light-sensitive material.*

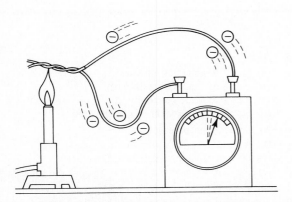

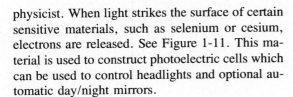

FIGURE 1-12 *Electron flow is produced by heating the connection of two different metals.*

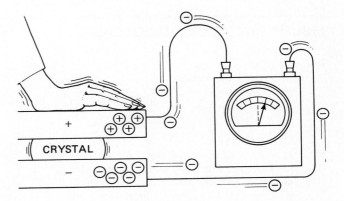

FIGURE 1-13 *Electron flow is produced by pressure on certain crystals.*

physicist. When light strikes the surface of certain sensitive materials, such as selenium or cesium, electrons are released. See Figure 1-11. This material is used to construct photoelectric cells which can be used to control headlights and optional automatic day/night mirrors.

3. *Thermoelectric.* Electron movement can be created by heating the connection of two dissimilar metals. If the two metals are connected to a voltage-sensitive gauge, such as a galvanometer, an increase in the temperature of the wire junction will increase the voltage reading. See Figure 1-12. This type of thermometer is called a thermoelectric *pyrometer.* A pyrometer is often used to measure exhaust temperature on diesel trucks.

4. *Piezoelectric.* Certain crystals, such as Rochelle salt and quartz, become electrically charged when pressure is applied to the crystals. The difference in potential produced increases with increased pressure. See Figure 1-13. A phonograph pickup crystal is an example of piezoelectric principles changing the varying grooves of a record into electrical pulses. Piezoelectric units are used in detonation (knock) sensors on computer-operated automotive systems.

5. *Electromagnetic induction.* A current can be created in any conductor that is moved through a magnetic field. The conductor can be stationary and the magnetic field moved. The voltage induced is increased with the speed of the movement and the number of conductors that are cut. All alternators, starters, and ignition systems work as a result of electromagnetism. See Chapter 4 for detailed explanations.

SUMMARY

1. Electricity is the movement of free electrons from one atom to another.
2. Electricity can be measured.
3. If 6 billion billion electrons flow past a certain point, 1 *ampere* of current is flowing.
4. Electrical pressure is measured in *volts*.
5. The resistance that the current must flow through is measured in *ohms*.

STUDY QUESTIONS

1-1. What are free electrons?

1-2. Describe how electrical current flows through a conductor.

1-3. What is a coulomb?

1-4. Describe the meaning of volts, amperes, and ohms.

1-5. List five methods that can be used to create electricity.

MULTIPLE-CHOICE QUESTIONS

1-1. An electron has:
(a) a positive (+) charge.
(b) a negative (−) charge.
(c) a neutral charge.
(d) no charge.

1-2. An insulator has:
(a) eight or more free electrons.
(b) more than four electrons in its outer orbit.
(c) exactly four electrons in its outer orbit.
(d) fewer than four electrons in its outer orbit.

1-3. An ampere is:
(a) a unit of electrical pressure.
(b) a unit of electrical resistance.
(c) a unit of static electricity.
(d) a unit of the amount of current flow.

1-4. A volt is:
(a) a unit of electrical pressure.
(b) a unit of electrical resistance.
(c) a unit of static electricity.
(d) a unit of the amount of current flow.

1-5. An ohm is:
(a) a unit of electrical pressure.
(b) a unit of electrical resistance.
(c) a unit of static electricity.
(d) a unit of the amount of current flow.

1-6. As temperature increases:
(a) the resistance of a conductor decreases.
(b) the resistance of a conductor increases.
(c) the resistance of a conductor remains the same.
(d) the conductor should be longer.

1-7. Piezoelectric crystals can create a voltage whenever:
(a) pressure is put on the crystals.
(b) the crystals are heated.
(c) the crystals are cooled.
(d) the crystals are put into water.

1-8. Materials that release electrons when exposed to light are called:
(a) thermoelectric.
(b) selenium or cesium.
(c) piezoelectric.
(d) chemical.

2

Basic Electrical Circuits

THE BASIC ELECTRICITY TERMS DEFINED IN CHAPTER 1 ARE APPLIED TO electrical circuits in this chapter. Basic circuits and circuit analysis are explained and applied to automotive electrical systems. The topics covered in this chapter include:

1. Parts of a circuit
2. Open circuits
3. Short circuits
4. Ground circuits
5. Series circuits
6. Parallel circuits
7. Series–parallel circuits
8. General automotive circuits and safety considerations

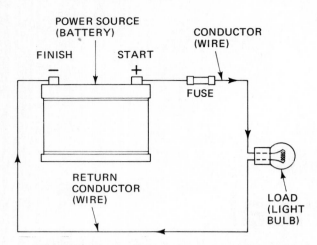

FIGURE 2-1 *All complete circuits must have a power source, a power path, an electrical load (light bulb), and a return path back to the power source.*

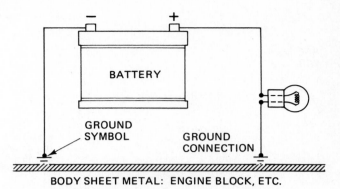

FIGURE 2-2 *The return path back to the battery can be any electrical conductor, such as the metal frame or body of the vehicle.*

CIRCUITS

A *circuit* is the path that electrons travel from a power source (such as a battery) through a resistance (such as a light bulb) and back to the power source. It is called a circuit because the current must start and finish at the same place (power source). See Figure 2-1.

For *any* electrical circuit to work at all, it must be continuous from the battery, through all the wires and components, and back to the battery. A circuit that is continuous throughout is said to have *continuity*.

PARTS OF A COMPLETE CIRCUIT

Every *complete circuit* contains the following parts:

1. *power source,* such as a car's battery.

2. *Protection* from harmful overloads (excessive current flow). Fuses, circuit breakers, and fusible links are examples of electrical circuit protection devices.

3. A *path* for the current to flow through from the power source *to* the resistance. This path from a power source to the resistance (a light bulb in this example) is usually an insulated copper wire. See Figure 2-2.

4. The *electrical "load"* or resistance—that which the electrical current is operating or lighting.

5. A *return path* for the electrical current from the load back to the power source so that there is a

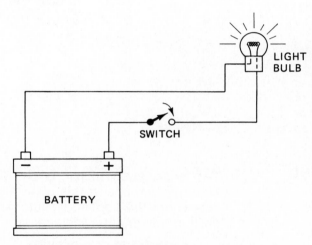

FIGURE 2-3 *An electrical switch "opens" the circuit and no current flows. The switch could also be on the return path wire.*

complete circuit. This return path is usually the metal body, frame, and engine block of the vehicle. This is called the *ground return path.* See Figure 2-2.

6. *Switches and controls* to turn the circuit on and off. See Figure 2-3.

OPEN CIRCUITS

An *open circuit* is any circuit that is *not* complete and lacks continuity. *No current at all* will flow through an incomplete circuit. An open circuit may be created by a break in the circuit or a switch that "opens" (turns off) the circuit and prevents the flow of current. A light switch in a home or the headlight switch in a car are examples of devices that open a circuit to control its operation.

OPEN CIRCUITS

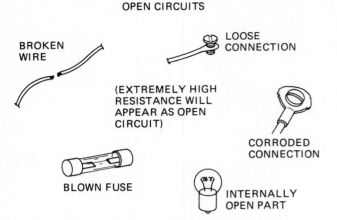

FIGURE 2-4 *Examples of common causes of open circuits. Some of these causes are often difficult to find.*

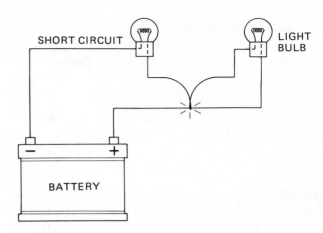

FIGURE 2-5 *A short circuit permits electrical current to bypass some or all of the resistance in a circuit.*

OPEN IS A FOUR-LETTER WORD

When a circuit is open, no current flows and the electrical units in the circuit do not operate. See Figure 2-4. Trying to find and correct an open circuit in a car can cause the use of many other four-letter words.

SHORT CIRCUITS

A *short circuit* is a circuit where current flows, but bypasses some or all of the resistance in the circuit. In other words, a short circuit means that the current takes a short-cut back to the return path of the circuit. A short circuit can be further defined as:

1. A connection that involves a "copper-to-copper" connection. A short involves the power ("hot")

side of the circuit or a "short to voltage." A short to voltage usually causes both affected circuits to malfunction. See Figure 2-5.

2. A short circuit allows *more* current to flow through the circuit than normal because the short has bypassed some of the resistance in the circuit. This additional current flow could overheat the wires and cause a fire.

3. To prevent the possibility of a fire caused by excessive current flow because of a short circuit, all circuits are protected by fuses or circuit breakers. These devices act to open the circuit to prevent current flow when a certain maximum number of amperes flows through the circuit. See Chapter 6 for details on fuses.

4. If a short occurs in a circuit protected by a fuse, *a short causes an open,* and no current flows through an open circuit. See Figure 2-6.

FIGURE 2-6 *A fuse or circuit breaker "opens" the circuit to prevent possible overheating damage in the event of a short circuit.*

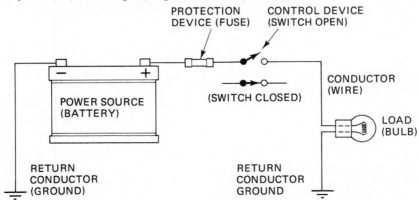

SHORT TO GROUND

A *short to ground* is a type of short circuit wherein the current bypasses part of the normal circuit and flows directly to ground. Since the ground return circuit is metal (car frame, engine, or body), this type of circuit is identified as having current flowing from "copper to steel." A defective component or circuit that is shorted to ground is commonly called "grounded."

EARTH GROUND, CHASSIS GROUND, AND FLOATING GROUND

A *ground* represents the lowest possible voltage potential in a circuit. There is actually more than one ground. The "earth" is the most "grounded" ground. A car with rubber tires is insulated "above" ground. The battery and all electrical components are connected to the chassis or frame of the car, and this type of ground is called *chassis ground*. There is another type of ground which is called *floating ground*. This ground acts as the return circuit for an electrical component in a car where no part of the circuit connects to the chassis. A floating type of ground is commonly used for radio and speaker connections.

Whenever "ground" is used throughout this book, it will mean chassis ground unless stated otherwise.

> **NOTE:** The *SAE Handbook* specifies that the negative side of the battery should be grounded. Before the early 1960s, some cars grounded the positive battery post.

ONE-WIRE SYSTEMS

A *one-wire system* is used for most automotive electrical circuits where the ground return path is provided by connecting the circuit return path to the battery through the metal of the vehicle's body and/or frame. See Figure 2-7.

FIGURE 2-7 *The metal components of the vehicle provide the ground return path for most automotive electrical components.*

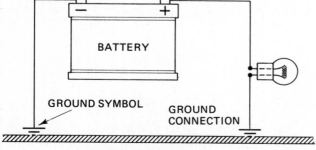

THE ONE-WIRE SYSTEM CAN CAUSE PROBLEMS

Because *all* the current has to flow through the vehicle's metal frame and/or body to return to the negative (−) side of the battery, there are many locations and connections that can cause high resistance for the entire circuit.

Most cars have ground wires connecting the negative (−) side of the battery to the car's *body*. Remember, all current flowing to any part of the body will ground to the body, and since the negative (−) battery cable connects to the engine, there *must* be an electrical connection between the engine block and the body. The body is usually electrically insulated from the engine because of the following rubber insulators:

1. The engine is mounted in rubber motor mounts.
2. The exhaust system is insulated from the body with rubber hangers.
3. The drive train and suspension are insulated from the body by rubber spring insulators and bushings.

Lights, horns, or other electrical accessories may not operate correctly (or not at all) if body ground wires are loose or corroded. Transmission and other drive-line problems can also be caused by loose or missing engine-to-body ground wires. Current normally travels through the body ground wires. If these wires are not properly connected, current could flow through dirt and moisture under the car and then through the transmission as it tries to return to the negative (−) side of the battery.

This current (up to 35 A) flows through the drive shaft and transmission. This current can arc or spark across the transmission bearings and U-joints. Always check the condition and tightness of all ground wires and do not fail to reconnect the wires attached to the valve covers and other engine and body locations.

DISCONNECT ONLY THE NEGATIVE BATTERY CABLE

If work is being performed on any automotive electrical component, the negative battery cable should be disconnected. This creates an open circuit and is safe because:

1. Removing *either* battery cable from the battery disconnects (opens) the entire car's electrical system.
2. The negative cable is connected to the engine block. Ground wires electrically connect the engine block to the car's body and frame. If the wrench used to remove the battery cable from the

battery accidentally touched the car's body or frame, nothing would happen. If, however, the positive cable were being removed and the wrench happened to touch the car body, a spark would occur which could cause a battery explosion. A spark could occur because current could flow through the wrench from the positive post of the battery directly to ground.

Always remember that the negative (−) cable should be the *first* cable removed and the *last* cable installed whenever electrical work is being performed.

SERIES CIRCUITS

A *series circuit* is a complete circuit with two or more resistances connected so that the current has to go through one resistance to go through the next. See Figure 2-8. A series circuit can have any number of resistances in the circuit. The resistance can be any of the following:

1. Resistors
2. Light bulbs
3. Horn
4. Electric motors
5. Coils
6. Relays
7. Solenoids
8. Heating elements (cigar lighter)
9. Connectors/junctions
10. A length of wire or conductor

In a series circuit, the voltage varies across each resistance, but the current flow in amperes is constant throughout the entire circuit.

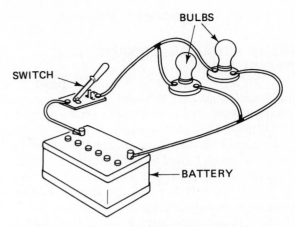

FIGURE 2-9 *Parallel circuit with two resistances. Electrical current from the battery can flow through either bulb.*

PARALLEL CIRCUITS

A *parallel circuit* is a type of complete circuit in which the current flows through the circuit by more than one path—similar to the driver of a car going through a city. The driver could travel straight through the city fighting heavy traffic (high resistance), or go around the city using a long bypass. Since both paths are available to all traffic, each road carries fewer vehicles.

In picture form, a parallel circuit looks as shown in Figure 2-9. In a parallel circuit, the voltage in each "leg" of the circuit is the same, but the current flow in amperes varies according to the resistance in each leg.

HOW TO DETERMINE A PARALLEL CIRCUIT

To test if a circuit is truly a parallel circuit, pretend to cut one wire to one of the light bulbs. See Figure 2-10. If the other bulbs are still connected to both a power source *and* a ground, the circuit is still complete and current can still flow. The circuit is therefore a parallel circuit.

FIGURE 2-10 *If the wire is cut going to bulb 2, the bulb will not light because of the open circuit created. Bulb 1 is still connected to a complete circuit and will operate normally.*

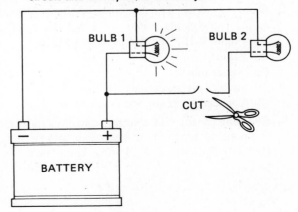

FIGURE 2-8 *Series circuit with three resistances. All of the current must flow through all of the bulbs.*

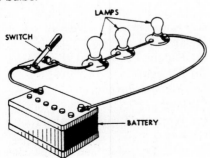

WHERE PARALLEL CIRCUITS ARE USED IN A CAR

Parallel circuits are used in almost every automotive electrical component. The exterior lights are all controlled by the headlight switch and are wired in parallel. If wired in series and one bulb burned out, *all* lights would go out due to the *open circuit* caused by the one defective bulb. This does not occur with a parallel circuit. If any bulb is defective, current can still flow through the other resistances (bulbs) as if nothing happened.

SERIES–PARALLEL CIRCUITS

A *series–parallel circuit* is any type of circuit containing resistances in both series and parallel in one circuit. Series–parallel circuits are also called "combination," or "compound" circuits. A series–parallel circuit is the most commonly used type of automotive circuit.

THE FOUR BASIC AUTOMOTIVE CIRCUITS

All automotive electrical components are divided into four operating circuits and all are series–parallel circuits. These circuits, all of which use the battery, are classified by function. During troubleshooting, a technician must know what components are connected together and their function in each specific circuit. Then by using a systematic testing procedure, the defective component can be located. These four circuits are as follows:

1. *Ignition circuit.* The components in the ignition circuit are designed to generate and deliver a high-voltage spark at the exact time necessary to fire the spark plugs in the correct order. See Figure 2-11. The components include:
   (a) Battery
   (b) Ignition coil
   (c) Distributor (including all electrical components)
   (d) Spark plugs
   (e) Spark plug wires
2. *Cranking circuit.* The cranking circuit includes all of the components needed to crank the engine. See Figure 2-12. The components include:
   (a) Battery
   (b) Starter motor
   (c) Starter solenoid/relay
   (d) Connecting cables and connections
3. *Charging circuit.* The charging circuit includes all components required to keep the battery fully charged. See Figure 2-13. The components include:

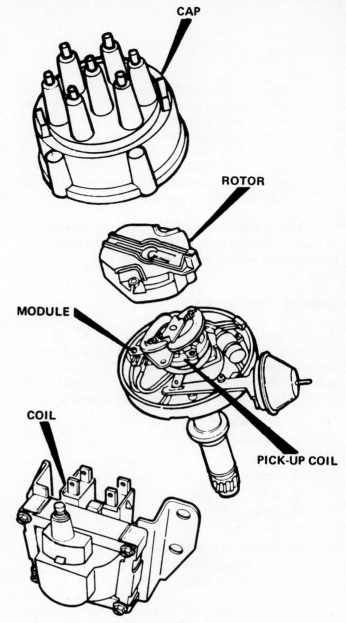

FIGURE 2-11 *Major components of the ignition circuit. (Courtesy of General Motors Corporation.)*

   (a) Battery
   (b) Alternator/generator
   (c) Voltage regulator
   (d) Connecting wires and connections
4. *Lighting and accessory circuits.* The lighting and accessory circuits include all other circuits (see Figure 2-14):
   (a) Battery
   (b) All lights
   (c) All dash instruments
   (d) Horn
   (e) Windshield wipers/washers
   (f) Radio
   (g) All other safety and convenience items

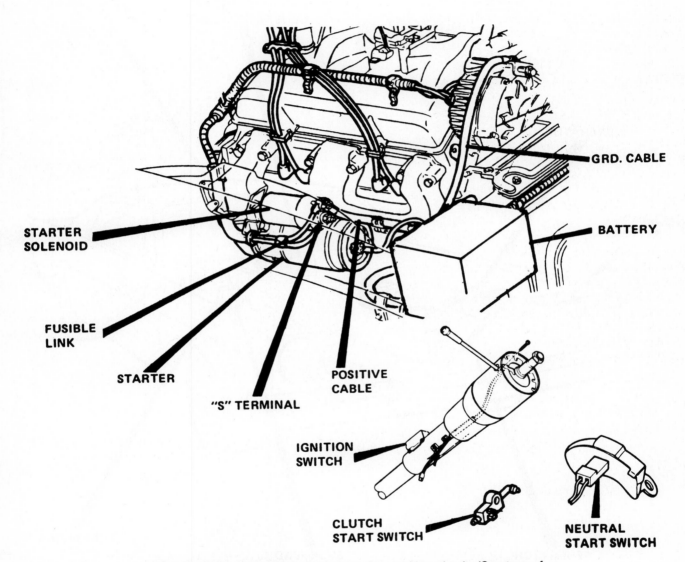

GRD. CABLE

BATTERY

STARTER
SOLENOID

FUSIBLE
LINK

STARTER

"S" TERMINAL

POSITIVE
CABLE

IGNITION
SWITCH

CLUTCH
START SWITCH

NEUTRAL
START SWITCH

FIGURE 2-12 *Major components of the cranking circuit. (Courtesy of General Motors Corporation.)*

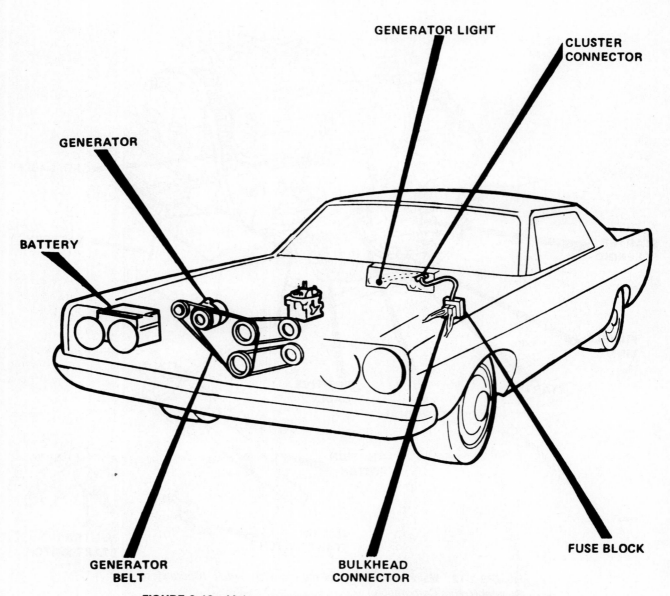

FIGURE 2-13 *Major components of the charging circuit. (Courtesy of General Motors Corporation.)*

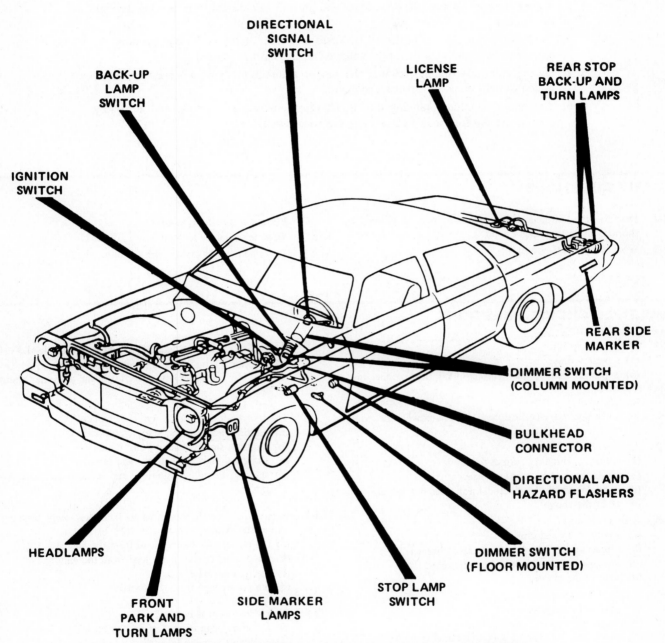

DIRECTIONAL
SIGNAL
SWITCH

BACK-UP
LAMP
SWITCH

LICENSE
LAMP

REAR STOP
BACK-UP AND
TURN LAMPS

IGNITION
SWITCH

REAR SIDE
MARKER

DIMMER SWITCH
(COLUMN MOUNTED)

BULKHEAD
CONNECTOR

DIRECTIONAL AND
HAZARD FLASHERS

DIMMER SWITCH
(FLOOR MOUNTED)

HEADLAMPS

FRONT
PARK AND
TURN LAMPS

SIDE MARKER
LAMPS

STOP LAMP
SWITCH

FIGURE 2-14 *The lighting and accessory circuits include all of the other automotive electrical wiring and components. (Courtesy of General Motors Corporation.)*

SUMMARY

1. No electrical current will flow unless the electrical circuit is complete.
2. A break or open in any part of a circuit will prevent any current from flowing through the circuit.
3. Automotive circuits are protected by fuses and will "blow," thereby opening the circuit in the event of a "short to voltage" or a "short to ground."
4. During electrical troubleshooting, the proper automotive "circuit" must be determined to help pinpoint the exact problem.
5. The ignition circuit, cranking circuit, charging circuit, and lighting and accessory circuits are the four major automotive electrical circuits.

STUDY QUESTIONS

2-1. Describe the differences between an open, a closed, a shorted, and a grounded circuit.

2-2. What is a short to ground?

2-3. What is a floating ground?

2-4. What is meant by a "one-wire system"?

2-5. Describe a series circuit and a parallel circuit.

MULTIPLE-CHOICE QUESTIONS

2-1. All circuits require:
  (a) a battery only.
  (b) a power source, an electrical load, and connecting wires.
  (c) a load only.
  (d) a battery, an electrical load, and a power-side wire only.

2-2. An open circuit:
  (a) is one in which no current flows.
  (b) can be created by a blown fuse.
  (c) is one in which the electrical unit does not operate.
  (d) all of the above.

2-3. A short circuit is:
  (a) a type of normal circuit.
  (b) copper-to-copper type of defective circuit.
  (c) copper-to-steel type of defective circuit.
  (d) none of the above.

2-4. A ground circuit is:
  (a) a type of normal circuit.
  (b) copper-to-copper type of defective circuit.
  (c) copper-to-steel type of defective circuit.
  (d) both (a) and (c) could be correct.

2-5. The major parts of any circuit include:
  (a) a power source, a power lead, an electrical load, and a return lead.
  (b) a power source, a return lead, and a ground lead.
  (c) a battery, a light, and a ground return lead.
  (d) a power source, a hot power lead, and a ground return lead.

2-6. The four basic automotive circuits include:
  (a) lighting, accessories, starter, and battery.
  (b) ignition, starter, battery, and radio.
  (c) charging, starting, battery, and ignition.
  (d) starting, charging, ignition, and lighting and accessories.

2-7. What should be disconnected to facilitate safely working on the electrical system of any car?
  (a) the power-side battery lead from the battery.
  (b) the ground (return path) lead from the battery.
  (c) the ignition wire.
  (d) the power lead from the starter.

2-8. Which statement is correct?
  (a) a "short" causes an "open" by blowing a fuse.
  (b) a "short" is an electrical connection that is from "copper to copper."
  (c) a "short to ground" is an electrical connection that is from "copper to steel."
  (d) all of the above.

3

Ohm's Law and Circuit Analysis

OHM'S LAW IS USED IN THIS CHAPTER TO EXPLAIN THE BEHAVIOR AND characteristics of electricity through series and parallel circuits. This knowledge can help a technician troubleshoot a problem by understanding the way a circuit is supposed to work. The topics covered in this chapter include:

1. Ohm's law
2. Watts
3. Ohm's law and series circuits
4. Kirchhoff's laws
5. Voltage drops
6. Ohm's law and parallel circuits
7. Ohm's law and series-parallel circuits
8. Mathematical calculations for all types of circuits

OHM'S LAW

A German physicist, Georg Simon Ohm (1787–1854), established that electric pressure (EMF) in volts, electrical resistance in ohms, and the amount of current in amperes flowing through any circuit are all related. See Figure 3-1. According to *Ohm's law,* it requires "1 volt to push 1 ampere through 1 ohm of resistance." This means that if the voltage is doubled, the number of amperes of current flowing though a circuit will also double if the resistance of the circuit remains the same.

Ohm's law can also be stated as a simple formula that can be used to calculate one value of an electrical circuit if the other two are known:

$$I = \frac{E}{R}$$

where

> I = current in amperes (A)
> E = electromotive force (EMF) in volts (V)
> R = resistance in ohms (Ω)

1. Ohm's law can determine the resistance if the volts and amperes are known $\left(R = \dfrac{E}{I} \right)$

2. Ohm's law can determine the *voltage* if the resistance (ohms) and amperes are known ($E = I \times R$).

3. Ohm's law can determine the amperes if the resistance and voltage are known $\left(I = \dfrac{E}{R} \right)$

OHM'S LAW APPLIED TO SIMPLE CIRCUITS

If a battery with 12 V is connected to a light bulb with a resistance of 4 Ω as shown in Figure 3-2, how many amperes will flow through the circuit? Using Ohm's law, we can calculate the number of amperes that will flow through the wires and the bulb. Remember, if two factors are known (volts and ohms in this example), the remaining factor (amperes) can be calculated using Ohm's law.

$$I = \frac{E}{R}$$
$$= \frac{12 \text{ V}}{4 \text{ }\Omega}$$

(the values for the voltage (12) and the resistance (4) were substituted for the letters E and R)

I is thus 3 A (12 ÷ 4 = 3).

If we wanted to connect a light bulb to a 12-V battery, we now know that this simple circuit requires 3 A to operate. This may help us for two reasons:

1. We can now determine the wire diameter that we will need based on the number of amperes flowing through the circuit.

2. The correct fuse rating can be selected to protect the circuit.

FIGURE 3-2 *Closed circuit, including a power source, power-side wire, resistance (bulb), and return path wire.*

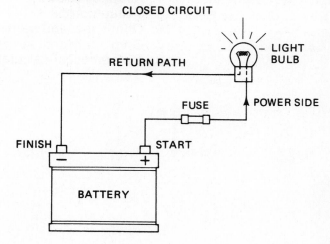

FIGURE 3-1

OHM'S LAW IN TRIANGLE EXPRESSION

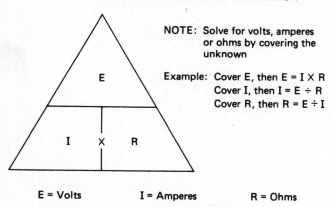

NOTE: Solve for volts, amperes or ohms by covering the unknown

Example: Cover E, then E = I X R
Cover I, then I = E ÷ R
Cover R, then R = E ÷ I

E = Volts   I = Amperes   R = Ohms

WATTS

A *watt* is the electrical unit for *power*, the capacity to do work. It is named after a Scottish inventor, James Watt (1736–1819). The symbol for power is *P*. Electrical power is amperes × volts:

$$P \text{ (power)} = I \text{ (amperes)} \times E \text{ (volts)}$$

This can easily be remembered by thinking of lowercase letters and what this spells—"pie." For example: How many watts are used to run an electric motor on 110 V using 2 A?

$$P = I \times E$$
$$= 2 \text{ A} \times 110 \text{ V}$$

P is, therefore 220 W. 220 W could also be expressed in kilowatts: 1000 watts equals 1 kilowatt (kw); therefore, 220 W equals 0.22 kW. A watt is also the metric standard for engine power. One horsepower equals 746 watts. Therefore, an automotive engine with 150 hp would be rated at 111,900 W (150 × 746 = 111,900) or 111.9 kW.

OHM'S LAW AND SERIES CIRCUITS

A series circuit is a circuit containing more than one resistance in which all of the current must flow through all of the resistances in the circuit. Ohm's law can be used to calculate the value of one unknown (voltage, resistance, amperes), if the other two values are known. Since there is more than one resistance in a series circuit, the *total* resistances of the entire circuit must be used.

Since *all* of the current flows through *all* of the resistances, the total resistance is the sum (addition) of all of the resistances. See Figure 3-3. The total resistance of the circuit shown here is 6 Ω (1 Ω + 2 Ω + 3 Ω). The formula for total resistance for a series circuit is

$$\text{total resistance } (R_t) = R_1 + R_2 + R_3 + \cdots$$

Using Ohm's law to find the current flow, we have

$$I \text{ (amperes)} = \frac{E \text{ (volts)}}{R \text{ (ohms)}}$$
$$= \frac{12 \text{ V}}{6 \text{ Ω}}$$
$$= 2 \text{ A}$$

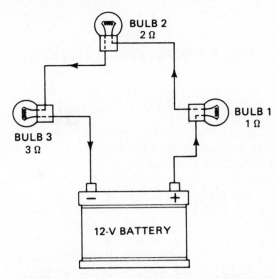

FIGURE 3-3 *Series circuit with three bulbs. All of the current flows through all of the resistances (bulbs).*

Therefore, with a total resistance of 6 Ω using a 12-V battery in the series circuit shown above, 2 A of current will flow through the entire circuit. If the amount of the resistance of a circuit is reduced, more current will flow.

In Figure 3-4, one resistance has been eliminated and now the total resistance is 3 Ω (1 Ω + 2 Ω). Using Ohm's law to calculate current flow yields

$$I \text{ (amperes)} = \frac{E \text{ (volts)}}{R \text{ (ohms)}}$$
$$= \frac{12 \text{ V}}{3 \text{ Ω}}$$
$$= 4 \text{ A}$$

Notice that the current flow is double (4 A instead of 2 A) when the resistance was cut in half (from 6 Ω to 3 Ω).

FIGURE 3-4 *Series circuit with two bulbs.*

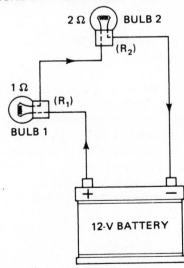

ELECTRICITY IS FARSIGHTED

Electricity almost seems to act as if it knows what resistances are ahead on the long trip through a circuit. If the trip has many high-resistance components in the circuit, very few electrons (amperes) will choose to attempt to make the trip. If a circuit has little or no resistance (for example, a short circuit), as many electrons (amperes) as possible attempt to flow through the complete circuit. If the flow exceeds the capacity of the fuse or the circuit breaker, the circuit is "opened" and all current flow stops.

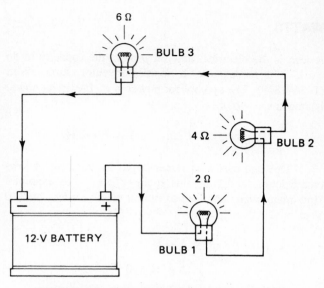

FIGURE 3-5 *In a series circuit, the voltage is dropped or lowered by each resistance in the circuit. The higher the resistance, the greater the drop in voltage.*

SERIES CIRCUITS AND VOLTAGE DROPS

The voltage that flows through each resistor in a series circuit "drops" in a way similar to the way the strength of an athlete drops each time a strenuous physical feat is performed. The *greater* the resistance, the *greater* the drop in voltage.

KIRCHHOFF'S VOLTAGE LAW

A German physicist, Gustav Robert Kirchhoff (1824–1887), developed laws about electrical circuits. *Kirchhoff's second law* (called the *voltage law*) concerns voltage drops and states: "The voltage around any closed circuit is equal to the sum (total) of the voltage drops across the resistances."

APPLYING KIRCHHOFF'S VOLTAGE LAW

Kirchhoff states in his second law that the voltage will drop in proportion to the resistance and that the total of all voltage drops should equal the applied voltage. Using Figure 3-5, the total resistance of the circuit can be determined by adding together the individual resistances ($2 \Omega + 4 \Omega + 6 \Omega = 12 \Omega$). The current through the circuit is determined by using Ohm's law, $I = E/R = 12 \text{ V}/12 \Omega = 1$ A. Therefore, in the circuit above, the following values are known:

$$\text{resistance} = 12 \, \Omega$$

$$\text{voltage} = 12 \text{ V}$$

$$\text{current} = 1 \text{ A}$$

Everything is known *except* the voltage drop caused by each resistance. The *voltage drop* can be determined by using Ohm's law and calculating for voltage (E) using the value of the resistance individually:

$$E = I \times R$$

where

E = voltage

I = current in the circuit (remember, the current is constant in a series circuit; only the voltage varies)

R = resistance of only one of the resistances

Therefore, the voltage drops are as follows:

Voltage drop for bulb 1:

$$E = I \times R$$
$$= 1 \text{ A} \times 2 \, \Omega$$
$$= 2 \text{ V for bulb 1}$$

Voltage drop for bulb 2:

$$E = I \times R$$
$$= 1 \text{ A} \times 4 \, \Omega$$
$$= 4 \text{ V for bulb 2}$$

Voltage drop for bulb 3:

$$E = I \times R$$
$$= 1 \text{ A} \times 6 \, \Omega$$
$$= 6 \text{ V for bulb 3}$$

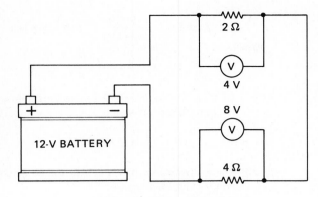

A. I = E/R (TOTAL "R" = 6 Ω)
    = 12/6 = 2 A

B. E = IR (VOLTAGE DROP)
    AT 2 Ω RESISTANCE =
    E = 2 × 2 = 4 V
    AT 4 Ω RESISTANCE =
    E = 2 × 4 = 8 V

C. 4 + 8 = 12 V
    SUM OF VOLTAGE DROP
    EQUALS APPLIED VOLTAGE

FIGURE 3-6

According to Kirchhoff, the sum (addition) of the voltage drops should equal the applied voltage (battery voltage):

Total of voltage drops = 2 V + 4 V + 6 V = 12 V = battery voltage

This proves Kirchhoff's second (voltage) law. Another example is illustrated in Figure 3-6.

THE USE OF VOLTAGE DROPS

The voltage drops due to built-in resistance are used in automotive electrical systems to drop the voltage in the following examples:

1. *Dash lights.* Most cars are equipped with a method of dimming the brightness of the dash lights by turning a variable resistor. This type of resistor can be changed (variable) and therefore varies the voltage to the dash light bulbs. A high voltage to the bulbs causes them to be bright and a low voltage results in a dim light.

2. *Blower motor* (heater or air-conditioning fan). Speeds are usually controlled by a fan switch sending current through high-, medium-, or low-resistance wire resistors. The highest resistance would drop the voltage the most and the motor would run at the lowest speed. The highest speed of the motor would occur when *no* resistance is in the circuit and full battery voltage is switched to the blower motor.

VOLTAGE DROPS AS A TESTING METHOD

Any resistance in a circuit causes the voltages to drop in proportion to the amount of the resistance. Since a high resistance will drop the voltage more than a lower resistance, we can use a voltmeter to measure resistance. Voltage-drop testing for determining high resistance in wiring or connections is discussed in detail in Chapter 11.

CURRENT FLOW THROUGH PARALLEL CIRCUITS

A parallel circuit is a complete circuit where the current has more than one path to travel to complete the circuit. A break or open in one leg or section of a parallel circuit does not stop the current flow through the remaining legs of the parallel circuit.

KIRCHHOFF'S CURRENT LAW

Kirchhoff's first law (called the *current law*) states: "The current flowing into any junction of an electrical circuit is equal to the current flowing out of that junction."

In Figure 3-7, Kirchhoff's current law can be illustrated using Ohm's law. Kirchhoff's law states that the amount of current flowing into junction A should equal the current flowing out of junction A.

FIGURE 3-7 *The amount of current flowing into junction point A equals the total amount of current flowing out of the junction.*

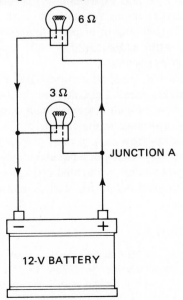

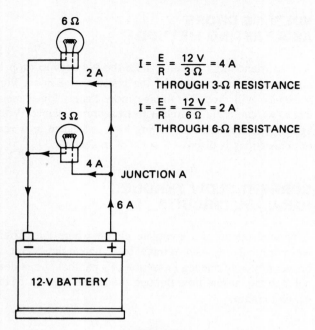

$$I = \frac{E}{R} = \frac{12 \text{ V}}{3 \, \Omega} = 4 \text{ A}$$

THROUGH 3-Ω RESISTANCE

$$I = \frac{E}{R} = \frac{12 \text{ V}}{6 \, \Omega} = 2 \text{ A}$$

THROUGH 6-Ω RESISTANCE

FIGURE 3-8

Using Ohm's law, we can calculate the currents in each leg as shown in Figure 3-8.

Since the 6 Ω leg requires 2 A and the 3 Ω resistance leg requires 4 A, it is necessary that the wire from the battery to junction A be capable of handling 6 A. Also notice that the sum of the current flowing *out* of a junction (2 + 4 = 6 A) is equal to the current flowing *into* the junction (6 A), proving Kirchhoff's current law.

OHM'S LAW AND PARALLEL CIRCUITS

According to Kirchhoff's current law, the current that flows through a parallel circuit "splits" into each leg or path in the circuit. The number of amperes (current) in each individual path, when added together, will equal the current leaving the power source.

To illustrate Kirchhoff's current law and the flow of current through a parallel circuit, Ohm's law can be used in the same way as with a series circuit except when calculating total resistance in the circuit.

The current flow in a parallel circuit can be determined by treating each leg of a parallel circuit as a separate simple or series circuit, *or* the total resistance of the circuit must be determined before the total current can be calculated.

EXAMPLE 3-1

Determine current flow in a parallel circuit by treating it as several simple circuits. See Figure 3-9.

    (a) The 6 Ω resistance bulb is connected to a 12-V battery. Using Ohm's law, we have

$$I \text{ (amperes)} = \frac{E \text{ (volts)}}{R \text{ (ohms)}} \qquad \begin{aligned} I &= \text{unknown} \\ E &= 12 \text{ V} \\ R &= 6 \, \Omega \end{aligned}$$

Dividing R (6 Ω) into E (12 V), the current flowing through the 6 Ω bulb is 2 A.

$$I = \frac{12}{6} = 2 \text{ A}$$

    (b) The 3 Ω resistance bulb's current can be determined in a similar manner:

$$I = \frac{E}{R}$$

$$= \frac{12 \text{ V}}{3 \, \Omega}$$

$$= 4 \text{ A}$$

Notice that the current flow is exactly *double* the amperes compared to the resistance, which is exactly one-half. Therefore, more current flows through the part of the circuit with the lower amount of resistance.

The total current in a parallel circuit is the total of all the separate current flows in each leg of the circuit. Therefore, the greater the numbers of legs in a parallel circuit, the greater the total current that could flow from the power source ($I = 4 + 2$; $I = 6$ A).

Homes and factories are wired in *parallel* and separated into several different parallel circuits. Each circuit is designed with the proper-size wire (conductor) and fused

FIGURE 3-9

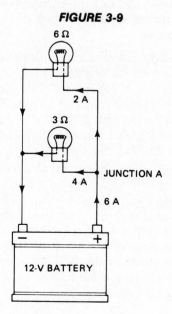

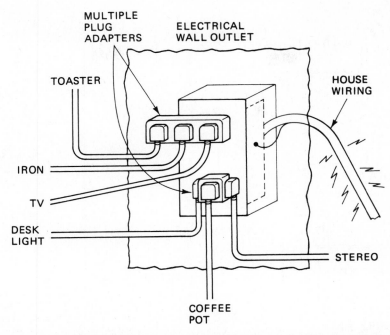

MULTIPLE PLUG ADAPTERS

ELECTRICAL WALL OUTLET

TOASTER

HOUSE WIRING

IRON

TV

DESK LIGHT

STEREO

COFFEE POT

FIGURE 3-10 *Household wiring is connected in parallel. The more items that are plugged into an outlet, the greater the current flow. Excessive current flow can cause overheating of the wiring and possible fire.*

for the proper amount of current. When additional electrical units are plugged into an outlet, each unit is actually adding another path for current to flow. This can cause excessive current to flow through the wires leading *to* the outlet. This excessive current can open a fuse or circuit breaker or cause the wires to overheat, which could cause a fire. See Figure 3-10.

DETERMINING TOTAL RESISTANCE IN PARALLEL CIRCUITS

To determine the *total* current flow in a parallel circuit, the total resistance of the entire circuit must be calculated first and then the value of this total resistance (R_t) must be used in Ohm's law to determine total current (I). However, the total resistance of a parallel circuit is *not* the addition of the resistances. Because not *all* of the current flows through each resistance *and* because the *most* current flows through the *smallest* resistance, the total resistance in a parallel circuit is *always less than the smallest resistance* in each leg of the circuit. There are two formulas that can be used to calculate the total resistance of a parallel circuit:

Formula 1: $\dfrac{1}{R_t} = \dfrac{1}{R_1} + \dfrac{1}{R_2} + \dfrac{1}{R_3} + \dfrac{1}{R_4} + \cdots$

Formula 2: $R_t = \dfrac{R_1 \times R_2}{R_1 + R_2}$

If more than two resistances are used, formula 2 can still be used. Calculate the total resistance for two resistances (R_t), then use that figure as R_1 and use the value of the third resistance as R_2. Repeat the formula.

EXAMPLE 3-2: Parallel Circuit

Using the example in Figure 3-11, the total current flow can be calculated by using Ohm's law and using total resistance:

$$\text{total resistance} = R_t = \frac{R_1 \times R_2}{R_1 + R_2}$$

$$= \frac{3 \times 6}{3 + 6} = \frac{18}{9}$$

$$= 2\ \Omega$$

Notice that the total resistance (2 Ω) is less that the smallest resistance (3 Ω). Now using Ohm's law we have,

$$I = \frac{E}{R}$$

$$= \frac{12\ \text{V}}{2\ \Omega}$$

$$= 6\ \text{A}$$

The total current was the same as that we determined by the method used earlier, where this same parallel circuit was treated as two separate series circuits.

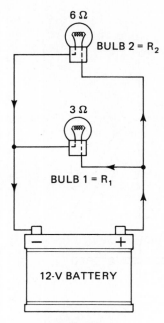

FIGURE 3-11

EXAMPLE 3-3: Parallel Circuit

In this example both methods will be used to determine current flow through the circuit shown in Figure 3-12.

Method 1: Treat all parallel circuits as separate simple circuits.

$$I = \frac{E}{R} = \frac{12\ V}{1\ \Omega} = 12\ A \text{ through the 1-}\Omega \text{ resistance}$$

$$I = \frac{E}{R} = \frac{12\ V}{2\ \Omega} = 6\ A \text{ through the 2-}\Omega \text{ resistance}$$

$$I = \frac{E}{R} = \frac{12\ V}{3\ \Omega} = 4\ A \text{ through the 3-}\Omega \text{ resistance}$$

$$\text{total current} = 22\ A$$

Notice that the most current (12 A) is flowing through the smallest resistance (1 Ω).

ELECTRICITY IS SMART, BUT LAZY

As in Example 3-3, it was determined that the higher the resistance, the less current flows through that part of the circuit. Electricity is smart, but lazy. It is smart because electricity will not leave home (the battery) without a way back home (a complete circuit). Electricity is also lazy because it *always* tries to take the path of least resistance.

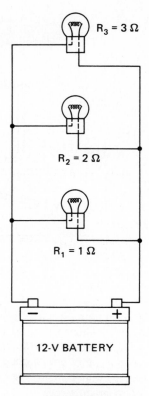

FIGURE 3-12

Method 2: Method 2 will solve for total current flow (I) by first determining total resistance using the formula:

$$R_t = \frac{R_1 \times R_2}{R_1 + R_2}$$

$$= \frac{1 \times 2}{1 + 2} = \frac{2}{3}$$

$$= 0.66\ \Omega$$

Now use 0.66 Ω as one of the two resistances in the formula and solve for R_t again using the third resistance (R_3) as R_2:

$$R_t = \frac{R_1 \times R_2}{R_1 + R_2}$$

$$= \frac{0.66 \times 3}{0.66 + 3} = \frac{1.98}{3.66}$$

$$= 0.54\ \Omega$$

The total resistance in this complex circuit is 0.54 Ω. Notice that the total resistance is *always* less than the smallest resistance in the circuit.

Now to calculate for total current, use Ohm's law:

$$I = \frac{E}{R}$$

$$= \frac{12}{0.54}$$

$$= 22 \text{ A}$$

the same result as that obtained in method 1.

OHM'S LAW AND SERIES–PARALLEL CIRCUITS

A series–parallel circuit is any type of circuit containing components (resistances) in both series and parallel in one circuit. A series–parallel circuit is the most common type of automotive circuit. It is also the most complex. Because a circuit is large does not mean that it is complicated. The circuit in Figure 3-13 looks at first to be difficult to calculate, but if the following steps are taken, even the most complex circuit can be simplified into separate series circuits and parallel circuits.

To calculate the total resistance and the current flow through the circuit in Figure 3-13, the resistances can be combined to simplify the circuit.

The two 4-Ω resistances can be combined because in the circuit these would act the same as one 8-Ω resistance. See Figure 3-14.

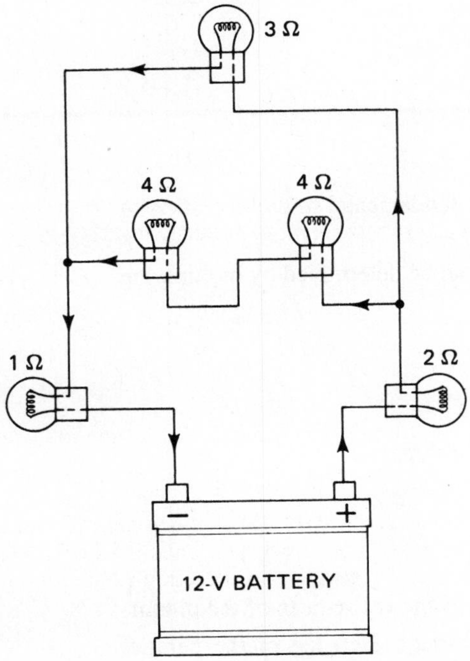

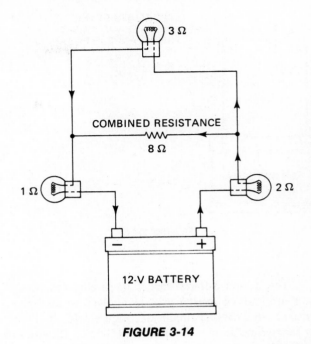

FIGURE 3-14

The 1- and 2-Ω resistances can be combined to simplify the circuit further. Even though the 2-Ω resistance is on the power side of the circuit and the 1-Ω resistance on the ground (return) side of the circuit, they are both in series because *all* of the current in the circuit *has* to flow through each. See Figure 3-15.

FIGURE 3-13

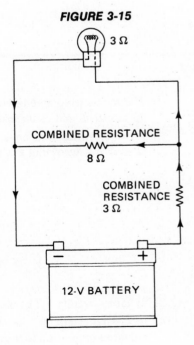

FIGURE 3-15

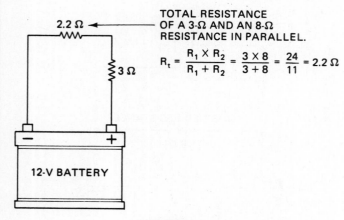

TOTAL RESISTANCE OF A 3-Ω AND AN 8-Ω RESISTANCE IN PARALLEL.

$$R_t = \frac{R_1 \times R_2}{R_1 + R_2} = \frac{3 \times 8}{3 + 8} = \frac{24}{11} = 2.2\ \Omega$$

FIGURE 3-16

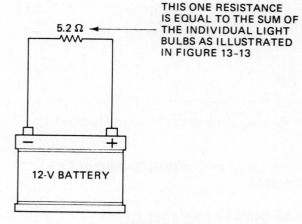

THIS ONE RESISTANCE IS EQUAL TO THE SUM OF THE INDIVIDUAL LIGHT BULBS AS ILLUSTRATED IN FIGURE 13-13

FIGURE 3-17

The 3- and 8-Ω resistances in parallel can now be combined into one resistance by using the formula for total resistance in a parallel circuit. See Figure 3-16. By replacing the two resistances in parallel (8 Ω and 3 Ω) with one 2.2 Ω resistance, the circuit is finally becoming a simple-looking circuit.

Figure 3-16 can be simplified even further to a simple circuit with only one resistance of 5.2 Ω (3 Ω + 2.2 Ω) because the two resistances are in series and *all* of the current in the circuit *has* to flow through both resistances. The resistances can be added together and represented as one resistance of 5.2 Ω. See Figure 3-17. Using Ohm's law, the total current in the circuit can be calculated since the battery voltage (12 V) and the total resistance (5.2 Ω) are known.

$$I = \frac{E}{R}$$

$$= \frac{12\ V}{5.2\ \Omega}$$

$$= 2.3\ A$$

By calculating the current flow in the complex circuit, the proper-size conductor wire can be selected to carry the current safely and the circuit can be fused correctly.

SUMMARY

1. The relationship among volts, amperes, and ohms is fundamental to the operation of all electrical and electronic units.

2. If two electrical units are known, the remaining unit can be determined by multiplying or dividing according to the following equations:

$$I\ (amperes) = \frac{E\ (volts)}{R\ (ohms)}$$

$$E\ (volts) = I\ (amperes) \times R\ (ohms)$$

$$R\ (ohms) = \frac{E\ (volts)}{I\ (amperes)}$$

3. In series circuits, all of the current flows through all of the resistances of the circuit.

4. In a parallel circuit, the current splits and travels through each leg of the parallel circuit in proportion to the amount of resistance in each leg; the greater the resistance, the lower the current.

STUDY QUESTIONS

3-1. State Ohm's law.

3-2. What is a watt?

3-3. What does Kirchhoff's current law state?

3-4. What does Kirchhoff's voltage law state?

3-5. Why is a voltage drop equal to resistance?

MULTIPLE-CHOICE QUESTIONS

3-1. Ohm's law states that:
(a) It takes 1 Ω of pressure to push 1 A through 1 V.
(b) It takes 1 V of pressure to push 1 A through 1 Ω of resistance.
(c) It takes 1 A of pressure to push 1 Ω through 1 V.
(d) none of the above.

3-2. A watt is:
(a) amperes times ohms.
(b) amperes times volts.
(c) ohms times volts.
(d) amperes divided by volts.

3-3. In a series circuit:
(a) all the current flows through every resistor.
(b) some of the current flows through every resistor.
(c) none of the current flows through every resistor.
(d) only about one-half of the current flows through any single resistor.

3-4. The higher the resistance:
(a) the smaller the battery has to be.
(b) the greater the voltage drop.
(c) the lower the voltage drop.
(d) the voltage remains the same.

3-5. In a parallel circuit:
(a) only a part of the total current flows in each leg of the circuit.
(b) all of the current flows through all of the resistors.
(c) one-half of all the current flows through all of the resistors.
(d) the voltage will increase as the current flows through each resistor.

3-6. How many amperes will flow through the circuit shown in Figure 3-18?
(a) 12 A.
(b) 4 V.
(c) 4 A.
(d) 3 V.

3-7. How much will the voltage be dropped by the 3 Ω resistor in Figure 3-18?
(a) 3 V.
(b) 4 V.
(c) 12 V.
(d) 0 V.

3-8. What is the total resistance of the circuit shown in Figure 3-19?
(a) 2 Ω.
(b) 4 Ω.
(c) 18 Ω.
(d) 12 Ω.

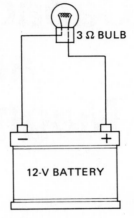

FIGURE 3-18

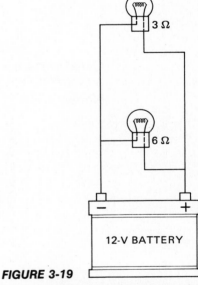

FIGURE 3-19

3-9. In Figure 3-19, what is the total current through the entire circuit?
(a) 6 A.
(b) 12 A.
(c) 2 A.
(d) 22 A.

3-10. How much current flows through the 3 Ω resistor in the circuit shown in Figure 3-19?
(a) 1 V.
(b) 1 Ω.
(c) 12 A.
(d) 4 A.

4

Capacitance and Magnetism

CAPACITANCE AND MAGNETISM ARE INTRODUCED IN THIS CHAPTER WITH an explanation of terms, units of measurement, and uses in automotive applications. The topics covered in this chapter include:

1. Capacitance (capacitors/condensers)
2. Basics of magnetism
3. Electromagnetism
4. Magnetic field strength
5. Electromagnetic induction
6. Electromagnetic switches (relays/solenoids)
7. Residual magnetism
8. Self-induction
9. Mutual induction
10. Symbols and abbreviations

CAPACITORS/CONDENSERS

Capacitors (also called *condensers*) are electrical components that can be used to perform a variety of functions. Electrons can be "stored" on the inside of a capacitor on two or more conductor plates separated by an insulator called a *dielectric*. The dielectric material in a condenser can be paper, mica (a type of silicate rock in thin layers), or air. See Figure 4-1. The greater the dielectric strength of a material, the greater the resistance of the material to voltage penetration.

If a capacitor is connected to a battery or another electrical power source, it is capable of storing the electrons from the power source. See Figure 4-2. This storing capacity is called *capacitance* and is measured in the unit called *farad*, named for Michael Faraday (1791–1867), an English physicist. A farad is the capacity to store 1 coulomb of electrons at 1 volt of potential difference between the plates of the capacitor. This is a very large number, so most capacitors for automotive use list values measured in microfarads (one millionth of a farad).

A capacitor will accept electrons when connected to a power source until the capacitor's maximum charge is reached. Electrons do not flow through a capacitor because of the insulating strength of the dielectric between the plates. The charge is stored in the capacitor until the plates are connected to a lower-voltage circuit. This will cause

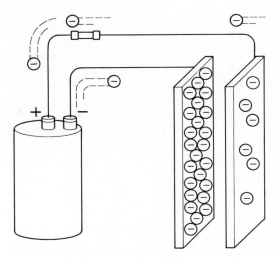

FIGURE 4-2 *Simple condenser. Air is the dielectric between two conductor plates. Notice that when connected to a battery, the electrons tend to "pile up" on the negative plate of the condenser. (Courtesy of Ford Motor Company.)*

the stored electrons to flow out of the capacitor and into a conductor which has a lower voltage. A capacitor can pass current that is constantly changing its direction of flow (alternating current—ac), but blocks the flow of direct current (dc).

COMPARISON BETWEEN A CAPACITOR AND A WATER TOWER

A capacitor can store electrons similar to the way a water tower can store water for use at a later time. The comparisons are listed below:

FIGURE 4-1 *A foil and paper condenser can store electrons on the surface of the foil. (Courtesy of Chrysler Corporation.)*

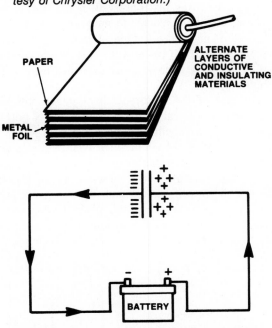

THE CAPACITOR CHARGES TO THE SOURCE VOLTAGE

Water Tower	Capacitor
1. Water can be pumped into a water tower if the water pressure is high enough to make the water flow into the tower.	1. Electrons can flow into a capacitor if connected to a power source such as a battery.
2. As the water level gets higher, the pressure of the water in the tower increases, and therefore a greater pressure is required to maintain water flow into the tower.	2. As the capacitor builds up a charge, the voltage increases in the capacitor, and higher and higher voltages are required to maintain current flow *into* the capacitor.

Comparison between a capacitor and a water tower

Water Tower	**Capacitor**
3. If a valve is opened at the base of a water tower, water will flow quickly out of the tower, but only for a short time because of the limited amount of water ''stored'' in the tower.	3. If a conductor is connected to both sides of a capacitor, current will flow quickly out of the capacitor but only for a short time because of the limited amount of stored electrons.

Purposes of a capacitor compared to a water tower

Water Tower	**Capacitor**
1. A water tower is used to store water when more water is available than can be used immediately (which could cause flooding), but which could be used later during a dry season.	1. A capacitor can temporarily store electrons (electrical energy) when too much energy could damage electrical components. An example of this purpose is the capacitor (condenser in a point-type ignition system) used to protect the ignition points from damage due to arcing by the high voltage induced in the ignition coil when the points open.
2. A water tower acts as a ''surge tank'' which tends to smooth out pulses of a water pump. A large tank can absorb the very high pressures, and slowly and evenly release the water when the pulses of the pump are weak.	2. A capacitor acts as a ''surge tank'' to voltages connected to the capacitor, which vary quickly from very high voltage to very low voltage. A capacitor can absorb electrons during the time when the voltage is high and release the electrons when the voltage is low. As a result of this ''smoothing out'' of the voltage, radio interference is

Water Tower	**Capacitor**
	reduced which would normally be produced by the rapidly changing voltages. A capacitor used for this purpose is called a *filter capacitor* and is used on alternators and ignition systems to reduce radio-frequency interference.

MAGNETISM

Like electricity, magnetism is sometimes difficult to visualize. Although electricity and magnetism cannot be seen, the *effects* can be both seen and felt.

DID YOU KNOW?

Magnetism is extremely important to automotive applications because everything electrical in the automobile, except the lights and the cigar lighter, work as a result of magnetism.

Magnetism was first observed in the way a natural stone called *lodestone* reacted to metal objects. Lodestone is a variety of magnetite (a type of iron ore) that attracts pieces of iron and will point to the earth's magnetic north pole if a long piece of this ore is suspended from a string. See Figure 4-3.

The end of the lodestone that pointed toward the earth's north pole is called the north pole or N pole. The opposite end is the south or S pole. These ''poles'' of a magnetic substance act similarly to electrostatic charges: Like poles repel each other, while opposite poles are attracted. Since a magnet also shows attraction for metal products such as tacks, nails, and iron filings, it is clear that a force surrounds the magnetic material. Magnetic lines of force are invisible, but when iron filings are placed on a piece of paper held above a magnet, the filings move and

FIGURE 4-3 *Lodestone (magnetite) is a variety of a natural magnet.*

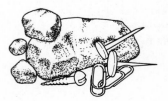 

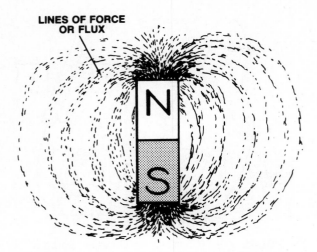

FIGURE 4-4 *The magnetic lines of force leave the north pole and enter the south pole. (Courtesy of Chrysler Corporation.)*

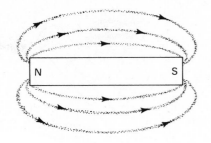

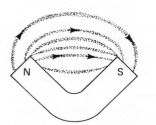

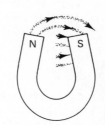

FORMING A HORSESHOE MAGNET

FIGURE 4-6 *(Courtesy of General Motors Corporation.)*

become stationary along a definite pattern between and around both the N and S poles. This pattern indicates parallel (side by side and not touching) lines of magnetic force that leave the N pole and enter the S pole. See Figure 4-4.

WHAT MAKES A MAGNET MAGNETIC

The most accepted theory (scientific explanation) indicates that the random positions of the atoms in a magnetic material are all aligned in one direction so that their combined forces generate the magnetizing force. The strength of a magnet varies greatly with the type of material used. See Figure 4-5.

CLASSIFICATION OF MAGNETIC MATERIALS

There are *natural* magnets such as lodestone and many other materials which can be *made* magnetic. If iron, for example, is rubbed by a strong magnet, the magnetic properties will be transferred to the iron. This transfer of magnetic properties is called *magnetic induction*. Magnetic

FIGURE 4-5 *(Courtesy of Chrysler Corporation.)*

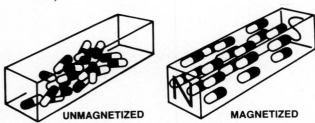

UNMAGNETIZED   MAGNETIZED

induction creates new *permanent magnets* if the material is of the proper type. The best materials for permanent magnets are hard metals or metal alloys. Alloys are a combination of two or more metals. Common alloys include:

1. Permalloy (permanent alloy), made from nickel and iron
2. Alnico, an alloy of *al*uminum, *ni*ckel, and *co*balt
3. Cunife, an alloy of copper (Cu), nickel (Ni), and iron (Fe)
4. Magnequench, made of neodymium, iron, and boron, a powerful alloy developed in the mid-1980s by General Motors for initial use in permanent-magnet starter motors.

Permanent magnets can be made into many shapes, including bar and horseshoe shapes. See Figure 4-6.

ELECTROMAGNETISM

It was not until about 1820 that it was discovered that a wire carrying an electrical current had an effect on a compass. See Figure 4-7. Until that time, the only thing that affected a compass was a magnetic field. Further study revealed that a magnetic field surrounds any conductor (wire) which carries an electrical current. A magnetic field created by current flow is called *electromagnetism*. The characteristics of electromagnetism can be summarized as follows:

FIGURE 4-7 *(Courtesy of General Motors Corporation.)*

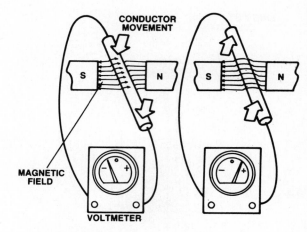

FIGURE 4-9 *(Courtesy of Chrysler Corporation.)*

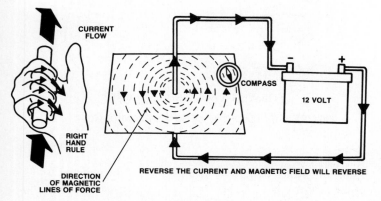

FIGURE 4-8 *(Courtesy of Chrysler Corporation.)*

1. The direction of the lines of force is determined by the *right-hand rule*. Place your right hand around the wire with your thumb pointing toward the direction of conventional current flow (positive to negative) and your fingers will point in the direction of the magnetic lines of force. See Figure 4-8.

2. The magnetic lines of force do *not* move except to progress farther away from the conductor with greater current flow.

3. The density and strength of the magnetic lines of force increase directly with increased current flow (in amperes) through the conductor.

MAGNETIC FIELDS IN A COIL OF WIRE

If a wire (conductor) is coiled and current is sent through the wire, the same magnetic field that surrounds a straight wire combines to form one larger magnetic field with true north and south poles. The north pole can be determined by the right-hand rule for *coils* (see Figure 4-9): Grasp the coil with the fingers pointed in the direction of current flow [conventional current flow (+ to −)] and the thumb will point toward the north pole of the coil. Current flowing through a coil of wire creates a useful magnetic field and is the principle used in countless electrical components. However, a more powerful magnetic field can be generated by placing an iron core in the center of the coil of wire.

The iron in the center of the coil provides an excellent conductor for the magnetic field which travels through the center of the wire coil. The measurement of a material's ability to conduct magnetic lines of force is called *permeability*. The permeability of various materials is based on a rating of 1 for air, which is a poor conductor of magnetic lines of force. The permeability of iron is 2000 and that of some alloy steels can be 50,000 or more. The increase in magnetic field strength is the reason that most coils and electromagnets contain an iron core. See Figure 4-10.

FIGURE 4-10 *(Courtesy of General Motors Corporation.)*

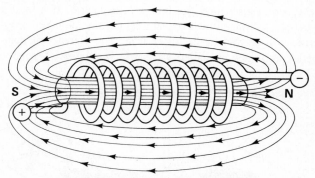

IRON CORE INCREASES FIELD STRENGTH

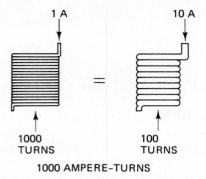

1000 AMPERE-TURNS

FIGURE 4-11 *Ampere-turns is the unit used to measure electromagnetic field strength. (Courtesy of General Motors Corporation.)*

MAGNETIC FIELD STRENGTH

The strength of the magnetic field surrounding a coil is increased by one or both of the following factors:

1. An increased number of turns of wire in a coil.
2. An increased amount of current flow through the coil, measured in amperes. See Figure 4-11.

The magnetic field strength for automotive use is measured in *ampere-turns*. The magnetic field strength is measured by multiplying the current flow in amperes through a coil by the number of complete turns of wire in the coil. For example, a 1000-turn coil with 1 A of current would generate 1000 ampere-turns (At). This 1000-turn coil would have the same field strength as that of a similar coil of 100 turns × 10 A or 1000 At. The ampere-turn strength of a coil also depends on the resistance to the magnetic lines of force.

RELUCTANCE

The resistance to the movement of magnetic lines of force is called *reluctance*. Reluctance (resistance) is reduced by using highly permeable materials for the cores of coils, and it is increased by placing air gaps in the coil cores. (Remember, air is a poor conductor of magnetic lines of force.) See Figure 4-12. Therefore, most automotive coils use an iron core because the magnetic field strength is increased for coils with cores compared to a similar coil without a core.

FIGURE 4-12 *Increasing the air gap greatly decreases the magnetic field strength. (Courtesy of General Motors Corporation.)*

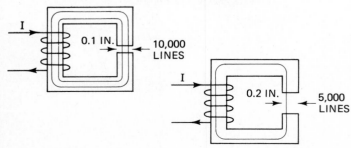

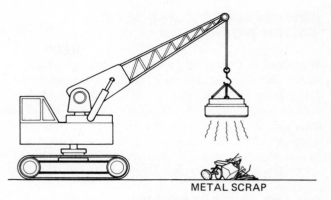

METAL SCRAP

FIGURE 4-13 *Electromagnetic crane. When electrical current is sent to large electromagnets, the crane can pick up ferrous metal products of any shape.*

ELECTROMAGNETIC USES

The relationship between electricity and magnetism is very important since many electrical components on an automobile use electromagnetism. An electromagnet can be made by wrapping an iron bar (core) with a coil of wire connected to a battery or another electrical power source. An example of an electromagnet is a scrap metal crane. See Figure 4-13. Whenever current is flowing through the coil of an electromagnet, it is magnetized with a force proportional to the ampere-turns of the magnet.

ELECTROMAGNETIC SWITCHES

Electromagnets are widely used in automotive electrical systems in the form of electromagnetic switches. An electromagnetic switch is one that opens or closes electrical contacts using an electromagnet. See Figure 4-14.

A low-current electromagnet switch is usually used to control (open or close) a high-current circuit. For example, an ignition switch circuit (low current) can control a high-current starter motor circuit by using an electromagnetic switch.

FIGURE 4-14 *Electromagnetic switch. A light current (low amperes) produces an electromagnet and causes the contact points to close. The contact points then conduct a heavy current (high amperes) to an electrical unit.*

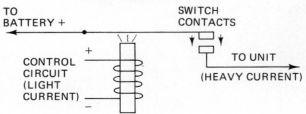

HOW AN ELECTROMAGNETIC SWITCH WORKS

When the electromagnetic wires are connected to a power source, the resulting magnetic "pull" on the upper movable contact forces the switch into contact with the lower contact point. These contact points complete (close) another circuit. Because the electromagnetic switch controls a higher current than the control current, it is often called a *relay* because it "relays" heavy current in the circuit.

RELAYS AND SOLENOIDS

If an electromagnetic switch uses a movable arm, it is called a *relay*. If an electromagnetic switch uses a movable iron core, it is called a *solenoid*. A solenoid, besides operating as a switch, can also use a movable core to perform mechanical work, such as engaging a starter gear. Solenoids are usually constructed to transfer heavier current than a movable arm relay. A solenoid may, therefore, be called a relay when used to transfer heavy current such as that used in diesel engine glow plug circuits. The proper term to use is determined by the individual automobile manufacturer as specified in the service literature. See Figure 4-15.

FIGURE 4-15 *Example of a typical electromagnetic switch found on a Ford starting circuit. The proper term for this switch is a starter solenoid because it contains a movable core.*

RESIDUAL MAGNETISM

Residual magnetism is the term used to describe the magnetism remaining in the core of an electromagnet *after* the magnetizing electrical current is shut off. The cause of some residual magnetism is the fact that the magnetizing force "lags behind" the applied current. This lag is called *hysteresis,* from the Greek word *hysterein,* which means "to be behind, to lag." To reduce residual magnetism, "soft iron" (a pure iron with little carbon content) is used as a core because it can be magnetized easily but cannot remain magnetized after the current is removed. If hard iron were used, some magnetizing power would remain. In some cases residual magnetism is an unwanted side effect of an electromagnet. In other applications, residual magnetism may be useful.

ELECTROMAGNETIC INDUCTION

In 1831, Michael Faraday (1791–1867) discovered that electrical energy can be induced from one circuit to another by using magnetic lines of force. When a conductor is moved through a magnetic field, a difference of potential is set up between the ends of the conductor and a voltage is induced. This action is called *electromagnetic induction.* See Figure 4-16. This voltage exists only when the magnetic field *or* the conductor is in motion.

To induce 1 V, 100,000,000 (100 million) magnetic lines of force must be cut *per second.* The induced voltage can be increased by increasing the *speed* with which the magnetic lines of force cut the conductor, or by increasing the *number of conductors* that are cut. Electromagnetic induction is the principle behind the operation of all ignition systems, starter motors, generators, alternators, and relays.

A *coil* is often referred to as an *inductor.* As the current increases through the coil, the magnetic strength eventually reaches a leveling-off point where an additional increase of the magnetizing force current no longer increases the magnetic field strength. This condition is called

FIGURE 4-16 *A wire (conductor) moving through a magnetic field generates a voltage.*

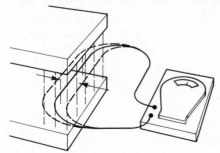

saturation. The magnetic lines of force represent stored energy. If the applied voltage is removed, the lines of force collapse, returning the energy back into the wire.

SELF-INDUCTION

Self-induction is the generation of an electric current in the wires of the coil itself when a current is *first* connected or disconnected. This induced current is in the *opposite* direction of the applied current and tends to reduce the magnetizing force. This is the reason that coils become fully saturated only after a slight delay. Self-induction also tends to maintain the applied voltage when the circuit is opened by a switch, because the energy in the coil becomes a source of voltage. Self-induction, first observed in 1834 by Heinrich Lenz, a German physicist (1804–1865), is called Lenz's law, which states: An induced current is in such direction that its magnetic effect opposes the change by which the current is induced. It is the *change* in current, not the current itself, that is opposed by the induced EMF. Self-induction is a generally *undesirable* electrical characteristic because whenever any automotive circuit containing an inductor is switched off, self-induction attempts to continue to supply current in the same direction as the applied (original) current. This causes an electrical arc to occur in the switch of any electrical circuit. Self-induction is commonly found in the following:

1. Ignition circuit (ignition coil)
2. Air conditioning (electromagnetic clutch coil)
3. Blower motors (motor field and armature windings) (see Figure 4-17)
4. Any other component containing a coil or an electric motor

To reduce or eliminate this high-voltage arc, capacitors (condensers) and/or diodes are commonly connected in the circuit to absorb or direct the high voltage created by the switching on and off of inductive circuits.

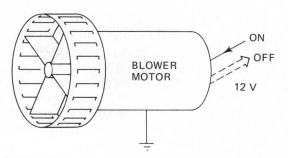

FIGURE 4-17 *Every electric motor contains magnetic windings. Whenever electrical power is shut off, self-induction produces a current in the opposite direction from the applied current flow needed to turn the fan motor.*

MUTUAL INDUCTION

Mutual induction is the *desirable* induction of a voltage in a conductor (coil) due to a changing magnetic field of an adjacent coil. Mutual induction is used in ignition coils where a rapidly changing magnetic field in the primary winding of the coil creates a voltage in the secondary winding of the coil. See Figure 4-18. The voltage induced in the secondary windings is basically determined by the number of turns of wire in both windings. The operation of the ignition coil and the ignition circuit is discussed in Chapter 18.

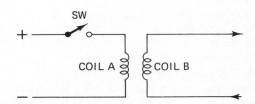

FIGURE 4-18 *Mutual induction can create a current in coil B if the current flow through coil A is switched on and off.*

SUMMARY

1. Capacitors (condensers) are used in numerous automotive applications. Because of their ability to block direct current and pass alternating current, they are used to control radio-frequency interference (RFI) and are installed in various electronic circuits to protect and control changing current.
2. Most automotive electrical components use magnetism, and the strength of the magnetism depends on both the amount of current (amperes) and the number of turns of wire of each electromagnet.

3. The strength of electromagnets is increased by using a soft-iron core.
4. The amount of voltage that can be induced from one circuit to another can be increased by the following:
   (a) Increasing the speed with which the magnetic lines cut across the conductors.
   (b) Increasing the number of the conductors that are cut.

   Self-induction, however, is undesirable because it can create extremely high voltage surges (7000 V or more) throughout the entire automotive electrical systems. Computer systems are subject to damage if these high-voltage spikes are not controlled or prevented.

STUDY QUESTIONS

4-1. What is a dielectric?

4-2. Capacitance is measured in what unit?

4-3. List three functions that a capacitor/condenser can perform in an electrical circuit.

4-4. Name four permanent-magnet alloys.

4-5. What does "permeability" mean when applied to electromagnets?

4-6. Explain how an electrical current can be created by a moving magnetic field.

4-7. What is the difference between self-induction and mutual induction?

MULTIPLE-CHOICE QUESTIONS

4-1. Capacitance is measured in units of:
   (a) ohms.
   (b) coulombs.
   (c) amperes.
   (d) farads.

4-2. Electromagnetism is measured in units of:
   (a) ampere-turns.
   (b) farads.
   (c) volt-turns.
   (d) coulombs.

4-3. Electromagnetism is used in almost every automotive circuit except:
   (a) lights and horn.
   (b) horn and cigarette lighter.
   (c) radio and horn.
   (d) lights and cigarette lighter.

4-4. A coil of wire wrapped around a soft-iron core will form a _____ if the wires are connected to a power source.
   (a) coil.
   (b) electromagnet.
   (c) solenoid.
   (d) electric switch.

4-5. A capacitor (condenser) can:
   (a) prevent high-voltage surges.
   (b) reduce radio interference.
   (c) prevent ignition-point arcing.
   (d) all of the above.

4-6. The major difference between a solenoid and a relay is:
   (a) a relay uses a movable core.
   (b) a solenoid uses a movable core.
   (c) a solenoid uses a movable arm.
   (d) none of the above.

4-7. A natural magnetic material is:
   (a) lodestone.
   (b) permalloy.
   (c) alnico.
   (d) neodymium–iron–boron.

4-8. The name of the electromagnetic switch that can conduct a high current flow is:
   (a) a relay.
   (b) a solenoid.
   (c) an ignition switch.
   (d) a permalloy switch.

HOW TO TEST DIODES AND TRANSISTORS

Diodes and transistors can best be tested with an ohmmeter. The diode or transistor being tested should be *disconnected* from the circuit for best results. Use an ohmmeter and set it to low ohms or "diode check" if using a digital ohmmeter. See Figure 5-12.

Diodes. A good diode should read high ohms with the test leads attached to each lead of the diode and low ohms when the leads are reversed.

1. A low ohm reading with the ohmmeter leads attached both ways across a diode means that the diode is *shorted* and must be replaced.

2. A high ohm reading with the ohmmeter leads attached both ways across a diode means that the diode is *open* and must be replaced.

Transistors. A good transistor should show continuity between the emitter (*E*) and the base (*B*) and between the base (*B*) and the collector (*C*) with an ohmmeter connected one way, and high ohms when the ohmmeter test leads are reversed. There should be a high ohmmeter reading (no continuity) in either direction when a transistor is tested between the emitter (*E*) and the collector (*C*). See Figure 5-13. A transistor tester can also be used if available.

SUMMARY

1. Diodes and transistors use P- and N-type materials, in combination, to provide an electrical conductor or insulator, depending on the polarity of the control current.

2. Diodes permit current flow in only one direction, except for zener diodes, which permit reverse-bias current flow, but only after reaching a certain breakdown voltage.

3. Transistors are two diodes back to back that can conduct current only after a specific current is sent to the center section (base) of the transistor.

4. If current can flow from the base to the emitter, heavier current will flow from the collector to the emitter.

5. All semiconductors must be kept cool to prevent damage to the junction between the layers of semiconductor materials.

STUDY QUESTIONS

5-1. What is "doping"?

5-2. Explain how current can flow across the junction of a diode.

5-3. What is "reverse bias"?

5-4. How does a zener diode differ from a standard diode?

5-5. Name the three parts of a transistor and explain how a transistor works.

5-6. What is a Darlington pair?

5-7. Explain how to test diodes and transistors.

MULTIPLE-CHOICE QUESTIONS

5-1. A semiconductor is a material with:
   (a) fewer than four electrons in the outer orbit of its atoms.
   (b) more than four electrons in the outer orbit of its atoms.
   (c) exactly four electrons in the outer orbit of its atoms.
   (d) other factors beside the number of electrons.

5-2. Two types of semiconductors are:
   (a) holes and blocks.
   (b) P and N.
   (c) white and black.
   (d) none of the above.

5-3. The arrow of a semiconductor:
 (a) points toward the negative.
 (b) points away from the negative.
 (c) is attached to the emitter.
 (d) both (a) and (c) are correct.

5-4. A transistor is controlled by the polarity and current at:
 (a) the collector.
 (b) the emitter.
 (c) the base.
 (d) both the collector and the emitter.

5-5. A transistor can:
 (a) switch "on" and "off."
 (b) amplify.
 (c) "throttle."
 (d) all of the above.

5-6. What happens when a transistor or diode "blows"?
 (a) The N-type material becomes P.
 (b) The P-type materials becomes N.
 (c) Both (a) and (b).
 (d) The junction between the N- and the P-type material is destroyed.

5-7. Zener diodes are used:
 (a) in voltage regulators.
 (b) to control radio static.
 (c) to improve radio reception.
 (d) both (b) and (c).

5-8. The voltage required to turn a transistor on is:
 (a) 0.7 V for silicon transistors.
 (b) 0.7 V for germanium transistors.
 (c) 0.3 V for silicon transistors.
 (d) none of the above.

6

Wiring
and
Circuit Diagrams

THIS CHAPTER TAKES THE READER ONE STEP AT A TIME THROUGH THE explanation, symbols, codes, and markings used in automotive electrical and electronic circuit diagrams and schematics. Information on electrical wiring, gauge sizing, metric sizing, automotive fuses, fusible links, and circuit breakers is included. The topics covered in this chapter include:

1. Wire gauge sizes/terminals and connectors
2. Soldering procedures
3. Aluminum wire repair
4. Fuses
5. Circuit breakers
6. Fusible links
7. Wiring diagrams

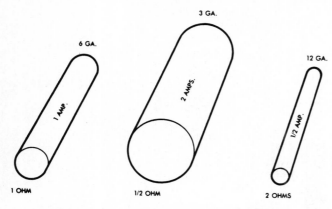

FIGURE 6-1 *(Courtesy of Chrysler Corporation.)*

AUTOMOTIVE WIRING

Most automotive wire is made from strands of copper covered by insulating plastic. Copper is an excellent conductor of electricity, reasonably priced, and very flexible. Even copper can break when moved repeatedly, and therefore, most copper wiring is constructed of multiple small strands that allow for repeated bending and moving without breaking. Solid copper wire is generally used for components such as starter armature and alternator stator windings which do not bend or move during normal operation.

GAUGE SIZE

Wiring is sized and purchased according to *gauge* size as assigned by the American wire gauge (AWG) system. AWG numbers can be confusing because as the gauge number *increases*, the size of the conductor wire *decreases*. Therefore, a 14-gauge wire is smaller than a 10-gauge wire. The *greater* the amount of current (in amperes), the *larger* the conducting wire (the smaller the gauge number) required. See Figure 6-1.

EXAMPLES OF WIRE-GAUGE-SIZE APPLICATIONS

Listed below are general applications for the most commonly used wire gauge sizes. Always check installation instructions or the manufacturer's specifications for wire gauge size before replacing any automotive wiring.

20–22 gauge: radio speaker wires

18 gauge: small bulbs and short leads

16 gauge: taillights, gas gauge, turn signals, windshield wipers

14 gauge: horn, radio power lead, headlights, cigar lighter, brake lights

12 gauge: headlight switch to fuse box, rear window defogger, power windows/locks

10 gauge: ammeter, generator/alternator to battery

Some manufacturers indicate on the wiring diagrams the wire sizes measured in square millimeters of cross-sectional area. Below is the conversion comparison size equivalent between the metric and the AWG gauge sizes. Notice that the metric wire size increases with size (area), whereas the AWG gauge size gets smaller with larger size wire.

Metric Size (mm²)	AWG Gauge
0.22	24
0.35	22
0.5	20
0.8	18
1.0	16
2.0	14
3.0	12
5.0	10
8.0	8
13.0	6
19.0	4
32.0	2

The gauge number should be decreased (wire size increased) with increased lengths of wire. See Figure 6-2. For example, a trailer may require 14-gauge wire to light

FIGURE 6-2

Wire Gauge Selection Chart

1. Measure wire length. When a return ground wire is used, measure both wires.
2. Find the correct wire gauge by matching the amperage, at the correct voltage, with the wire length or the next larger footage on the chart.

NOTE: This chart is based upon a 10% maximum voltage drop, for 5% voltage drop, use double the measure wire length.

12 Volt Amps	*WIRE GAUGE REQUIRED—BY LENGTH OF CIRCUIT IN FEET*														
	3′	5′	7′	10′	15′	20′	25′	30′	40′	50′	60′	70′	80′	90′	100′
1	20	20	20	20	20	20	20	20	20	20	20	20	20	20	20
1.5	20	20	20	20	20	20	20	20	20	20	20	20	18	18	18
2	20	20	20	20	20	20	20	20	20	20	18	18	16	16	16
3	20	20	20	20	20	20	20	20	18	18	16	16	14	14	14
4	20	20	20	20	20	20	20	18	16	16	14	14	14	14	12
5	20	20	20	20	20	20	18	18	16	14	14	14	12	12	12
6	20	20	20	20	20	18	18	16	14	14	14	12	12	12	10
7	20	20	20	20	20	18	16	16	14	14	12	12	12	10	10
8	20	20	20	20	18	16	16	14	14	12	12	12	10	10	10
10	20	20	20	20	18	16	14	14	12	12	10	10	10	10	8
12	20	20	20	18	16	14	14	14	12	10	10	10	8	8	8
15	20	20	20	18	16	14	12	12	10	10	10	10	8	8	6
20	20	20	18	16	14	12	12	10	10	8	8	8	6	6	6
24	20	18	16	14	14	12	10	10	8	8	8	6	6	6	4
30	18	16	16	14	12	10	10	10	8	8	6	6	6	4	4
36	16	14	14	14	12	10	10	8	8	6	6	4	4	4	4
50	14	14	14	12	10	8	8	8	6	4	4	4	2	2	2
100	14	12	10	8	8	6	6	4	4	2	2	1	0	0	2/0
150	12	10	8	6	6	4	4	2	2	1	0	2/0	2/0	3/0	3/0
200	10	8	8	6	4	2	2	2	1	0	2/0	3/0	4/0	4/0	4/0

— WHEN MECHANICAL STRENGTH IS A FACTOR, USE NEXT LARGER WIRE GAUGE

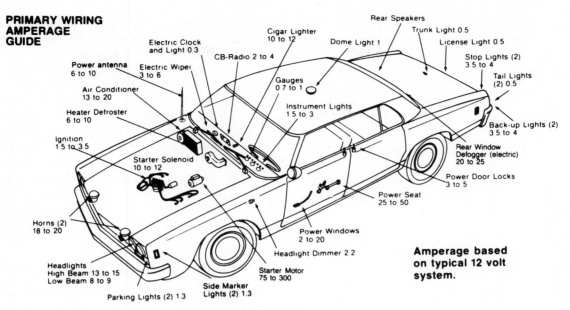

PRIMARY WIRING AMPERAGE GUIDE

Power antenna 6 to 10
Electric Wiper 3 to 6
Air Conditioner 13 to 20
Heater Defroster 6 to 10
Ignition 1.5 to 3.5
Starter Solenoid 10 to 12
Horns (2) 18 to 20
Headlights High Beam 13 to 15 Low Beam 8 to 9
Parking Lights (2) 1.3
Side Marker Lights (2) 1.3
Starter Motor 75 to 300
Headlight Dimmer 2.2
Power Windows 2 to 20
Power Seat 25 to 50
Electric Clock and Light 0.3
CB-Radio 2 to 4
Cigar Lighter 10 to 12
Gauges 0.7 to 1
Instrument Lights 1.5 to 3
Dome Light 1
Rear Speakers
Trunk Light 0.5
License Light 0.5
Stop Lights (2) 3.5 to 4
Tail Lights (2) 0.5
Back-up Lights (2) 3.5 to 4
Rear Window Defogger (electric) 20 to 25
Power Door Locks 3 to 5

Amperage based on typical 12 volt system.

FIGURE 6-3

all the trailer lights, but if the wire required is over 25 ft long, 12-gauge wire should be used. If the length is over 50 ft the wire size should be increased further, to 10 gauge, to prevent excessive voltage drops due to the connecting wires. Most automotive wire, except for spark plug wire, is often called *primary wire* because it is designed to operate at or near battery voltage (named for the voltage range used in the primary ignition circuit). See Figure 6-3.

SOLDERING AND WIRE CONNECTORS

Solderless connectors can also be used to connect two wires. Special crimping pliers are available to provide the necessary high-pressure crimps (two crimps on terminal connectors, four crimps on butt connectors). See Figure 6-4 for wire repair procedures.

ALUMINUM WIRE REPAIR

Since the mid-1970s, many automobile manufacturers have used plastic-coated solid aluminum wire for some body wiring. Since aluminum wire is brittle and could break due to vibration, it is only used where there is no possible movement of the wire, such as along the floor or sill area. This section of wire is stationary and changes back to copper wire at a junction terminal after the wiring reaches the trunk or rear section of the vehicle, where movement of the wiring may be possible.

If any aluminum wire must be repaired or replaced,

the following procedure should be used to be assured of a proper repair. The aluminum wire is usually found protected in a plastic conduit. This conduit is normally slit and the wires can easily be removed for repair.

FIGURE 6-4 *Wire repair procedure.*

STEP 1   Remove insulation as required.

STEP 2   Twist wires

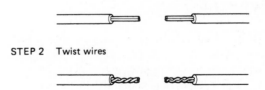

STEP 3   Twist wires together to form a strong mechanical connection that can't be pulled apart.

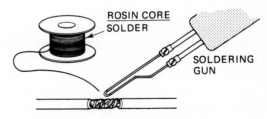

ROSIN CORE SOLDER
SOLDERING GUN

USE SOLDERING GUN TO HEAT WIRE (NOT THE SOLDER). TOUCH SOLDER TO HOT TWISTED WIRE TO MELT SOLDER.

STEP 4   Use vinyl electrical tape or shrink tubing to insulate and protect splice.

HINT: If splice is located where physical damage may cut through insulation, slip a length of rubber vacuum hose over one wire before connecting the two wires. After soldering, slip hose over splice and tape in place.

Step 1. Carefully strip only about ¼ in. (0.6 cm) from the aluminum wire, being careful not to nick or damage the aluminum wire case.

Step 2. Use a crimp connector to join two wires together. Do *not* solder an aluminum wire repair. Solder will not readily adhere to aluminum because the heat causes an oxide coating on the surface of the aluminum.

Step 3. The spliced, crimped connection must be coated with petroleum jelly to prevent corrosion.

Step 4. The coated connection should then be covered with shrinkable plastic tubing (installed over the wire before crimping the connector) or wrapped with electrical tape to seal out moisture and insulate the splice.

BATTERY CABLES

Battery cables are the largest wires used in the automotive electrical system. The cable gauge numbers are usually 4-gauge, 2-gauge, or 1-gauge wires. Wires larger than 1 gauge are called 0 gauge (pronounced ought). Larger cables are labeled 2/0 (called double ought) and 3/0 (triple ought). Six-volt electrical systems require battery cables two sizes larger than those for 12-V electrical systems.

JUMPER CABLES

Jumper cables are 4 to 2/0 gauge electrical cables with large clamps attached used to connect a vehicle that has a discharged battery to a vehicle that has a good battery. Good-quality jumper cables are necessary to prevent excessive voltage drops due to the cable's resistance. Aluminum wire jumper cables should not be used because even though aluminum is a good electrical conductor (although not as good as copper), it is less flexible and can crack and break when bent or moved repeatedly. The size should be 4 gauge or larger. "Ought"-gauge welding cable can be used to construct an excellent set of jumper cables using welding clamps on both ends. Welding cable is usually constructed of many very fine strands of wire which allow for easier bending of the cable because the strands of fine wire can slide against each other inside the cable.

FUSES

Fuses should be used in every circuit to protect the wiring from overheating and damage caused by excessive current flow as a result of a short circuit or other malfunction.

The symbol for a fuse is a wavy line between two points: ◠◡. A fuse is constructed of a fine tin conductor inside a glass, plastic, or ceramic housing. The tin is designed to melt and open the circuit if excessive current flows through the fuse. Each fuse is rated according to its maximum current-carrying capacity.

Many fuses are used to protect more than one circuit of the automobile. See Figure 6-5. A typical example is the fuse for the cigar lighter. The cigar lighter fuse may also be used to protect the courtesy lights, clock, and other circuits. Therefore, a fault in one circuit could cause the fuse to melt, which will prevent the operation of all other circuits that are protected by the fuse.

WHAT SIZE FUSE SHOULD BE USED?

The maximum anticipated current flow through a circuit should be approximately 80% of the fuse rating for that circuit. For example, if an automotive circuit has a maximum designed current of 12 A, a 15-A fuse should be used to protect the circuit (80% of 15 A is 12 A: 15 × 0.80 = 12). Circuit breakers also usually adhere to the 80% rule.

STANDARD FUSES

Standard glass tube or ceramic fuses are rated according to maximum current and size. Typical glass-tube fuses are generally all ¼ in. in diameter with lengths of 1¼, 1, ⅞, ¾, and ⅝ in. There are many different sizes and time-delay factors designed into each of the various manufacturers' fuses. Fuses manufactured under the standards established by the Society of Fuse Engineers (SFE) vary in length according to amperage rating. SFE fuses usually are longer as amperage rating increases. See Figure 6-6. The fuses manufactured by the Bussmann Manufacturing Division of McGraw-Edison Company are the same length for each type regardless of amperage rating. Some common Buss fuse types include:

AGA	formerly 1AG
AGW	formerly 7AG
AGC	formerly 3AG
AGY	formerly 9AG
AGX	formerly 8AG

The letters used are a Bussmann identification code and have no other meaning. Certain fuses are designed to

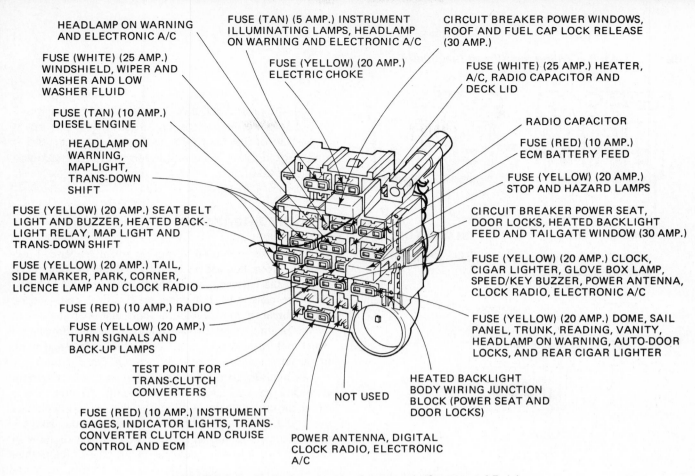

HEADLAMP ON WARNING AND ELECTRONIC A/C

FUSE (WHITE) (25 AMP.) WINDSHIELD, WIPER AND WASHER AND LOW WASHER FLUID

FUSE (TAN) (10 AMP.) DIESEL ENGINE

HEADLAMP ON WARNING, MAPLIGHT, TRANS-DOWN SHIFT

FUSE (YELLOW) (20 AMP.) SEAT BELT LIGHT AND BUZZER, HEATED BACK-LIGHT RELAY, MAP LIGHT AND TRANS-DOWN SHIFT

FUSE (YELLOW) (20 AMP.) TAIL, SIDE MARKER, PARK, CORNER, LICENCE LAMP AND CLOCK RADIO

FUSE (RED) (10 AMP.) RADIO

FUSE (YELLOW) (20 AMP.) TURN SIGNALS AND BACK-UP LAMPS

TEST POINT FOR TRANS-CLUTCH CONVERTERS

FUSE (RED) (10 AMP.) INSTRUMENT GAGES, INDICATOR LIGHTS, TRANS-CONVERTER CLUTCH AND CRUISE CONTROL AND ECM

FUSE (TAN) (5 AMP.) INSTRUMENT ILLUMINATING LAMPS, HEADLAMP ON WARNING AND ELECTRONIC A/C

FUSE (YELLOW) (20 AMP.) ELECTRIC CHOKE

NOT USED

POWER ANTENNA, DIGITAL CLOCK RADIO, ELECTRONIC A/C

CIRCUIT BREAKER POWER WINDOWS, ROOF AND FUEL CAP LOCK RELEASE (30 AMP.)

FUSE (WHITE) (25 AMP.) HEATER, A/C, RADIO CAPACITOR AND DECK LID

RADIO CAPACITOR

FUSE (RED) (10 AMP.) ECM BATTERY FEED

FUSE (YELLOW) (20 AMP.) STOP AND HAZARD LAMPS

CIRCUIT BREAKER POWER SEAT, DOOR LOCKS, HEATED BACKLIGHT FEED AND TAILGATE WINDOW (30 AMP.)

FUSE (YELLOW) (20 AMP.) CLOCK, CIGAR LIGHTER, GLOVE BOX LAMP, SPEED/KEY BUZZER, POWER ANTENNA, CLOCK RADIO, ELECTRONIC A/C

FUSE (YELLOW) (20 AMP.) DOME, SAIL PANEL, TRUNK, READING, VANITY, HEADLAMP ON WARNING, AUTO-DOOR LOCKS, AND REAR CIGAR LIGHTER

HEATED BACKLIGHT BODY WIRING JUNCTION BLOCK (POWER SEAT AND DOOR LOCKS)

FIGURE 6-5 *Typical automotive fuse panel. (Courtesy of Buick Motor Division, GMC.)*

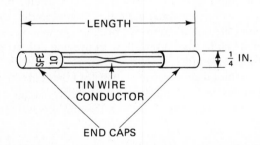

LENGTH

¼ IN.

TIN WIRE CONDUCTOR

END CAPS

FIGURE 6-6 *Fuses are all the same diameter, but the length varies depending on the type of fuse.*

allow an overload for varying lengths of time. Some fuses are designed to delay "blowing" to allow for the starting current, usually accompanying the starting of electric motors such as the blower motor or the windshield wiper motor. Other fuses are designed to be "quick blow" and immediately open the circuit due to any current flow higher than the fuse rating. Because each fuse is selected to meet a wide range of operating conditions, temperatures, shocks, and vibrations, it is important to replace a blown fuse with the correct type (for example, SFE, AGA, AGC) and amperage rating.

BLADE FUSES

Colored blade-type fuses have been used since 1977. The color of the plastic of blade fuses indicates the maximum current flow measured in amperes. The following chart lists the color and the amperage rating.

Amperage Rating	Color
1	Dark green
2	Gray
2.5	Purple
3	Violet
4	Pink
5	Tan
6	Gold
7.5	Brown
9	Orange
10	Red
14	Black
15	Light blue
20	Yellow
25	White
30	Light green

All blade fuses are the same size regardless of amperage rating. See Figure 6-7.

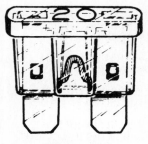 

GOOD FUSE   BLOWN FUSE

FIGURE 6-7 *Blade fuse. (Courtesy of Buick Motor Division, GMC.)*

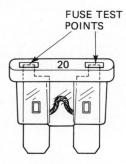

FUSE TEST POINTS

FIGURE 6-8

TESTING FUSES

It is important to test the condition of a fuse if the circuit being protected by the fuse does not operate. Most blown fuses can be detected quickly because the center conductor is melted. Fuses can also fail and open the circuit because of a poor connection in the fuse itself or in the fuse holder. Therefore, just because a fuse "looks okay" does not mean that it *is* okay. All fuses should be tested with a test light. The test light should be connected to first one side of the fuse and then the other. A test light should "light" on both sides. See Figure 6-8.

THE 7-HOUR FUSE TEST

A technician spent 7 hours troubleshooting an older-model Toyota on which the "charge" light remained on whenever the engine was running. Other technicians had replaced both the alternator and the voltage regulator. After hours of troubleshooting, the problem was discovered to be the fuse for the charging circuit. The fuse had been checked by removing the fuse and checking it with an ohmmeter. It did have continuity. However, when placed back in the circuit, the fuse would not conduct enough amperes for the circuit to operate. The fuse was corroded

---

inside the end caps, yet looked perfect on the outside. Replacing the fuse restored proper operation of the alternator, voltage regulator, and charge light. *Always* test fuses in the fuse panel with a test light.

CIRCUIT BREAKERS

Circuit breakers are used to prevent harmful overload (excessive current flow) in a circuit by opening the circuit and stopping the current flow to prevent overheating and possible fire caused by hot wires or electrical components. Circuit breakers are mechanical units that open the circuit by the heating effect of two different metals (bimetallic) which deform and open a set of contact points that work in the same manner as an "off" switch. See Figure 6-9.

Circuit breakers, therefore, are reset when the current stops flowing, which causes the bimetallic strip to cool and closes the circuit again. A circuit breaker is used in circuits that could affect the safety of the passengers if a conventional nonresetting fuse were used. The headlight circuit is an excellent example of the use of a circuit breaker rather than a fuse. A short or grounded circuit anywhere in the headlight circuit could cause excessive current flow, and therefore, the opening of the circuit. A sudden loss of headlights at night could have disastrous results. A circuit breaker, however, would open and close the circuit rapidly, thereby protecting the circuit from overheating and also being able to provide sufficient current flow to provide at least partial headlight operation.

Circuit breakers are also used for other circuits where conventional fuses could not provide for the surges of high current commonly found in the following accessories:

1. Power seats
2. Power door locks
3. Power windows

FUSIBLE LINKS

A fusible link is a type of fuse that consists of a short length of standard copper-strand wire covered with a special nonflammable insulation. This wire is usually four wire sizes smaller than the wire of the circuits it protects. The special thick insulation over the wire may make the wire look larger than other wires of the same gauge number. See Figure 6-10.

If excessive current flow (caused by a short to ground or a defective component) occurs, the fusible link will melt in half and open the circuit to prevent a fire hazard. Some fusible links are identified with tags at the junction between the fusible link and the standard chassis wiring. These tags

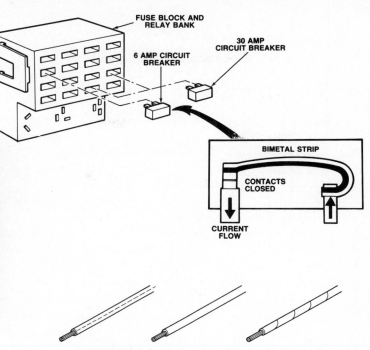

FIGURE 6-9 *Blade circuit breaker. (Courtesy of Chrysler Corporation.)*

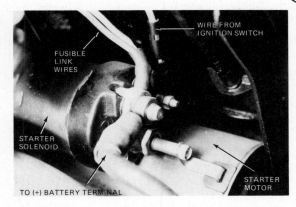

FIGURE 6-10 *Fusible links conduct electrical current from the battery or starter terminal (as illustrated) to the fuse panel. Notice that two wires are used. Each wire is four gauge sizes smaller (higher number) than the size used in the circuits they are protecting.*

FUSIBLE LINK INSTALLATION

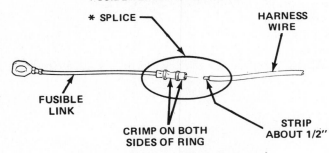

* **If original splice had 2 harness wires then 2 fusible links must be installed.**

FIGURE 6-11 *Fusible link replacement. (Courtesy of Buick Motor Division, GMC.)*

are labeled "fusible link" and represent only the junction. Fusible links are the backup system for circuit protection. All current except the current used by the starter motor flows through fusible links and then through individual circuit fuses. It is possible that a fusible link will melt and not blow a fuse. Fusible links are installed as close to the battery as possible so that they can protect the wiring and circuits coming directly from the battery. See Figure 6-11.

CIRCUIT INFORMATION

Many wiring diagrams contain numbers and letters near components and wires that may cause confusion to readers of the diagram. Most letters used near or on a wire are used to identify the color or colors of the wire. The first

FIGURE 6-12 *Solid color wire. For example, BRN = brown with tracer (either the left or right example) could be labeled BRN/WHT, indicating a brown wire with a white tracer.*

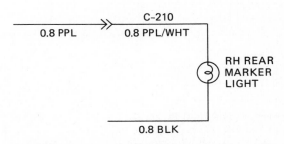

FIGURE 6-13 *Typical section of a wiring diagram.*

color or color abbreviation is the color of the wire insulation, and the second color (if mentioned) is the color of the strip or tracer on the base color. See Figure 6-12.

Wires with different color tracers are indicated by both colors with a slash (/) between them. For example, BRN/WHT means a brown wire with a white strip or tracer.

Figure 6-13 illustrates a rear-side marker bulb circuit.

Colors

BRN	=	brown	YEL	=	yellow
BLK	=	black	ORN	=	orange
GRN	=	green	DK. BLU	=	dark blue
WHT	=	white	LT. BLU	=	light blue
PPL	=	purple	DK. GRN	=	dark green
PNK	=	pink	LT. GRN	=	light green
TAN	=	tan	RED	=	red
BLU	=	blue	GRY	=	gray

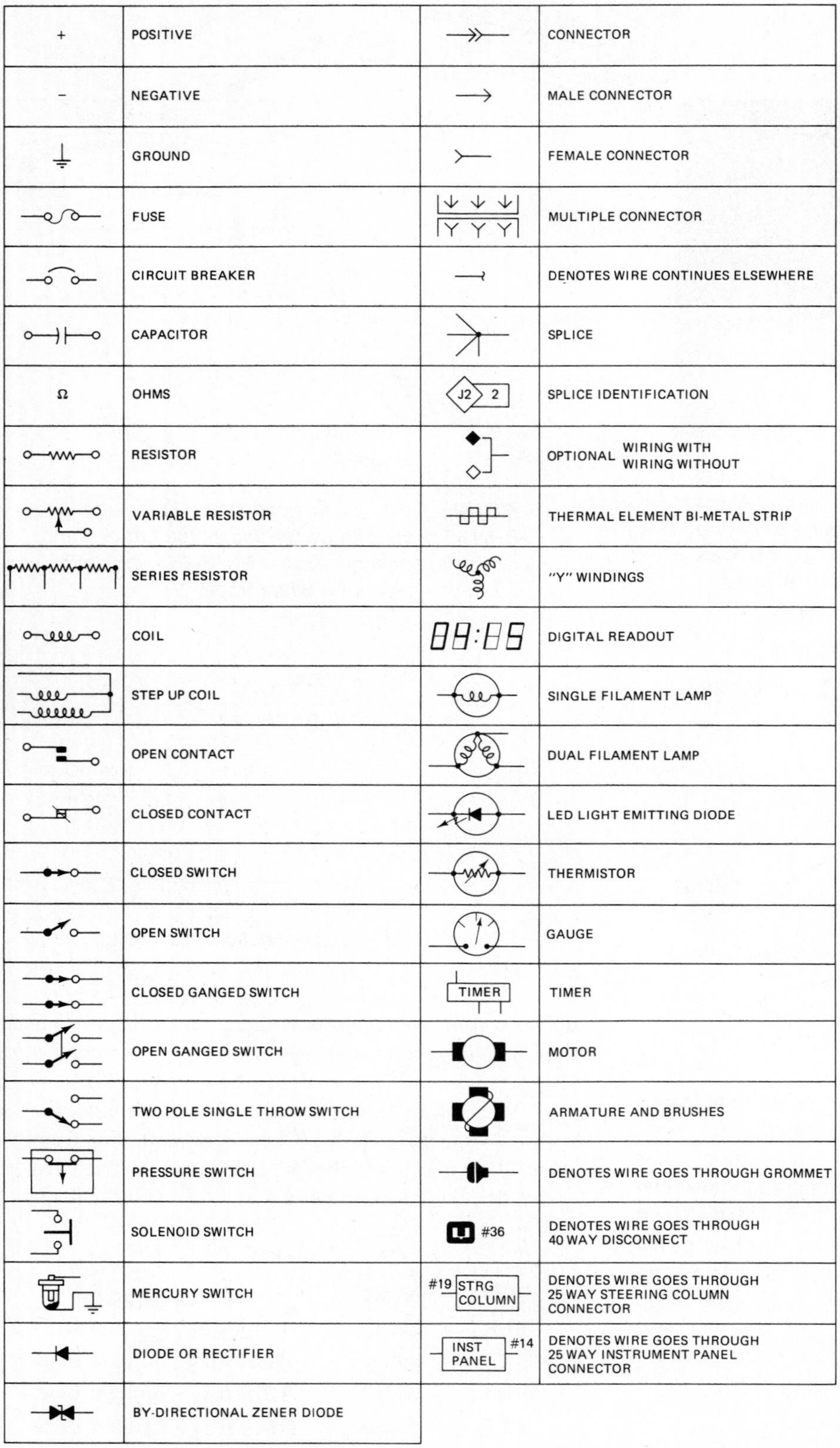

Symbol	Name	Symbol	Name
+	POSITIVE	⟶≫	CONNECTOR
–	NEGATIVE	⟶	MALE CONNECTOR
⏚	GROUND	⟩—	FEMALE CONNECTOR
—o⌢o—	FUSE	↓ ↓ ↓ / Y Y Y	MULTIPLE CONNECTOR
—o o—	CIRCUIT BREAKER	⟶	DENOTES WIRE CONTINUES ELSEWHERE
o—||—o	CAPACITOR	⟶	SPLICE
Ω	OHMS	J2 ⟩ 2	SPLICE IDENTIFICATION
o—www—o	RESISTOR	◆ / ◇	OPTIONAL WIRING WITH / WIRING WITHOUT
o—www—o	VARIABLE RESISTOR	⊓⊔⊓	THERMAL ELEMENT BI-METAL STRIP
wwwww	SERIES RESISTOR	ℓℓℓ	"Y" WINDINGS
o—ℓℓℓ—o	COIL	09:05	DIGITAL READOUT
STEP UP COIL	STEP UP COIL	⟨ℓℓℓ⟩	SINGLE FILAMENT LAMP
o—▪—o	OPEN CONTACT	⟨ℓℓ⟩	DUAL FILAMENT LAMP
o—▨—o	CLOSED CONTACT	⟶⊲	LED LIGHT EMITTING DIODE
—•o•—	CLOSED SWITCH	⟨www⟩	THERMISTOR
—•⁄ o—	OPEN SWITCH	⟨ ⟩	GAUGE
CLOSED GANGED SWITCH	CLOSED GANGED SWITCH	TIMER	TIMER
OPEN GANGED SWITCH	OPEN GANGED SWITCH	—●—	MOTOR
TWO POLE SINGLE THROW SWITCH	TWO POLE SINGLE THROW SWITCH	✦	ARMATURE AND BRUSHES
PRESSURE SWITCH	PRESSURE SWITCH	—◀▮—	DENOTES WIRE GOES THROUGH GROMMET
SOLENOID SWITCH	SOLENOID SWITCH	▪ #36	DENOTES WIRE GOES THROUGH 40 WAY DISCONNECT
MERCURY SWITCH	MERCURY SWITCH	#19 STRG COLUMN	DENOTES WIRE GOES THROUGH 25 WAY STEERING COLUMN CONNECTOR
—◀⊢	DIODE OR RECTIFIER	INST PANEL #14	DENOTES WIRE GOES THROUGH 25 WAY INSTRUMENT PANEL CONNECTOR
—◀▶⊢	BY-DIRECTIONAL ZENER DIODE		

FIGURE 6-14 *Standardized electrical and electronic component signs and symbols. (Courtesy of Chrysler Corporation.)*

In ".8 PPL," the ".8" indicates the metric wire gauge size in square millimeters and "PPL" indicates a solid purple wire.

The circuit illustrated in Figure 6-13 shows the color of the wire that changes at the number C210. This stands for "connection number 210" and is used for reference purposes. The symbol for the connection can vary as illustrated depending on the manufacturer. The color change from purple (PPL) to purple with a white tracer (PPL/WHT) is not important except for knowing where the wire changes color in the circuit. The wire gauge has remained the same on both sides of the connection (0.8 mm² or 18 gauge). The ground circuit is the ".8 BLK" wire.

CIRCUIT DIAGRAMS

A circuit diagram is a representation of actual electrical or electronic components using signs and symbols. These symbols (see Figure 6-14) are standardized by DIN (Deutsche Industrie Norm = German Industrial Standard) and the IEC (International Electrotechnical Commission).

ACTUAL DRAWING OF A COMPONENT AND ITS SYMBOL

A simple circuit consisting of a battery, light bulb, and connecting wires is illustrated in Figure 6-15. This shows how each component actually looks, with proper shape and detail.

Instead of drawing a complete battery which contains cells made from positive (+) and negative (−) plates, a drawing often uses just the symbol for a battery. Therefore, the same light bulb as drawn in Figure 6-15 could also be illustrated as shown in Figure 6-16. The battery illustrated shows six cells, with each cell (2 V each cell) represented

FIGURE 6-15 *Symbol drawing of a simple circuit.*

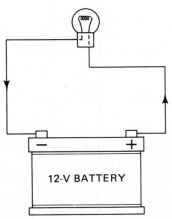

12-V BATTERY

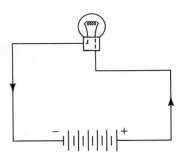

FIGURE 6-16 *The same circuit as Figure 6-15, except the battery cells are indicated.*

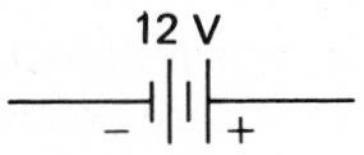

by (1/) indicating one negative plate and one positive plate. The *shorter* line indicates the negative (−) plate and the larger line indicates the positive (+) plate. However, even this illustration could get complicated and large for high-voltage batteries. Instead of drawing each cell of a battery, most diagrams indicate a battery with only a few cells, but label the voltage of the battery as indicated in Figure 6-17.

The light bulb shown in Figure 6-16 is also difficult to draw and would require a lot of space on a page. A light bulb "works" because of the resistance of the wire inside the bulb; the wire becomes hot and produces light. This resistance wire is called a *filament*. Many automotive bulbs contain two (dual) filaments in one bulb.

Since the wiring diagram is actually "shorthand" for electrical circuits, only the filament (resistance) is shown in Figure 6-18. This drawing does not look anything like the first drawing of a battery and a bulb (Figure 6-15). However, it represents the identical simple circuit. There can still be some reduction in the size and looks of this circuit to reduce the number of connecting wires. This is important because in large circuits with many wires and connections, a wiring diagram could get very crowded as each additional component is added to the drawing. The

12 V

FIGURE 6-17 *The battery cell symbol is shortened to indicate only two cells.*

FIGURE 6-18 *Simple circuit of a light bulb and a battery, with the R₁ symbol replacing the drawing of the light bulb.*

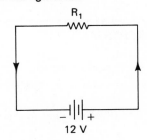

12 V

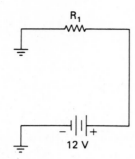

FIGURE 6-19 *Simple circuit showing ground symbols instead of the return path wire.*

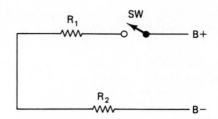

FIGURE 6-20 *Simple circuit showing a switch (SW), two bulbs (R_1 and R_2), and connections to a battery.*

last major reduction in the simple light bulb circuit shown in Figure 6-18 is to eliminate the ground return wire and replace it with the symbol ⏚ indicating that the current will flow back to the negative (−) side of the battery through the metal vehicle frame and/or body. The negative (−) side of the battery is also connected to the metal block and therefore is also drawn with the ground symbol connected to the battery. Either ⏚ or ⏚ may be found to represent ground on automotive wiring diagrams. In Figure 6-19, the simple light bulb circuit is reduced to symbols alone.

Since the resistance symbol could represent many different bulbs in an automotive circuit, most wiring diagrams abbreviate the exact location and purpose of the bulb. The

battery is often eliminated from the diagram and labeled so that the power side goes to the positive (+) of the battery. Often the term "B+" is used on drawings to indicate the positive (+) of the battery and "B−" is used to indicate the negative (−) of the battery.

In Figure 6-20 there is a switch added to the single bulb circuit. Figure 6-21 is an example of a typical wiring diagram, including a drawing of the actual component. Note the use of a pictorial rather than the symbol of a battery, starter, and alternator.

Figure 6-22 illustrates a complete electrical diagram. Complete body lighting diagrams can be understood easily if the reader looks carefully and traces each circuit from the power source through the bulb and back to the power source through the ground return path.

SUMMARY

1. All circuits *must* be protected by fusible links, circuit breakers, or fuses.
   (a) Fusible links are often smaller-gauge wires than those used for the rest of the circuit, causing them to become hot and melt in the event of a serious short circuit.
   (b) Circuit breakers are electrical units that can open an electrical circuit if excessive current flows through the circuit. A circuit breaker can be reset or will reset itself when it cools.
   (c) A fuse contains a tin strip that will melt and open the circuit if excessive current flow, above the rating of the fuse, occurs. All fuses should be tested using a test light.
2. All circuit diagrams illustrate wire-gauge number, wire color, circuit number, and other circuit information. By checking a wiring diagram, other electrical components which may share the same components, fuse, or ground connection can be identified. An electrical malfunction can be traced to its source.

STUDY QUESTIONS

6-1. Explain the AWG wire size system.

6-2. Explain the proper procedure for making an aluminum wire repair.

6-3. Explain the 80% rule regarding fuse and circuit breaker protection.

6-4. Explain the proper procedure for testing fuses.

6-5. What is a fusible link?

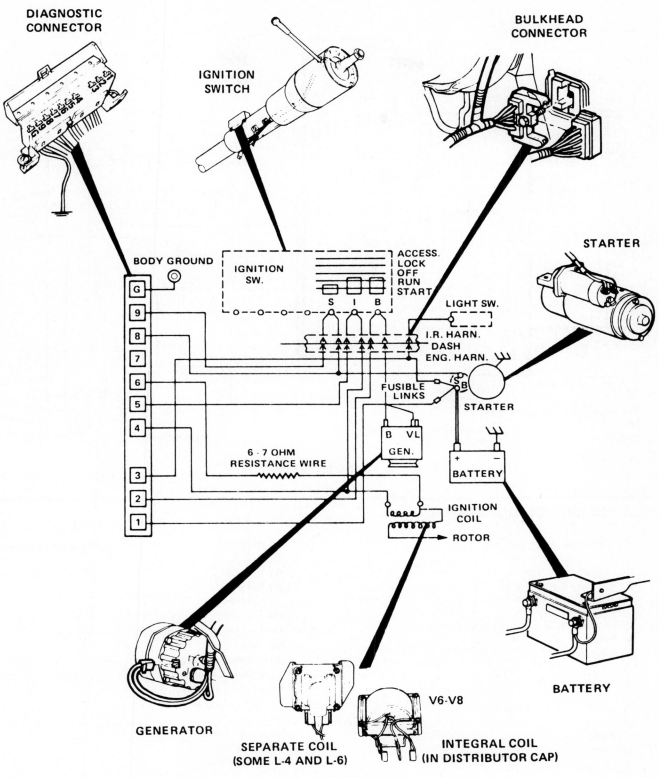

DIAGNOSTIC CONNECTOR

IGNITION SWITCH

BULKHEAD CONNECTOR

STARTER

BODY GROUND

IGNITION SW.

ACCESS.
LOCK
OFF
RUN
START

S I B

LIGHT SW.

I.R. HARN.
DASH
ENG. HARN.

FUSIBLE LINKS

STARTER

B VL
GEN.

6 - 7 OHM
RESISTANCE WIRE

BATTERY

IGNITION COIL

ROTOR

BATTERY

GENERATOR

**SEPARATE COIL
(SOME L-4 AND L-6)**

V6-V8

**INTEGRAL COIL
(IN DISTRIBUTOR CAP)**

IGNITION COIL

FIGURE 6-21

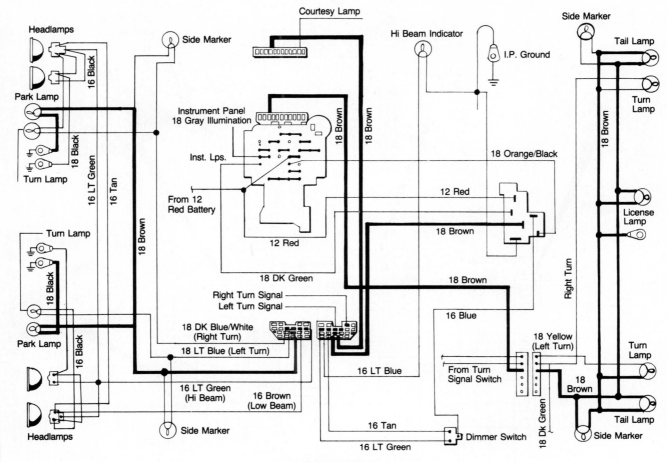

FIGURE 6-22

MULTIPLE-CHOICE QUESTIONS

6-1. The higher the gauge number:
(a) the smaller the wire.
(b) the larger the wire.
(c) the same the wire size.
(d) the thicker the wire.

6-2. Metric wire size is measured in units of:
(a) meters.
(b) cubic centimeters.
(c) liters.
(d) square millimeters.

6-3. The metal in fuses is made from:
(a) tin.
(b) lead.
(c) copper.
(d) silver.

6-4. Jumper cables should be made from:
(a) copper, at least 8 gauge.
(b) aluminum or copper, at least 6 gauge.
(c) copper, at least 4 gauge.
(d) none of the above.

6-5. 18-gauge wire is illustrated in the metric system on a wiring diagram as:
(a) 0.5 mm².
(b) 0.8 mm².
(c) 1.0 mm².
(d) 2.0 mm².

6-6. Most automotive "primary" wire is made from:
(a) solid copper.
(b) stranded copper.
(c) solid aluminum.
(d) stranded aluminum.

6-7. All fuses should be tested with:
(a) a fuse tester.
(b) an ohmmeter.
(c) a voltmeter.
(d) a test light.
(e) (a) and (d) only.

6-8. Blade fuses are:
(a) all the same amperage rating.
(b) color coded according to amperage rating.
(c) color coded according to time delay.
(d) none of the above.

7

Lighting and Signaling Circuits

THIS CHAPTER INTRODUCES THE OPERATION AND TROUBLESHOOTING OF lighting and signaling circuits found on all cars. The proper operation of these circuits is critical for safe vehicle operation. The topics covered in this chapter include:

1. Bulb trade numbers
2. Headlight circuit operation and troubleshooting
3. Brake lights
4. Directional (turn) signal operation and troubleshooting
5. Hazard flasher operation
6. Courtesy lights
7. Illuminated entry
8. Fiber optics

TYPICAL AUTOMOTIVE LIGHT BULBS

Trade Number	Design Volts	Design Amperes	Watts: $P = I \times E$	Candlepower	Average Life (hrs)
37	14.0	0.09	1.3	0.5	2500
37E	14.0	0.09	1.3	0.5	2500
51	7.5	0.22	1.7	1.0	1000
53	14.4	0.12	1.7	1.0	1000
55	7.0	0.41	2.9	2.0	500
57	14.0	0.24	3.4	2.0	500
57X	14.0	0.24	3.4	2.0	500
63	7.0	0.63	4.4	3.0	1000
67	13.5	0.59	8.0	4.0	5000 +
68	13.5	0.59	8.0	4.0	5000 +
70	14.0	0.15	2.1	1.5	100
73	14.0	0.08	1.1	0.3	7000
74	14.0	0.10	1.4	0.7	1000
81	6.5	1.02	6.6	6.0	500
88	13.0	0.58	7.5	6.0	750
89	13.0	0.58	7.5	6.0	750
90	13.0	0.58	7.5	6.0	750
93	12.8	1.04	13.3	15.0	700
94	12.8	1.04	13.3	15.0	700
158	14.0	0.24	3.4	2.0	500
161	14.0	0.19	2.7	1.0	4000
168	14.0	0.35	4.9	3.0	1500
192	13.0	0.33	4.3	3.0	1000
194	14.0	0.27	3.8	2.0	2500
194E-1	14.0	0.27	3.8	2.0	2500
194NA	14.0	0.27	3.8	1.5	2500
209	6.5	1.78	11.6	15.0	100
211-2	12.8	0.97	12.4	12.0	1000
212-2	13.5	0.74	10.0	6.0	2000
214-2	13.5	0.52	7.0	4.0	1000
561	12.8	0.97	12.4	12.0	1000
562	13.5	0.74	10.0	6.0	2000
563	13.5	0.52	7.0	4.0	1000
631	14.0	0.63	8.8	6.0	1000
880	12.8	2.10	27.0	43.0	300
881	12.8	2.10	27.0	43.0	300
906	13.0	0.69	9.0	6.0	1000
912	12.8	1.00	12.8	12.0	1000
1003	12.8	0.94	12.0	15.0	200
1004	12.8	0.94	12.0	15.0	200
1034	12.8	1.80/0.59	23.0/7.6	32.0/3.0	200/5000 +
1073	12.8	1.80	23.0	32.0	200
1076	12.8	1.80	23.0	32.0	200
1129	6.4	2.63	16.8	21.0	200
1133	6.2	3.91	24.2	32.0	200
1141	12.8	1.44	18.4	21.0	1000
1142	12.8	1.44	18.4	21.0	1000
1154	6.4	2.63/.75	16.8/4.5	21.0/3	200/1000
1156	12.8	2.10	26.9	32.0	1200
1157	12.8	2.10/0.59	26.9/7.6	32.0/3.0	1200/5000 +
1157A	12.8	2.10/0.59	26.9/7.6	32.0/3.0	1200/5000 +
1157NA	12.8	2.10/0.59	26.9/7.6	32.0/3.0	1200/5000 +
1176	12.8	1.34/0.59	17.2/7.6	21.0/6.0	300/1500
1195	12.5	3.00	37.5	50.0	300
1196	12.5	3.00	37.5	50.0	300
1445	14.4	0.135	1.9	0.7	2000
1816	13.0	0.33	4.3	3.0	1000
1889	14.0	0.27	3.8	2.0	2000
1891	14.0	0.24	3.4	2.0	500
1892	14.4	0.12	1.7	0.75	1000

TYPICAL AUTOMOTIVE LIGHT BULBS *(cont'd)*

Trade Number	Design Volts	Design Amperes	Watts: $P = I \times E$	Candlepower	Average Life (hrs)
1893	14.0	0.33	4.6	2.0	7500
1895	14.0	0.27	3.8	2.0	2000
2057	12.8	2.10/0.48	26.9/6.1	32.0/2.0	1200/5000
2057NA	12.8	2.10/0.48	26.9/6.1	32.0/2.0	1200/5000
P25–1	13.5	1.86	25.1	36.6	250
P25–2	13.5	1.86	25.1	35.0	250
R19/5	13.5	0.37		4.0	400
R19/10	13.5	0.74		9.94	400
W10/3	13.5	0.25		1.75	1000

LIGHTING

Exterior lighting is controlled by the headlight switch, which is connected directly to the battery. Therefore, the lights can be left on and drain the battery. The headlight switch controls the following lights on most cars:

1. Headlights
2. Tail lights
3. Side marker lights
4. Front parking lights
5. Dash lights
6. Interior (dome) light(s)

BULB NUMBERS

The number used on automotive bulbs is called the bulb *trade number,* as recorded with the *American National Standards Institute* (ANSI), and the number is the same regardless of manufacturer. Amber-colored bulbs which use natural amber glass are indicated with an NA (meaning "natural amber") at the end of the number (for example, 1157NA). A less expensive amber bulb with a painted glass bulb is labeled with only the letter A for amber (for example, 1157A).

The trade number also identifies the size, shape, number of filaments, and amount of light produced. The amount of light produced is measured in *candlepower.* For example, the candlepower of an 1156 bulb commonly used for backup lights is 32. A 194 bulb, commonly used for dash or side marker lights, is rated at only 2 candlepower. The amount of light produced by a bulb is determined by the resistance of the filament wire, which also affects the amount of current (in amperes) required by the bulb.

It is important that the correct trade number bulb always be used for replacement to prevent circuit or component damage. See Figure 7-1 for a typical service manual bulb chart as specified for one car.

HEADLIGHT SWITCHES

The headlight switch operates the exterior and interior lights of most cars. The headlight switch is connected directly to the battery through a fusible link and has power to it at all times. This is called "hot all the time." The interior dash lights can be dimmed manually by rotating the headlight switch knob, which controls a variable resistor (called a *rheostat*) built into the headlight switch. See Figure 7-2.

The rheostat drops the voltage sent to the dash lights. Whenever there is a voltage drop (resistance), there is heat. This coiled resistance wire is built into a ceramic holder which is designed to insulate the heat from the rest of the switch and allow heat to escape. Continual driving with the dash lights dimmed can result in the headlight switch knob getting hot to the touch. This is normal and the best prevention is to increase the brightness of the dash lights to reduce the amount of heat generated in the switch. The headlight switch also contains a built-in circuit breaker that will rapidly turn the headlights on and off in the event of a short circuit. This prevents a total loss of headlights. If the headlights are rapidly flashing on and off, check the entire headlight circuit for possible shorts. The circuit breaker controls only the headlights. The other lights controlled by the headlight switch (taillights, dash lights, and parking lights) are fused separately. Flashing headlights may also be caused by a failure in the built-in circuit breaker, requiring replacement of the switch assembly.

REMOVING A HEADLIGHT SWITCH

Most dash-mounted headlight switches can be removed by first removing the dash panel. However, to get the dash panel off, the headlight switch knob usually has to be removed. Some knobs can be removed by depressing a small clip in a notch in the knob itself. See Figure 7-3. Other headlight switch knobs are removed by depressing a spring-loaded release, which allows for removal of the entire

CUTLASS BULB CHART

LAMP USAGE	QUANTITY	TRADE NO.	CANDLE POWER
EXTERNAL			
Back Up	2	1156	32
Front-Park-Turn (exc. Calais)	2	1157NA	1.5 & 24
Front Park-Turn (Calais Model)	2	2057	2.0 & 24
Headlamps, Dual— Low and High Beam	2	4652	50 & 60W
Headlamps, Dual— High Beam (Exc. Halogen)	2	4651	50W
Headlamps, Dual— High Beam (Halogen)	2	H4651	50W
License	2	194	2
Opera, Sail Panel	2	73	.03
Side Marker, Front	2	194	2
Side Marker, Rear	2	194	2
Tail-Stop-Turn (Coupe)	4	1157	3 & 32
Tail-Stop-Turn (Sedan)	2	1157	3 & 32
INTERIOR			
Ash Tray	1	1445	.7
Aux. Gen. Bulb— Diesel Only (Hidden)	1	168	3
Clock, Analog	2	194	2
Clock, Digital	1	168	3
Console Compartment	1	1891	2
Console, Shift Indicator	1	194	2
Courtesy, Front Floor— Right Side	1	89	6
Courtesy, Front Floor— Left Side	1	906	6
Courtesy, Sail Panel (Coupe with T-Tops)	2	562	16
Dome, (Standard)	1	561	12
Dome, (with Single Lens Reading Lamp)	1	562	6
Dome, (with Dual Lens Reading Lamps)	1	561	12
Dome, Rear— Station Wagon	1	561	12
Glove Box	1	1891	2
Heater and A/C Control	1	194	2
High Beam Indicator (Standard Cluster)	1	194	2
High Beam Indicator (Gage Package Cluster)	1	161	1
Radio Dial (Exc. Tapes)	1	216	1
Radio Dial (All Tapes)	1	1893	2
Reading Lamp, Single (with Combination Dome)	1	1004	15
Reading Lamps, Dual (with Combination Dome)	2	1004	15
Rear Window Defogger Switch Illum.	1	1893	2
Speedo Cluster (Standard)			
Upper Lighting	2	194	2
Lower Lighting	2	161	1
Speedo Cluster (with Gage Package)			
Upper Lighting	2	194	2
Gage Illumination	1	168	3
Fuel Gage Illumination	1	194	2
Lower Lighting and Shift Indicator	1	161	1
Telltale Indicator Lights			
"Oil/Choke"	1	161	1
"Temp."	1	161	1
"Fasten Belts"	1	161	1
"Water In Fuel"	1	194	2
"Brake"	1	161	1
"Charge"	1	168	3
"Check Engine"	1	161	1
"Wait"	1	161	1
Trunk Light	1	1003	15
Turn Signal Indicators (Standard Cluster)	2	194	2
Turn Signal Indicators (Gage Package Cluster)	2	161	1
Underhood Lamp	1	89	16
Visor Vanity Lighted Mirror	4	194	2

FIGURE 7-1 *Typical bulb usage chart. (Courtesy of Oldsmobile Division, GMC.)*

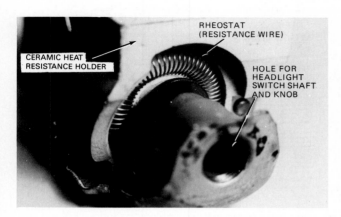

FIGURE 7-2 *Typical headlight switch. By rotating the headlight switch knob, the brightness of the dash lights can be varied.*

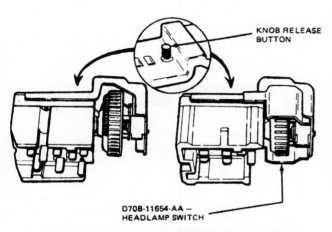

FIGURE 7-3 *Many headlight switch knobs can be removed by depressing a button on the top, side, or bottom of the headlight switch. (Courtesy of Ford Motor Company.)*

headlight switch knob and shaft. This will turn on the headlights. Therefore, to prevent a battery drain, disconnect the negative (−) battery cable before working on the headlight switch.

SEALED-BEAM HEADLIGHTS

Low-beam headlights contain two filaments: one for low beam and the other for high beam. High-beam headlights contain only one filament. Headlights are standardized so that they may be replaced by sealed-beam units that can be purchased at most auto-parts stores. Since low-beam headlights also contain a high-beam filament, the entire headlight assembly must be replaced if either filament is defective. See Figures 7-4 through 7-7.

A sealed-beam headlight can be tested with an ohmmeter. A good bulb should indicate low ohms between the

ground terminal and both power (''hot'')-side terminals. If either the high-beam or the low-beam filament is burned out, the ohmmeter will indicate infinity.

COMPOSITE HEADLIGHTS

Composite headlights are constructed using a replaceable bulb and a fixed lens cover that is part of the car. The replaceable bulbs are usually bright halogen bulbs. See Figure 7-8. Halogen bulbs get very hot during operation [between 500 and 1300°F (260 and 700°C)]. The glass of any halogen bulb must never be touched with bare fingers because the natural oils of the skin on the glass bulb can cause the bulb to break when it heats during normal operation.

HALOGEN SEALED-BEAM HEADLIGHTS

Halogen sealed-beam headlights are brighter and more expensive than normal headlights. Because of their extra brightness, it is common practice to have only two headlights on at any one time since the candlepower output would exceed the maximum U.S. federal standards when all four halogen headlights are on. Therefore, before trying to repair the problem that only two of the four lamps are on, check with the owner's manual or shop manual for proper operation.

> **HINT:** Often, halogen bulbs (replaceable bulbs and sealed beams) are equipped with both a bright and a low-intensity filament. If one filament burns out, the bulb (or sealed-beam) may be able to be switched since both bulbs are interchangeable.

DIMMER SWITCHES

The headlight switch controls the power or ''hot'' side of the headlight circuit. The current is then sent to the dimmer switch, which allows current to flow to either the high or the low filament of the headlight bulb. An indicator light also lights on the dash whenever the bright lights are selected.

The dimmer switch can be either foot-operated on the floor or hand-operated on the steering column. The popular steering column switches are actually attached to the *outside* of the steering column on most cars and are spring loaded. To replace most of these types of dimmer switches, the steering column needs to be lowered slightly to gain access to the switch itself, which is also adjustable for proper lever operation. See Figure 7-9. The dimmer switch used on the floor of many cars causes problems because of

HORIZONTAL 4-HEADLIGHT SYSTEM

VERTICAL 4-HEADLIGHT SYSTEM

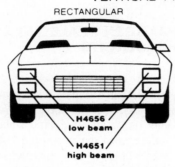

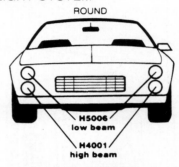

2-HEADLIGHT SYSTEM

THIS GE POWER PLUS™ HALOGEN HEADLAMP	REPLACES THIS STANDARD HEADLIGHT
H4001	4001 5001
H4651	4651
H4656	4652
H5006	4000
H6024 H6014	6014
H6054	6052

FIGURE 7-4   *Headlight trade numbers and locations. (Courtesy of General Electric Corporation.)*

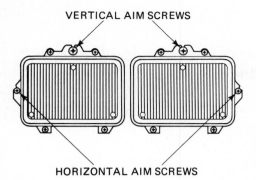

VERTICAL AIM SCREWS

HORIZONTAL AIM SCREWS

FIGURE 7-5 *Location of headlight aiming screws. (Courtesy of Pontiac Motor Division, GMC.)*

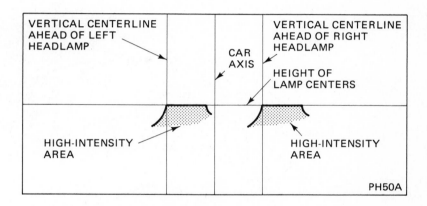

VERTICAL CENTERLINE AHEAD OF LEFT HEADLAMP

VERTICAL CENTERLINE AHEAD OF RIGHT HEADLAMP

CAR AXIS

HEIGHT OF LAMP CENTERS

HIGH-INTENSITY AREA

HIGH-INTENSITY AREA

PH50A

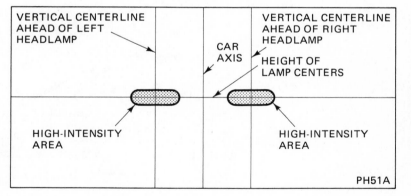

VERTICAL CENTERLINE AHEAD OF LEFT HEADLAMP

VERTICAL CENTERLINE AHEAD OF RIGHT HEADLAMP

CAR AXIS

HEIGHT OF LAMP CENTERS

HIGH-INTENSITY AREA

HIGH-INTENSITY AREA

PH51A

FIGURE 7-6 *The upper pattern represents what the headlights should look like 25 ft (7.6 m) from a wall on low beam. The lower pattern represents a properly aimed high-beam pattern. (Courtesy of Chrysler Corporation.)*

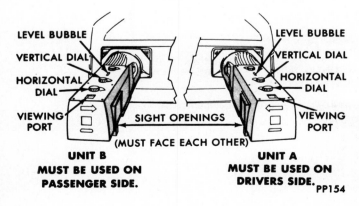

LEVEL BUBBLE

VERTICAL DIAL

HORIZONTAL DIAL

VIEWING PORT

SIGHT OPENINGS

LEVEL BUBBLE

VERTICAL DIAL

HORIZONTAL DIAL

VIEWING PORT

(MUST FACE EACH OTHER)

UNIT B MUST BE USED ON PASSENGER SIDE.

UNIT A MUST BE USED ON DRIVERS SIDE. PP154

FIGURE 7-7 *Typical mechanical headlight aimers. Always follow aiming equipment manufacturer's instructions for correct calibration and aiming procedures. (Courtesy of Chrysler Corporation.)*

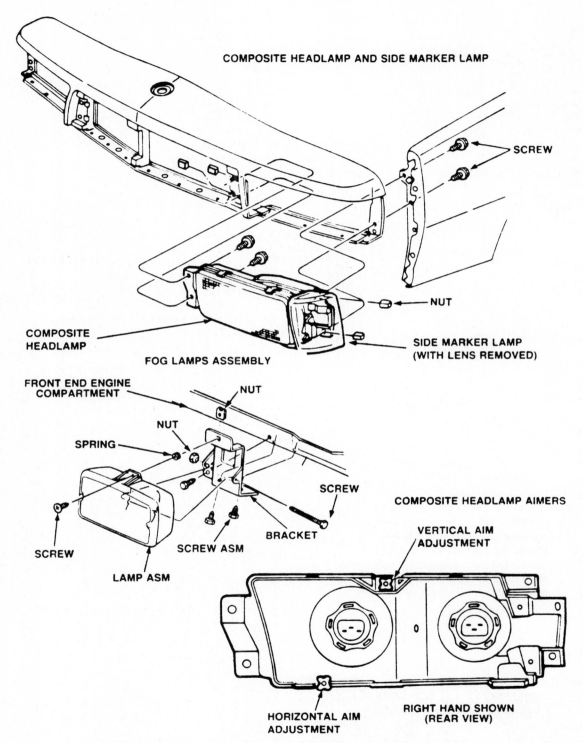

COMPOSITE HEADLAMP AND SIDE MARKER LAMP

SCREW

COMPOSITE
HEADLAMP

NUT

SIDE MARKER LAMP
(WITH LENS REMOVED)

FOG LAMPS ASSEMBLY

FRONT END ENGINE
COMPARTMENT

NUT

NUT

SPRING

SCREW

BRACKET

SCREW

SCREW ASM

LAMP ASM

COMPOSITE HEADLAMP AIMERS

VERTICAL AIM
ADJUSTMENT

HORIZONTAL AIM
ADJUSTMENT

RIGHT HAND SHOWN
(REAR VIEW)

FIGURE 7-8 *Typical composite headlights use replaceable halogen bulbs.*
(Courtesy of General Motors Corporation.)

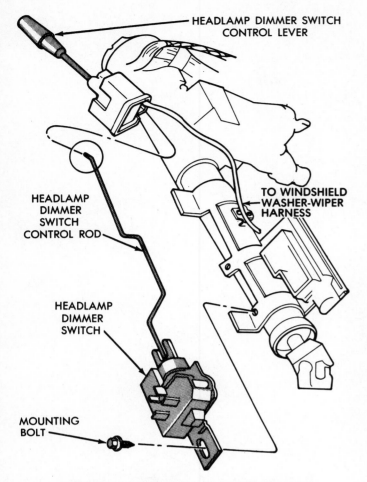

HEADLAMP DIMMER SWITCH
CONTROL LEVER

HEADLAMP
DIMMER
SWITCH
CONTROL ROD

TO WINDSHIELD
WASHER-WIPER
HARNESS

HEADLAMP
DIMMER
SWITCH

MOUNTING
BOLT

FIGURE 7-9 Typical headlight dimmer switch used on steering columns. Adjustment or replacement of this style of switch requires access to the steering column under the dash. Disassembly of the steering column is not necessary. (Courtesy of Chrysler Corporation.)

dirt and water corroding the switch. An open-circuit (defective) dimmer switch could cause *both* bright and/or low beams not to light. Some cars use a relay to control the high-beam lights. If the relay is defective, it could prevent proper headlight operation.

BRAKE LIGHTS

Brake lights can usually be operated without the ignition switch "on." Current from the fuse panel is sent to the high-intensity filament of the taillight bulbs whenever the brake pedal is depressed. The brake light switch can be located near the brake pedal arm under the dash or under on the master brake cylinder which operates the brake lights when brake fluid pressure is applied. This high-intensity filament in the taillight is the same filament that is operated for turn signals.

CENTER HIGH-MOUNTED STOP LAMPS

Center high-mounted stop lamps (CHMSLs) are required by U.S. federal law to be installed on all 1986 and newer cars. The additional stop light must be on the center line of the car and no lower than 3 in. below the bottom of the rear glass (less than 6 in. below the bottom of the rear window on convertibles). It may be illegal to mount any accessory, such as luggage rack, that would block the CHMSL. See Figure 7-10.

DIRECTIONAL (TURN) SIGNALS AND BRAKE LIGHTS ARE CONNECTED

The current *from* the brake light switch goes up the steering column where the directional signal switch is located. If neither directional signal is "on," the current from the brake light switch is sent to both brake lights.

If the left directional signal is "on," for example, current for the brake light is sent to the right brake light directly. The current for the left side is sent through the "flasher unit" and then to the left-side lights, which blink. Therefore, problems with brake lights could be caused by a defective directional signal switch located in the steering column.

DIRECTIONAL (TURN) SIGNALS

A directional (turn) signal flasher unit is a metal or plastic can containing a switch that opens and closes the directional signal circuit. This directional signal flasher unit is usually installed in a metal clip attached to the dash panel to allow the "clicking" noise of the flasher to be heard by the driver. The directional signal flasher is designed to transmit the current to only "light" the front and rear bulbs on one side at a time. When the directional signal flasher unit is old, both sides will flash slowly equally. The contact points inside the flasher unit may become corroded and pitted, requiring higher voltage to operate. To restore normal operation, replace the directional signal flasher unit. Other common directional signal problems and possible corrections include the following:

Problem 1:   Slow flashing on both sides equally.

Solution 1:   Replace the worn flasher unit. Check the battery and the charging voltage to

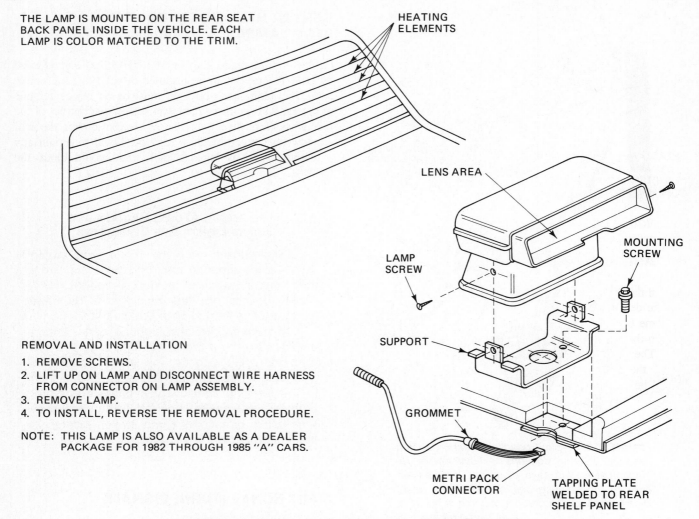

THE LAMP IS MOUNTED ON THE REAR SEAT
BACK PANEL INSIDE THE VEHICLE. EACH
LAMP IS COLOR MATCHED TO THE TRIM.

HEATING
ELEMENTS

LENS AREA

MOUNTING
SCREW

LAMP
SCREW

SUPPORT

GROMMET

METRI PACK
CONNECTOR

TAPPING PLATE
WELDED TO REAR
SHELF PANEL

REMOVAL AND INSTALLATION

1. REMOVE SCREWS.
2. LIFT UP ON LAMP AND DISCONNECT WIRE HARNESS
   FROM CONNECTOR ON LAMP ASSEMBLY.
3. REMOVE LAMP.
4. TO INSTALL, REVERSE THE REMOVAL PROCEDURE.

NOTE: THIS LAMP IS ALSO AVAILABLE AS A DEALER
       PACKAGE FOR 1982 THROUGH 1985 "A" CARS.

FIGURE 7-10 *Typical center high-mounted stop lamp (CHMSL).
(Courtesy of General Motors Corporation.)*

be certain that the charging circuit and
battery are supplying high-enough
voltage for proper operation of the directional signals. (See Chapter 11.)

Problem 2:   Slow or no flashing on one side only.

Solution 2:   Replace the defective bulb or clean
poor connections on the front or rear
bulbs on the side that does not work.

Problem 3:   Directional signals do not flash on
either side.

Solution 3:   The most likely cause is a defective
flasher unit, and replacement will be
necessary. However, defective bulbs
or connections on both sides could also
be the cause.

WHERE IS THE FLASHER UNIT LOCATED?

Most directional signal flasher units are
mounted in a metal clip that is attached to the
dash. The dash panel acts as a sounding board,
increasing the sound of the flasher unit. Most
four-way hazard flasher units are plugged into
the fuse panel. Some two-way directional signal
flasher units are also plugged into the fuse
panel. How do you know for sure where the
flasher unit is located? With both the directional
signal and the ignition on, listen and/or feel for
the clicking of the flasher unit. Some service
manuals also give general locations for the
placement of flasher units.

FIGURE 7-11 *Two styles of two-prong flasher units.*

HAZARD FLASHERS

Hazard flasher units are usually plugged into the fuse panel and are designed to flash four or more bulbs safely and at the same flashing speed regardless of the number of bulbs used in the lighting circuit.

Therefore, if trailer lights are connected to the taillights, the flasher unit for the four-way hazard flasher should be used in place of the standard directional signal flasher. However, the regular directional signal flasher *cannot* be used for the four-way hazard flashers. The result would be the very rapid flashing of the hazard flasher and damage to the flasher itself. See Figure 7-11.

FLASHER UNITS

Flasher units are cylindrical or rectangular in shape and can have either two or three plug-in prongs. Flasher units designed only for directional signals are usually thermal units that use the heat generated by the current flowing in the circuit to bend a bimetallic strip that opens a set of contact points. When the strip cools slightly, the contacts again conduct current to the directional signals. However, if more than a few bulbs are used in the circuit, the heating occurs too rapidly and the lights flash at over 100 times per minute. Hazard flashers (four-way flashers) eliminate this problem by using electromagnetic-type flasher units which are not sensitive to the number of bulbs in the circuit and will flash at a fixed rate of 50 to 100 times per minute regardless of the electrical load. To reduce inventory costs, many parts stores stock only heavy-duty (HD) four-way or hazard flasher units. These can safely be used for either hazard warning or directional (turn) signals. Lighting-circuit diagrams are shown in Figure 7-12.

FIGURE 7-12 *Lighting circuit diagnosis. (Courtesy of Oldsmobile Division, GMC.)*

DIAGNOSIS - LIGHTS AND LIGHTING CIRCUITS

Troubles in the lighting circuits are caused by loose connections, open or shorted wiring, burned out bulbs, failed switches, inadequate ground, or blown fuses. In each, trouble diagnosis requires following through the circuits until the source of difficulty is found. To aid in making an orderly check, refer to the wiring diagrams shown in the Electrical Troubleshooting Manual (ETM).

HEADLIGHT DIAGNOSIS

Condition	Possible Cause	Correction
One headlight inoperative or intermittent	1. Loose connection	1. Secure connections to sealed beam including ground (black wire)
	2. Sealed beam unit malfunction.	2. Replace sealed beam.
One or more headlights are dim.	1. Open ground connection at headlight.	1. Repair black wire connection between sealed beam and body ground.
	2. Black ground wire mislocated in head-light connector (three-wire, hi-lo, connector only)	2. Relocate black wire in connector.

HEADLIGHT DIAGNOSIS

Condition	Possible Cause	Correction
One or more headlights short life	1. Charge circuit problem.	1. Refer to Section 6D, charging system diagnosis.
All headlights inoperative or intermittent	1. Loose connection.	1. Check and secure connections at dimmer switch and light switch.
	2. Dimmer switch malfunction.	2. Check voltage at dimmer switch with test light. Refer to ETM for test points.
	3. Open wiring - light switch to dimmer switch.	3. Check yellow wire with test light. If bulb lights at light switch yellow wire terminal but not at dimmer switch, repair open wire.
	4. Open wiring - light switch to battery.	4. Check red wire terminal at light switch with test light. If bulb does not light, repair open red wire circuit to battery (possible open fusible link).
	5. Shorted ground circuit.	5. If, after a few minutes operation, headlights flicker "ON" and "OFF" and or a thumping noise can be heard from the light switch (circuit breaker opening and closing), repair short to ground in circuit between light switch and headlights. After repairing short, check for headlight flickering after one minute operation. If flickering occurs, the circuit breaker has been damaged and light switch must be replaced.
	6. Light switch malfunction.	6. Check red and yellow wire terminals at light switch with test light. If bulb lights at red wire terminal but not at yellow terminal, replace light switch.
Upper or lower beam will not light or intermittent.	1. Open connection or defective dimmer switch.	1. Check dimmer switch terminals with test light. If bulb lights at light green or tan wire terminals, repair open wiring between dimmer switch and headlights. If bulb will not light at either of these terminals,

FIGURE 7-12 *(cont'd)*

Condition	Possible Cause	Correction
	2. Loose connection or open circuit	2. Check all connectors. If OK, check continuity of circuit from fuse to light on either side of fuse, correct open circuit from battery to fuse.
	3. Blown fuse	3. Replace fuse. If new fuse blows, repair short to ground in circuit from fuse through gear selector or from fuse through gear selector or backup light switch to backup lights.
	4. Defective gear selector or backup light switch	4. With ignition "ON," check switch terminals in backup position with test light. If test bulb lights at dark blue wire terminal but not at light green wire terminal, replace light switch.
	5. Defective ignition switch	5. If test bulb lights at ignition switch battery terminal but not as output terminal, replace ignition switch.
Light will not turn off	1. Gear selector switch misadjusted (closed when shift lever is not in reverse position)	1. Readjust gear selector switch.

STOP LIGHTS

Condition	Possible Cause	Correction
One bulb inoperative	1. Bulb burned out.	1. Replace bulb.
One side inoperative (multi-bulb design)	1. Loose connection, open wiring or defective bulbs.	1. Turn on directional signal. If light does not operate, check bulbs. If bulbs are OK, check all connections. If light still does not operate, use test light and check for open wiring.
	2. Defective directional signal switch or cancelling cam	2. If light will operate by turning directional signal on, the switch is not centering properly during cancelling operation. Replace defective cancelling cam or directional signal switch.
All stop lights inoperative	1. Stop fuse blown	1. Replace fuse. If new fuse blows, repair short to ground in circuit between fuse and lights.
	2. Open in wire from fuse to stop-switch	2. Check for power at brown wire at stop-switch and at fuse. If there is power at fuse but

FIGURE 7-12 *(cont'd)*

FIGURE 7-12 *(cont'd)*

STOP LIGHTS

Condition	Possible Cause	Correction
		not at switch, check for open in brown wire.
	3. Stop-switch misadjusted or defective	3. With brake pedal depressed, check white wire terminal in steering column connector with test light. If bulb does not light, check stop switch for proper adjustment. If adjustment is OK, jumper stop switch. If stop lights operate, replace stop switch.
Will not turn off	1. Stop switch misadjusted or defective	1. Readjust switch. If switch still malfunctions, replace.

COURTESY LIGHTS

"Courtesy lights" is a generic term generally used for interior lights, including overhead (dome) and under-the-dash (courtesy) lights. These interior lights can be operated by rotating the headlight switch knob full counterclockwise (left) or by switches located in the door jams of the car doors. There are two types of circuits commonly used for these interior lights. Most manufacturers, except Ford, use the door switches to ground the courtesy light circuit. Ford uses the door switches to open and close the power side of the circuit.

ILLUMINATED ENTRY

Some cars are equipped with illuminated entry, where the interior lights are turned on for a given amount of time whenever the doors are locked and the outside door handle is operated. Most cars equipped with illuminated entry also light the exterior door keyhole to help assist unlocking the doors at night. Some cars equipped with body computers use the door handle electrical switch of the illuminated entry circuit to "wake up" the power supply for the body computer. See Chapter 27 for additional body computer information.

FIBER OPTICS

Fiber optics is the transmission of light through special plastic (polymethylmethacrylate) that keeps the light rays parallel even if the plastic is tied in a knot. These strands of plastic are commonly used in automotive use as indicators for the driver that certain lights are functioning. For example, some cars are equipped with fender-mounted units that light whenever the lights or turn signals are operating. Plastic fiber optic strands, which often look like standard electrical wire, transmit the light at the bulb to the indicator on the fender so that the driver can determine if a certain light is not operating. Fiber optics can also be run like wires and could indicate the operation of all lights on the dash or console. Fiber optics are also commonly used to light ashtrays, outside door locks, the other areas where a small amount of light is required. The source of the light can be any normally operating light bulb. A special bulb clip is usually used to retain the fiber optic plastic tube near the bulb.

SUMMARY

1. All electrical light bulbs use a trade number that identifies the bulb's candlepower, voltage, and wattage.
2. Headlights that use replaceable bulbs are called *composite* headlights.
3. Headlights also use a separate circuit breaker for protection, while other lighting circuits are usually grouped and use a fuse for protection.

4. Brake lights and directional (turn) lights are electrically connected at the directional signal switch, which is usually mounted in the steering column.

5. If directional signals fail to operate on one side only, the usual cause is a defective bulb or connection on the side that does not function. If both sides malfunction, the usual cause is a defective flasher unit.

6. Typical directional signal flasher units flash at a rate that depends on the amount of current flow through the flasher.

7. Heavy-duty or four-way hazard flashers can operate at a fixed flashing rate regardless of the number of bulbs in the circuit.

8. Courtesy lights are interior lights that are controlled by V-door jam switches, the headlight switch, or door handle switches.

STUDY QUESTIONS

7-1. What does "trade number" mean?

7-2. What is a rheostat, and where is it commonly used in automotive electrical systems?

7-3. What are composite headlights?

7-4. What is a CHMSL?

7-5. Explain the operation of directional (turn) signals, including the connection with the brake light circuit.

MULTIPLE-CHOICE QUESTIONS

7-1. Bulb numbers mean:
(a) the trade number.
(b) amperage required.
(c) candlepower.
(d) none of the above.

7-2. Low sealed-beam headlights:
(a) contain only a low beam.
(b) contain both a low beam and a high beam.
(c) use a two-wire connector.
(d) none of the above.

7-3. Technician A says that the brake lights operate the high-intensity filament of the taillight bulb. Technician B says that the current from the brake lights goes to the turn signal switch before going to the brake lights. Which technician is correct?
(a) A only.
(b) B only.
(c) both A and B.
(d) neither A or B.

7-4. A *slow* turn signal on one side only is most likely caused by:
(a) a defective flasher unit.
(b) the wrong flasher unit.
(c) a defective (open) bulb on the affected side.
(d) a poor ground connection on the affected side.

7-5. The headlight switch controls:
(a) the headlights, taillights, and dash lights.
(b) the headlights, taillights, and side marker lights.
(c) the dash lights, side marker lights, taillights, and headlights.
(d) the headlights, taillights, and dome light.
(e) all of the above could be correct.

7-6. Illuminated entry operates:
(a) only when the doors are locked.
(b) the headlights.
(c) the trunk light and keyhole light.
(d) by grounding the door jam switch.

7-7. The major difference between the flasher units commonly used for directional (turn) signals and four-way hazard flashers is:
(a) the size of the flasher unit.
(b) four-way flasher units are thermoelectric.
(c) two-way directional signal units are constant-rate flasher units.
(d) directional signal units are thermoelectric and four-way units are electromagnetic.

7-8. A defective directional (turn) signal flasher unit could cause:
(a) the turn signals to flash slowly in both directions.
(b) the turn signals not to operate in both directions.
(c) the hazard flashers and turn signals not to operate.
(d) both (a) and (b).

8

Analog
and Digital
Dash Instruments

DASH INSTRUMENTS SUCH AS SPEEDOMETERS AND FUEL-LEVEL GAUGES
are critical to the safe operation of any vehicle. This chapter explains the
operation of two types of commonly used analog (needle-type) dash instruments, engine warning lights, electronic digital dash instrument operation, and troubleshooting. The topics covered in this chapter include:

1. Thermoelectric gauge operation
2. Electromagnetic gauge operation
3. Dash warning lights (operation and troubleshooting), including:
   (a) Coolant temperature
   (b) Oil pressure
   (c) Charging
   (d) Brake warning
4. Digital dash operation:
   (a) LED displays
   (b) Liquid-crystal displays
   (c) Vacuum-tube fluorescent displays
   (d) CRT displays
5. Digital dash diagnosis and troubleshooting of the following
   gauges:
   (a) Speedometer
   (b) Tachometer
   (c) Oil pressure
   (d) Engine temperature
   (e) Fuel level
   (f) Charging system

DASH INSTRUMENTS

Dash instruments are either analog (needle-type) or digital. This section describes only the operation and testing of analog-type (also called needle-type) dash instruments. There are two basic types of gauges used: the electromagnetic and the thermoelectric. General Motors uses the electromagnetic type, whereas Ford, Chrysler, and American Motors usually use the thermoelectric type. The type used, if unknown, can usually be determined by looking at the fuel gauge with the ignition "off."

If the gauge reads the fuel level with the ignition "off," it is an electromagnetic gauge. If the gauge falls to empty with the ignition "off," the gauge is thermoelectric. All dash instruments work in exactly the same way; only the function being measured is different.

THERMOELECTRIC GAUGES

A thermoelectric gauge uses electrical current flow through the meter controlled by a *sending unit* or *sensor* to heat a curved bimetallic strip. As the current flow increases, the heat generated inside the gauge causes the indicator needle to swing toward the right.

This type of gauge moves very slowly, which is an advantage because turns and hills do not affect the readings of the fuel gauge, for example, and the needle tends to remain steady. However, a thermoelectric gauge is very sensitive to battery voltage variations. Therefore, to main-

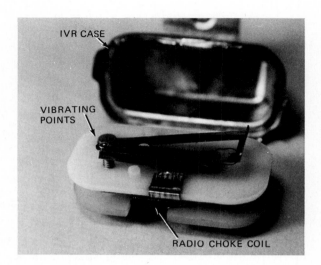

FIGURE 8-2 *Instrument voltage regulator. Vibrating points maintain current through the instruments at 5 V. The radio choke prevents radio interference created by the pulsing current flow.*

tain accuracy, thermoelectric-type gauges use a voltage regulator, commonly called an IVR (instrument voltage regulator). An IVR maintains instrument voltage at an average of 5 V. See Figure 8-1.

The regulator uses a bimetallic strip and an electric heating coil which will alternately open and close (pulses) a contact which produces the average 5 V for all instrument gauges. To prevent radio interference caused by the pulsation from the regulator, a small coil of wire called a *radio choke* is installed in the power lead going to the IVR. If *all* dash instruments are not functioning correctly, such as when all are reading high or low, the usual cause is the instrument voltage regulator located on the back of the instrument panel. See Figure 8-2.

ELECTROMAGNETIC GAUGES

Electromagnetic dash instruments use small electromagnetic coils that are connected to a sending unit for such things as fuel level, water temperature, and oil pressure. The resistance of the sensor varies with what is being measured. See Figure 8-3 for typical electromagnetic fuel gauge operations.

DASH INSTRUMENT DIAGNOSIS

With electromagnetic gauges, if the resistance of the sensor is low, the meter reads low. If the resistance of the sensor is high, the meter reads high.

FIGURE 8-1 *A thermoelectric fuel gauge has 5 V coming from the instrument voltage regulator (IVR). The IVR is shared by all other dash instruments. (Courtesy of Chrysler Corporation.)*

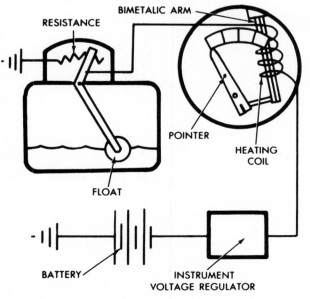

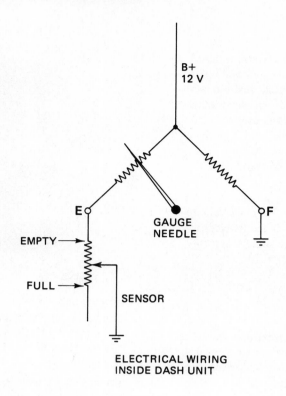

ELECTRICAL WIRING
INSIDE DASH UNIT

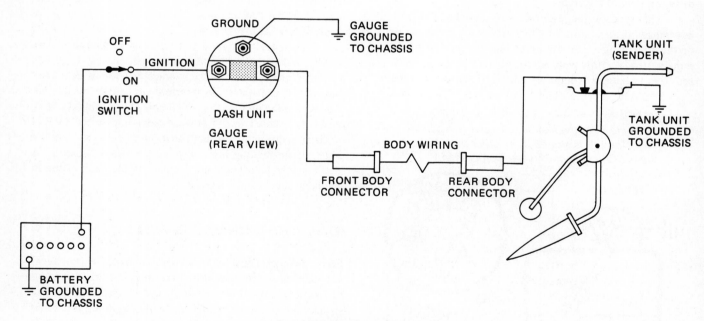

TYPICAL GAS GAUGE SYSTEM SCHEMATIC

FIGURE 8-3 *Electromagnetic fuel gauge wiring. If the sensor wire is unplugged and grounded, the needle should indicate "E" (empty). If the sensor wire is unplugged and held away from ground, the needle should indicate "F" (full).*

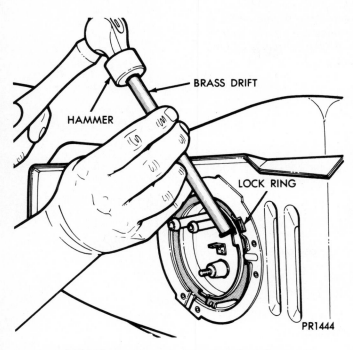

FIGURE 8-4 *Procedure that can be used to remove the lock ring to remove the gas tank sending unit from the gas tank. Some units can be removed only after detaching the gas tank from the vehicle. (Courtesy of Chrysler Corporation.)*

NOTE: Thermoelectric gauges are opposite from electromagnetic gauges and read low when resistance is high. The following procedures are given for electromagnetic gauges and should be reversed if working on thermoelectric gauges.

FUEL GAUGE DIAGNOSIS

When troubleshooting a fuel gauge, if the power wire is unplugged from the tank unit with the ignition "on," the dash unit should move toward "full" (high resistance). If the power lead is touched to a ground (low resistance), the fuel gauge should register "empty." The same operation can be used with oil pressure and water temperature gauges. See Figures 8-4 and 8-5.

TELL-TALE LIGHTS

Tell-tale lights (often called "idiot lights") warn the driver of system failure. Whenever the ignition is turned "on," all warning lights come on as a bulb check.

The charging system warning light may be labeled "CHARGE," "GEN," or "ALT" and will light if the

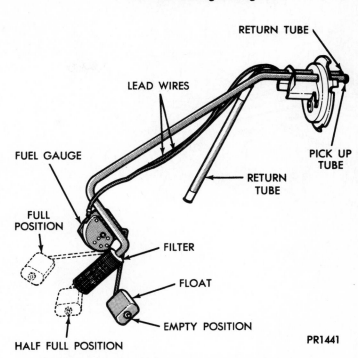

FIGURE 8-5 *A typical gas gauge sending unit includes the fuel pickup tube and filter. (Courtesy of Chrysler Corporation.)*

charging system voltage is lower than battery voltage. Complete operation of the charging system and the warning light circuit is discussed in Chapter 16.

The oil pressure light operates when an oil pressure sensor unit, which is screwed into the engine block, grounds the electrical circuit and lights the dash warning light in the event of low oil pressure [3 to 7 psi (20 to 50 kPa)]. Normal oil pressure is generally between 10 and 60 psi (70 and 400 kPa).

OIL PRESSURE LIGHT DIAGNOSIS

To test the operation of the oil pressure warning circuit, unplug the wire from the oil pressure sending unit, usually located near the oil filter, with the ignition switch "on." With the wire disconnected from the sending unit, the warning light should be off. If the wire is touched to a ground, the warning light should be on. If there is *any* doubt of the operation of the oil pressure warning light, always check actual engine oil pressure using a gauge that can be screwed into the opening left after unscrewing the oil pressure sending unit. To remove the sending unit, special sockets are available at most auto-parts stores or a 1- or 1^1/16-in. six-point socket may be used for most units. See Figure 8-6 for the location of typical oil pressure sending units.

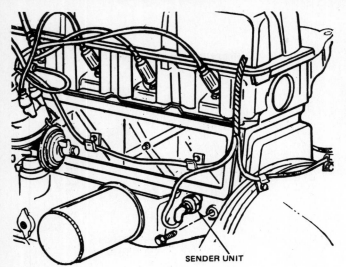

FIGURE 8-6 *Oil pressure switches or sender units are usually located near the oil filter. (Courtesy of Ford Motor Company.)*

SENDER UNIT

FIGURE 8-7 *(Courtesy of Buick Motor Division.)*

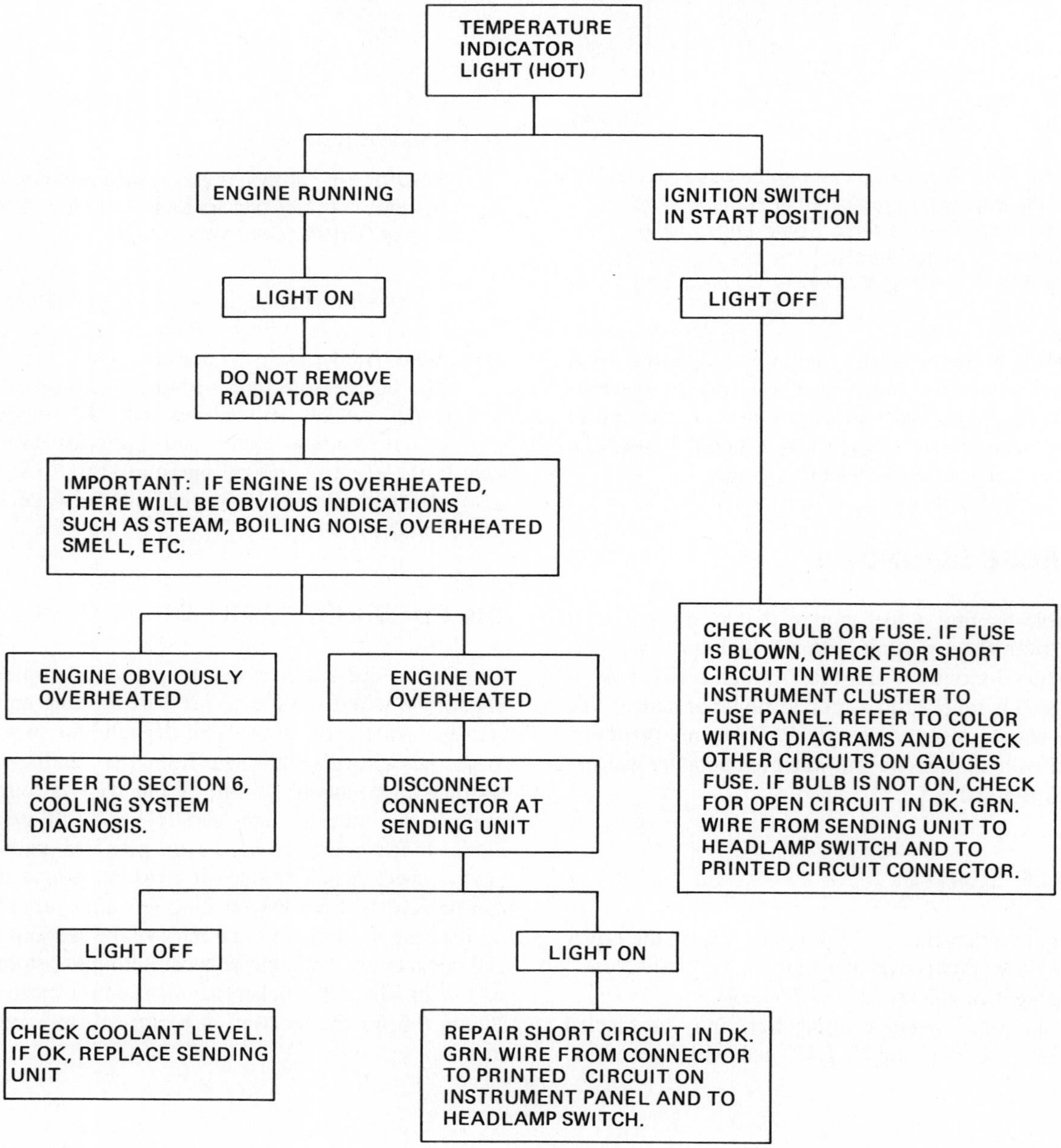

```
                    ┌─────────────────────┐
                    │ TEMPERATURE         │
                    │ INDICATOR           │
                    │ LIGHT (HOT)         │
                    └─────────────────────┘
              ┌─────────────┴──────────────────┐
      ┌───────────────┐              ┌───────────────────┐
      │ ENGINE RUNNING│              │ IGNITION SWITCH   │
      │               │              │ IN START POSITION │
      └───────────────┘              └───────────────────┘
              │                                │
       ┌──────────┐                     ┌──────────┐
       │ LIGHT ON │                     │ LIGHT OFF│
       └──────────┘                     └──────────┘
              │
     ┌─────────────────┐
     │ DO NOT REMOVE   │
     │ RADIATOR CAP    │
     └─────────────────┘
              │
```

IMPORTANT: IF ENGINE IS OVERHEATED, THERE WILL BE OBVIOUS INDICATIONS SUCH AS STEAM, BOILING NOISE, OVERHEATED SMELL, ETC.

ENGINE OBVIOUSLY OVERHEATED

ENGINE NOT OVERHEATED

REFER TO SECTION 6, COOLING SYSTEM DIAGNOSIS.

DISCONNECT CONNECTOR AT SENDING UNIT

CHECK BULB OR FUSE. IF FUSE IS BLOWN, CHECK FOR SHORT CIRCUIT IN WIRE FROM INSTRUMENT CLUSTER TO FUSE PANEL. REFER TO COLOR WIRING DIAGRAMS AND CHECK OTHER CIRCUITS ON GAUGES FUSE. IF BULB IS NOT ON, CHECK FOR OPEN CIRCUIT IN DK. GRN. WIRE FROM SENDING UNIT TO HEADLAMP SWITCH AND TO PRINTED CIRCUIT CONNECTOR.

LIGHT OFF

LIGHT ON

CHECK COOLANT LEVEL. IF OK, REPLACE SENDING UNIT

REPAIR SHORT CIRCUIT IN DK. GRN. WIRE FROM CONNECTOR TO PRINTED CIRCUIT ON INSTRUMENT PANEL AND TO HEADLAMP SWITCH.

TEMPERATURE LIGHT DIAGNOSIS

The ''hot'' light or engine coolant overheat warning light warns the driver whenever the engine coolant temperature is between 248 and 258°F (124°C). This temperature is just slightly below the boiling point of the coolant in a properly operating cooling system. To test the hot light, disconnect and ground the wire from the water-temperature sending unit. The hot light should come on. The sensor is located in the engine block, usually near the thermostat. Always check the cooling system operation *and* the operation of the warning light circuit whenever the hot light comes on during normal driving. See Figure 8-7.

BRAKE WARNING LIGHT DIAGNOSIS

All vehicles sold in the United States after 1967 must be equipped with a dual braking system and a dash-mounted warning light to signal the driver of a failure to one part of the hydraulic brake system. The switch that operates the warning light is called a pressure differential switch. This switch is usually the center portion of a multiple-purpose brake part called a combination valve. If there is unequal hydraulic pressure in the braking system, the switch usually *grounds* the 12-V lead at the switch and the light comes on. See Figure 8-8.

Unfortunately, the dash warning light is often the same light as that used to warn the driver that the parking light is ''on.'' The warning light is usually operated by using the parking brake lever or brake hydraulic pressure switch to complete the *ground* for the warning light circuit. If the warning light is on, first check if the parking brake is fully released. If the parking brake is fully released, the

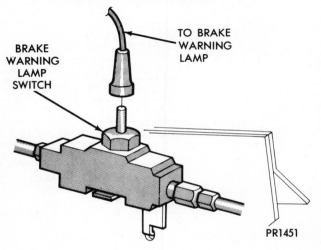

FIGURE 8-8 *Typical brake warning light switch located on or near the master brake cylinder. (Courtesy of Chrysler Corporation.)*

BRAKE WARNING LAMP SWITCH

TO BRAKE WARNING LAMP

PR1451

problem could be a defective parking brake switch or a hydraulic brake problem. To test which system is causing the light to remain on, simply unplug the wire from the valve or switch. If the wire on the pressure differential switch is disconnected and the warning light remains ''on,'' the problem is due to a defective or misadjusted parking brake switch. If, however, the warning light went out when the wire was removed from the brake switch, the problem is due to an hydraulic brake fault which caused the pressure differential switch to complete the warning light circuit. See Figure 8-9.

INTRODUCTION TO DIGITAL DASH OPERATION

Mechanical or electromechanical dash instruments use cables, mechanical transducers, and sensors to operate a particular dash instrument. Digital dash instruments use various electric and electronic sensors that activate segments or sections of an electronic display. Most electronic dash clusters use a computer chip and various electronic circuits to operate and control the internal power supply, sensor voltages, and display voltages. Electronic dash displays may use one or more of the several types of displays: LED, LCD, VTF, or CRT.

LED DIGITAL DISPLAYS

LED stands for *light-emitting diode*. All diodes emit some form of energy during operation and the LED is a semiconductor that is constructed to release energy in the form of light. Many colors of LEDs can be constructed, but the most popular are red, green, and yellow. See Figure 8-10. Red is difficult to see in direct sunlight; therefore, if a LED is used, most car manufacturers use yellow. Light-emitting diodes can be arranged in a group of seven. Seven segment LEDs can be used to indicate both numbers and letters. See Figure 8-11.

An LED display requires more electrical power than that required by other types of electronic displays. A typical LED display requires 30 mA for each *segment*; therefore, each number or letter displayed could require 210 mA.

LIQUID-CRYSTAL DISPLAYS

Liquid-crystal displays (LCDs) can be arranged into a variety of forms, letters, numbers, and bar-graph displays. LCD construction consists of a special fluid sandwiched between two sheets of polarized glass. The special fluid between the glass plates will permit light to pass if a small

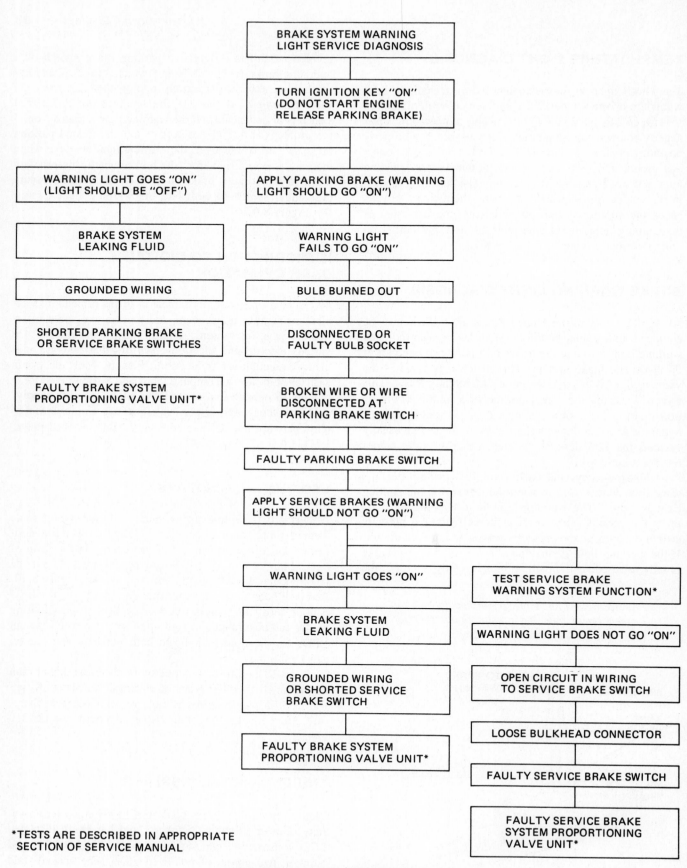

BRAKE SYSTEM WARNING
LIGHT SERVICE DIAGNOSIS

TURN IGNITION KEY "ON"
(DO NOT START ENGINE
RELEASE PARKING BRAKE)

WARNING LIGHT GOES "ON"
(LIGHT SHOULD BE "OFF")

BRAKE SYSTEM
LEAKING FLUID

GROUNDED WIRING

SHORTED PARKING BRAKE
OR SERVICE BRAKE SWITCHES

FAULTY BRAKE SYSTEM
PROPORTIONING VALVE UNIT*

APPLY PARKING BRAKE (WARNING
LIGHT SHOULD GO "ON")

WARNING LIGHT
FAILS TO GO "ON"

BULB BURNED OUT

DISCONNECTED OR
FAULTY BULB SOCKET

BROKEN WIRE OR WIRE
DISCONNECTED AT
PARKING BRAKE SWITCH

FAULTY PARKING BRAKE SWITCH

APPLY SERVICE BRAKES (WARNING
LIGHT SHOULD NOT GO "ON")

WARNING LIGHT GOES "ON"

BRAKE SYSTEM
LEAKING FLUID

GROUNDED WIRING
OR SHORTED SERVICE
BRAKE SWITCH

FAULTY BRAKE SYSTEM
PROPORTIONING VALVE UNIT*

TEST SERVICE BRAKE
WARNING SYSTEM FUNCTION*

WARNING LIGHT DOES NOT GO "ON"

OPEN CIRCUIT IN WIRING
TO SERVICE BRAKE SWITCH

LOOSE BULKHEAD CONNECTOR

FAULTY SERVICE BRAKE SWITCH

FAULTY SERVICE BRAKE
SYSTEM PROPORTIONING
VALVE UNIT*

*TESTS ARE DESCRIBED IN APPROPRIATE
SECTION OF SERVICE MANUAL

FIGURE 8-9   *(Courtesy of Chrysler Corporation.)*

LIGHT EMITTING DIODE

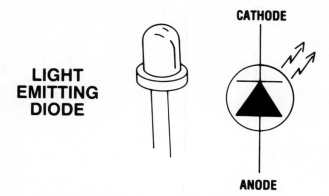

CATHODE

ANODE

FIGURE 8-10 *(Courtesy of General Motors Corporation.)*

SEVEN-SEGMENT LED

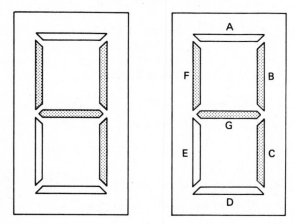

A

F B

G

E C

D

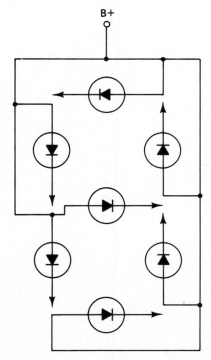

B+

FIGURE 8-11 *(Courtesy of General Motors Corporation.)*

LIQUID CRYSTAL DISPLAY

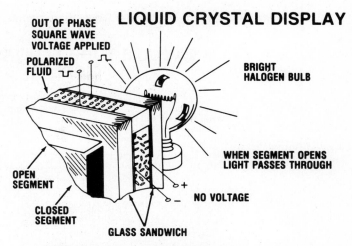

OUT OF PHASE SQUARE WAVE VOLTAGE APPLIED

POLARIZED FLUID

BRIGHT HALOGEN BULB

WHEN SEGMENT OPENS LIGHT PASSES THROUGH

NO VOLTAGE

OPEN SEGMENT

CLOSED SEGMENT

GLASS SANDWICH

FIGURE 8-12 *If voltage is applied to the special fluid inside the LCD display, the light from the halogen bulb passes through the glass sandwich and lights a small section (segment) of the display. (Courtesy of General Motors Corporation.)*

voltage is applied to the fluid through a conductive film laminated to the glass plates.

The light from a very bright halogen bulb behind the LCD shines through those segments of the LCD that have been polarized to let the light through, which then shows numbers or letters. See Figure 8-12. Color filters can be placed in front of the display to change the color of certain segments of the display, such as the maximum engine speed on a digital tachometer. LCD displays are used on newer model Chevrolet Corvettes and several other makes and models.

> **NOTE:** Be careful when cleaning a LCD display not to push on the glass plate covering the special fluid. If excessive pressure is exerted on the glass, the display may be permanently distorted. If the glass breaks, the fluid will escape and could damage other components in the vehicle due to its strong alkaline nature. Use only a soft damp cloth to clean these displays.

The major disadvantage of a LCD digital dash is that the numbers or letters are slow to react or change at low temperature.

VACUUM-TUBE FLUORESCENT DISPLAYS

The vacuum-tube fluorescent (VTF) display is a popular automotive and household appliance display because it is very bright and can easily be viewed in strong sunlight. The usual VTF display is green, but white is often used for home appliances. The VTF generates its bright display similar to a TV screen, where a chemical-coated light-emitting element called a *phosphor* is hit with high-speed elec-

VACUUM TUBE FLUORESCENT DISPLAY

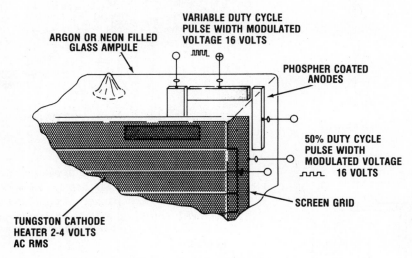

FIGURE 8-13 *The front of the display above is toward the top of the illustration. (Courtesy of General Motors Corporation.)*

trons. VTF displays are very bright and must be dimmed by dense filters or by controlling the voltage applied to the display. See Figure 8-13. A typical VTF dash is dimmed to 75% brightness whenever the parking or headlights are turned on. Some displays use a photocell to monitor and adjust the intensity of the display during daylight viewing. Most VTF displays are green for best viewing under most lighting conditions.

CRT DISPLAYS

A cathode ray tube (CRT) dash display was first used as standard equipment on the 1986 Buick Riviera. A CRT is similar to a television tube and permits the display of hundreds of controls and diagnostic messages in one convenient location. See Figure 8-14.

Using the touch-sensitive cathode ray tube, the driver or technician can select many different separate displays, including radio, climate, trip, and dash instrument information. All of the functions noted above can be accessed readily by the driver. Further diagnostic information can be displayed on the CRT if the proper combination of air-conditioning controls is touched. See Chapter 27 for operation and diagnosis of body computers.

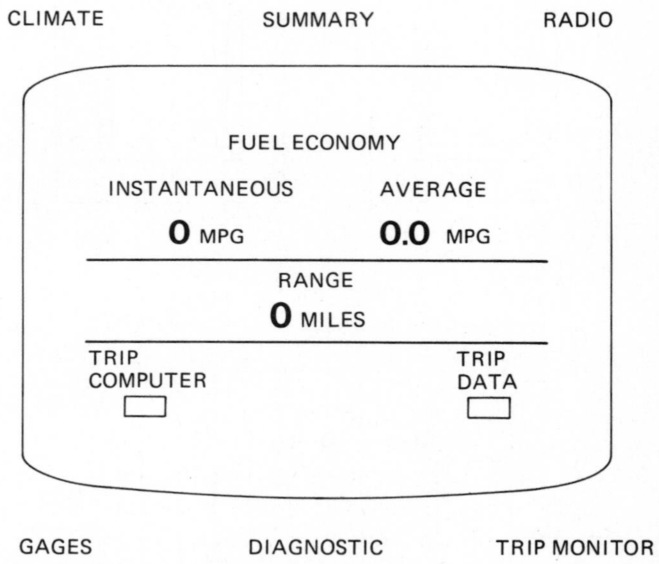

CLIMATE   SUMMARY   RADIO

FUEL ECONOMY

INSTANTANEOUS   AVERAGE

0 MPG   **0.0** MPG

RANGE

0 MILES

TRIP COMPUTER   TRIP DATA

GAGES   DIAGNOSTIC   TRIP MONITOR

FIGURE 8-14 *Typical CRT display as found on a newer-model Buick Riviera. (Courtesy of General Motors Corporation.)*

WHAT'S A "WOW!" DISPLAY?

When a car equipped with a digital dash is first started, all segments of the electronic display are turned on full brilliance for 1 or 2 seconds. This is commonly called the "WOW!" display and is used to show off the brilliance of the display. If numbers are part of the display, the number "8" is displayed, because this number uses all segments of a number display. Technicians can also use the "WOW" display to check if all segments of the electronic display are functioning correctly.

PM GENERATOR VEHICLE SPEED SENSOR

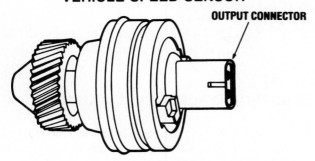

OUTPUT CONNECTOR

FIGURE 8-15 (Courtesy of General Motors Corporation.)

BUFFER CIRCUIT

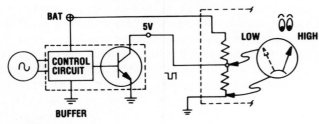

FIGURE 8-16 (Courtesy of General Motors Corporation.)

ELECTRONIC SPEEDOMETERS

Electronic dash displays usually use an electric vehicle speed sensor driven by a small gear on the output shaft of the transmission. These speed sensors contain a permanent magnet and generate a voltage in proportion to the vehicle speed. These speed sensors are commonly called *PM* (permanent magnet) *generators*. See Figure 8-15.

The output of a PM generator speed sensor is a voltage that varies in frequency and intensity with increasing vehicle speed. The PM generator speed signal is sent to the instrument cluster electronic circuits. These specialized electronic circuits include a buffer amplifier circuit that converts the variable sine-wave voltage from the speed sensor to an "on" and "off" signal that can be used by other electronic circuits to indicate a vehicle's speed. See Figure 8-16. The vehicle speed is then displayed by either an electronic needle-type speedometer or by numbers on a digital display.

ELECTRONIC ODOMETERS

An odometer is a dash display that indicates the total miles traveled by the vehicle. Some dash displays also include a trip odometer that can be reset and used to record total

STEPPER MOTOR ODOMETER

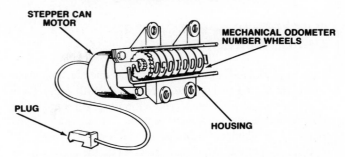

STEPPER CAN MOTOR

MECHANICAL ODOMETER NUMBER WHEELS

PLUG

HOUSING

FIGURE 8-17 (Courtesy of General Motors Corporation.)

miles traveled on a trip or the distance traveled between fuel stops. Electronic dash displays could use either an electrically driven mechanical odometer or a digital display odometer to indicate miles traveled. A small electric motor called a stepper motor is used to turn the number wheels of a mechanical-style odometer. A pulsed voltage is fed to this stepper motor, which moves in relation to the miles traveled. See Figure 8-17.

Digital odometers use LED, LCD, or VTF displays to indicate miles traveled. Since total miles must be retained when the ignition is turned off or the battery is disconnected, a special electronic chip must be used that will retain the miles traveled.

These special chips are called *nonvolatile random access memory* (NVRAM). "Nonvolatile" means that the information stored in the electronic chip is not lost when electrical power is removed. See Figure 8-18. Some cars use a chip called *electronically erasable programmable read-only memory* (EEPROM). Most digital odometers can read up to 999,999.9 km (621,388 miles); then the display indicates "error." If the chip is damaged or exposed to static electricity, it may fail to operate and "error" may appear.

FIGURE 8-18 (Courtesy of General Motors Corporation.)

NON-VOLATILE MEMORY ODOMETER

DIGITAL DISPLAY

NON-VOLATILE RAM

ELECTRONIC SPEEDOMETER AND ODOMETER SERVICE

If the speedometer and odometer fail to operate, the speed sensor should be the first item checked. With the vehicle safely raised off the ground and supported, disconnect the wires from the speed sensor near the output shaft of the transmission. Connect a multitester set on "AC volts" to the terminals of the speed sensor and rotate the drive wheels with the transmission in neutral. A good speed sensor should indicate approximately 2 V ac if the drive wheels are rotated by hand. If the speed sensor is working, check the wiring from the speed sensor to the dash cluster. If the wiring is good, the dash should be sent to a specialty repair facility. Consult your local dealer for the nearest authorized repair facility.

If the speedometer operates correctly but the mechanical odometer does not work, the odometer stepper motor, the number wheel assembly, or the circuit controlling the stepper motor is defective. If the digital odometer does not operate but the speedometer operates correctly, the dash cluster must be removed and sent to a specialized repair facility. If the odometer chip is defective, a replacement chip is available only through authorized sources and the original number of miles must be programmed into the replacement chip.

> **NOTE:** Some digital odometers only change (update) every 15 miles or whenever the ignition is turned off. Be certain to check for normal operation before attempting to repair the odometer.

Digital dash displays that use EEPROM odometer chips are the type most likely to update odometer readings periodically rather than continuously.

ELECTRONIC FUEL-LEVEL GAUGES

Electronic fuel-level gauges usually use the same fuel tank sending unit as that used on conventional fuel gauges. The tank unit consists of a float attached to a variable resistor. As the fuel level changes, the resistance of the sending unit changes. As the resistance of the tank unit changes, the dash-mounted gauge also changes. The only difference between a digital fuel-level gauge and a conventional needle-type is in the display. Digital fuel-level gauges can be either numerical (indicating gallons or liters remaining in the tank) or a bar-graph display. A bar graph consists of lights segments often corresponding to a gallon of fuel per segment. The electronic circuits inside the cluster light the corresponding number of gallons remaining or the number of segments, depending on the resistance of the tank sending unit. For example, a typical General Motors tank unit has 90Ω when the fuel tank is full and 0Ω when empty. Therefore, every decrease of 6Ω would decrease the display one segment if equipped with a 16-segment bar-graph fuel gauge. See Figure 8-19.

The diagnosis is the same as that described earlier for conventional fuel gauges. If the tests indicate that the dash unit is defective, usually the *entire* dash gauge assembly must be replaced.

DIGITAL FUEL MEASUREMENT

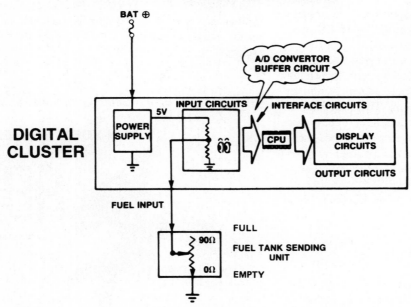

FIGURE 8-19 This analog-to-digital (A/D) converter is an electronic circuit designed to change analog (A) (changing) signals into digital (D) or "on" and "off" pulses which can then be processed by the circuits in the electronic display. (Courtesy of General Motors Corporation.)

COOLANT THERMISTOR

TEMPERATURE   **RESISTANCE**

FIGURE 8-20 *Typical water (coolant) tempera-ture sensor which uses a temperature-sensitive resistor (thermistor) to measure coolant temper-ature. (Courtesy of General Motors Corporation.)*

OIL PRESSURE SENDER

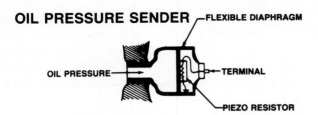

FIGURE 8-21 *(Courtesy of General Motors Corporation.)*

OTHER ELECTRONIC GAUGE DISPLAYS

Oil pressure, water temperature, and voltmeter reading are other commonly used electronic dash displays. Oil pressure is monitored by a variable-resistance sending unit threaded into an oil passage, usually near the oil filter. A typical oil pressure sending unit will have low resistance when the oil pressure is zero and higher resistance when the oil pressure is high. See Figures 8-20 and 8-21.

Water temperature is also sensed by a variable-re-sistance sending unit, usually located near the engine's thermostat. Similar to oil pressure, the higher the coolant temperature, the greater the number of segments that will be indicated based on the resistance of the coolant tempera-ture sensor.

> **NOTE:** The coolant temperature sensor for the dash display is usually a separate sensor from the cool-ant temperature sensor used by the engine com-puter.

A voltmeter is often included in a digital display and each segment of the display represents a particular voltage range. A warning light is often part of the electronic cir-cuits in the electronic display to warn the driver of high or low battery voltage.

WHAT ARE "SMART" DASH WARNING LIGHTS?

Most electronic dash clusters contain many electronic circuits for the processing of infor-mation from sensors and controlling the seg-ments or numbers displayed. Some clusters also receive signals about engine speed from the en-gine control computer. The electronics in the in-strument cluster compare engine speed with the oil pressure from the oil pressure sending unit. The "oil" tell-tale warning light will come on if the oil pressure is below acceptable limits for the speed of the engine. For example, an oil pressure of 12 psi may be acceptable at idle, but not acceptable if the engine is running above 2500 rpm. Following are typical "smart" low oil pressure warning light pressures and rpm's:

Warning Light "On"	Engine Speed (rpm)
4 psi or less	less than 1500
12 psi	2500
20 psi	4000
25 psi	7000

ELECTRONIC DASH INSTRUMENT DIAGNOSIS AND TROUBLESHOOTING

If one or more electronic dash gauges do not work cor-rectly, first check the "WOW" display that lights all seg-ments to full brilliance whenever the ignition switch is first switched on. If all segments of the display do *not* operate, the entire electronic cluster must be replaced in most cases. If all segments operate during the "WOW" display but do not function correctly after the "WOW" display, the prob-lem is most often due to a defective sensor or defective wiring to the sensor.

All dash instruments except the voltmeter use a var-iable-resistance unit as a sensor for the system being mon-itored. Most new-car dealers are required to purchase essential test equipment, including a test unit that permits the technician to insert various fixed resistance values in the suspected circuit. For example, if a 45-Ω resistance was put into the fuel gauge circuit that reads from 0 to 90 Ω, a properly operating dash unit should indicate one-half tank. The same tester can produce a fixed signal to test the opera-tion of the speedometer and tachometer. If this type of special test equipment is not available, the electronic dash instruments can be tested using the following procedure:

1. With the ignition switched off, unplug the wire(s) from the sensor for the function being tested. For example, if the oil pressure gauge is not functioning correctly, unplug the wire connector at the oil pressure sending unit.
2. With the sensor wire unplugged, turn the ignition switch on and wait until the "WOW" display stops. The display for the affected unit should indicate either fully lighted segments or no lighted segments, depending on the make of the vehicle and the type of sensor.
3. Turn the ignition switch "off." Connect the sensor wire lead to ground and turn the ignition switch "on." After the "WOW" display, the display should be opposite (either fully on or fully off) the results in step 2.

TESTING RESULTS

If the electronic display does function fully "on" and fully "off" with the sensor unplugged and then grounded, the problem is due to a defective sensor. If the electronic display fails to indicate fully "on" and fully "off" when the sensor wire(s) is opened and grounded, the problem is usually in the wiring from the sensor to the electronic dash or is a defective electronic cluster.

> **CAUTION:** Whenever working on or *near* any type of electronic dash display, always wear a wire attached to your wrist (wrist strap) connected to a good body ground, to prevent damaging the electronic dash with static electricity.

SUMMARY

1. Thermoelectric gauges require an instrument voltage regulator (IVR) to ensure accuracy. Electromagnetic gauges do not require an IVR.
2. Tell-tale lights warn the driver of high coolant temperature (248 to 258°F), low oil pressure (4 to 7 psi), or any other system selected.
3. All cars sold in the United States after 1967 are equipped with a dash warning light which lights in the event of a hydraulic failure in the braking system.
4. Electronic dash instruments may use LED, LCD, or VTF displays. LED displays require the most electrical power and are generally used for small displays. LCD displays use bright halogen bulbs behind the display to provide the necessary brightness for viewing in sunlight. VTF displays are very bright green and must be filtered or reduced in brightness for night time viewing.
5. Electronic speedometers use either an electronic needle or a digital display. Odometers on electronic dash displays use a stepper motor to move conventional odometer number wheels or an electronic display. If an electronic display is used for the odometer, a nonvolatile memory chip is used to retain the number of miles traveled.
6. Electronic displays for oil pressure, coolant temperature, and fuel level use the same basic sensors as those used in a conventional dash display. Most electronic dash clusters use a computer chip and other electronic circuits to control all aspects of dash operation. Most electronic dash displays are replaceable as an entire unit.

STUDY QUESTIONS

8-1. What is the difference between thermoelectric and electromagnetic gauges?

8-2. Explain how to diagnose dash instruments.

8-3. Explain the operation of LED, LCD, and VTF displays.

8-4. What is a "WOW" display?

8-5. Explain the difference between a stepper motor odometer and a digital electronic odometer display.

8-6. What are "smart" dash warning lights?

8-7. Explain digital dash diagnosis procedures.

MULTIPLE-CHOICE QUESTIONS

8-1. Technician A says that thermoelectric gauges require an instrument voltage regulator. Technician B says that if the power lead of the fuel gauge of an electromagnetic gauge is grounded, the dash unit reads empty. Which technician is correct?
(a) A only.
(b) B only.
(c) both A and B.
(d) neither A nor B.

8-2. The oil pressure light "lights" when oil pressure drops between 3 and 7 psi by:
(a) opening the circuit.
(b) shorting the circuit.
(c) grounding the circuit.
(d) conducting current to the dash light by oil.

8-3. The "hot" light comes on at about 250°F by:
(a) opening the circuit.
(b) shorting the circuit.
(c) grounding the circuit.
(d) conducting electricity to the dash light by steam.

8-4. The brightest electronic dash display is called:
(a) LED.
(b) VTF.
(c) LCD.
(d) analog.

8-5. Technician A says that LCD displays may be slow to work at low temperatures. Technician B says that LED displays require very low electrical power. Which technician is correct?
(a) A only.
(b) B only.
(c) both A and B.
(d) neither A nor B.

8-6. Technician A says that some electronic dashes use an analog speedometer. Technician B says that all electronic dash fuel gauges use a numerical display. Which technician is correct?

(a) A only.
(b) B only.
(c) both A and B.
(d) neither A nor B.

8-7. Digital odometers use _____ to retain the miles traveled.
(a) an EEPROM.
(b) a stepper motor.
(c) NVRAM.
(d) either (a) or (c).

8-8. When servicing an electronic dash:
(a) if the "WOW" does not display, the cluster must be replaced.
(b) individual gauges can be replaced if one is defective.
(c) the sensors used for an electronic dash are not replaceable.
(d) static electricity discharge can damage any electronic instrument cluster.

8-9. Technician A says that all halogen bulbs must not be touched with bare hands. Technician B says that it is normal for a VTF display to dim to 75% brightness whenever the headlights are "on." Which technician is correct?
(a) A only.
(b) B only.
(c) both A and B.
(d) neither A nor B.

8-10. Technician A says that the brake warning light is connected to the parking brake lever and the hydraulic portion of the brakes. Technician B says that the dash-mounted brake warning light comes on *only* if the brakes need to be relined. Which technician is correct?
(a) A only.
(b) B only.
(c) both A and B.
(d) neither A nor B.

9

Electrical Accessories

THIS CHAPTER INTRODUCES THE OPERATION, PARTS, AND BASIC DIAGNOSIS and troubleshooting of a number of electrical accessory systems. The systems covered in this chapter include:

1. Blower motors
2. Windshield wipers
3. Horns
4. Cruise control
5. Power windows
6. Power seats
7. Electric power door locks
8. Heated rear-window defogger
9. Radios and antenna
10. Power sunroofs
11. Power mirrors
12. Electronic leveling systems

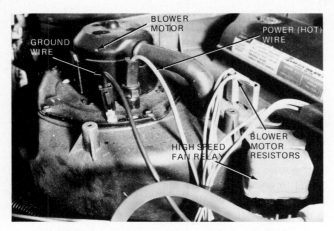

FIGURE 9-1 *Typical blower motor installation. The dropping resistors are always located near the blower motor to help keep the resistors cool.*

FIGURE 9-2 *Close-up view of blower motor resistors. If defective, replacement resistors are purchased as a unit, as shown.*

BLOWER MOTOR OPERATION

The same blower motor moves air inside the car for air conditioning, heat, and defrost. The fan switch controls the path that the current flows to the blower motor. The motor is usually a permanent-magnet, one-speed motor that operates at its maximum speed with full battery voltage. The switch gets current from the fuse panel with the ignition switch "on." The switch directs full battery voltage to the blower motor for high speed and through resistors for lower speeds. See Figures 9-1 and 9-2.

BLOWER MOTOR DIAGNOSIS

If the blower motor does not operate at any speed, the problem could be any of the following:

1. A defective ground wire or ground wire connection.
2. A defective blower motor (not repairable; must be replaced)
3. An open circuit in the power-side circuit, including fuse, wiring, or fan switch.

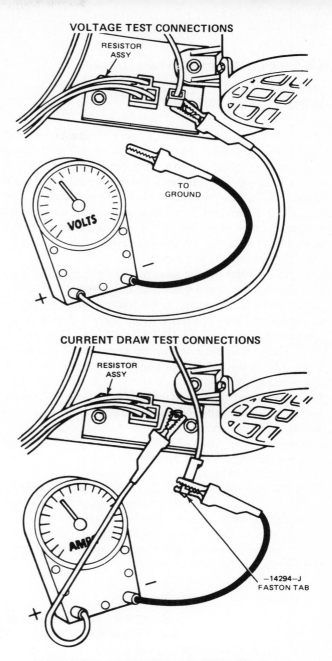

FIGURE 9-3 *Testing blower motor voltage and current draw. A typical blower motor should test approximately 3.5 A at 5 V on "low," 5.5 A at 7.5 V on "medium," and 8.5 A at 12.8 V on "high." (Courtesy of Ford Motor Company.)*

If the blower works on lower speeds but not on high speed, the problem is usually an "in-line" fuse or high-speed relay that controls the heavy current flow for high-speed operation. The high-speed fuse or relay usually fails as a result of internal blower motor bushing wear, which causes excessive resistance to motor rotation. At slow blower speeds, the resistance is not as noticeable and operates normally. The blower motor is a sealed unit and if defective, must be replaced as a unit. If the blower motor operates normally at high speed but not at any of the lower speeds, the problem could be melted wire resistors or a defective switch. See Figure 9-3.

WINDSHIELD WIPERS

The windshield wipers usually use a special two-speed electric motor. General Motors uses many different-shaped wiper motors. Most are compound-wound motors, which are a type of motor with both a series-wound field and a shunt field, which provides for two different speeds. The wiper switch provides the necessary electrical connections for either speed. Switches in the mechanical wiper motor assembly provide the necessary operation for "parking" and "concealing" of the wipers.

Other wiper motors usually use a permanent-magnet motor with a low-speed brush and a high-speed brush. The brush connects the battery to the internal windings of the motor and the two brushes provide for 2 different motor speeds.

The ground brush is directly opposite the low-speed brush. Off to the side of the low-speed brush is the high-speed brush. When current flows through the high-speed brush, the number of turns on the armature between the hot and ground brushes is less, and therefore the resistance is less. With less resistance, more current flows and the armature revolves faster. See Figures 9-4 or 9-5. See Figures 9-6 through 9-9 for a two-speed wiper motor using reverse-wound field coils.

Variable-delay wipers (also called "pulse wipers") use an electronic circuit with a variable resistor which controls the time of the charge and discharge of a capacitor. This charging and discharging of the capacitor controls the circuit for the operation of the wiper motors. See Figures 9-10 through 9-14.

HORNS

Automotive horns are usually wired directly to battery voltage from the fuse panel. The majority of automobiles, except most Fords, utilize a horn *relay*. With a relay, the horn button on the steering wheel or column completes a circuit to ground which closes a relay, and the heavy current flow required by the horn then travels from the relay to the horns. See Figure 9-15.

Horns are manufactured in several different tones or frequencies ranging between 1800 and 3550 Hz. Car manufacturers can select various horn tones for a particular car sound. When two horns are used, each has a different tone when operated separately, yet the sound combines when both are operated.

HORN DIAGNOSIS

To help determine the cause of a horn not operating, use a jumper wire and connect one end to the positive (+) post of the battery and the other end to the wire terminal of the horn itself. If the horn works, the problem is in the circuit supplying current to the horn. If the horn does not work, the horn itself could be defective or the mounting bracket may not be providing a good ground. See Figures 9-16 and 9-17.

If a replacement horn is required, attempt to replace it with a horn of the same tone as the original. The tone is usually indicated by a number or letter stamped on the body of the horn.

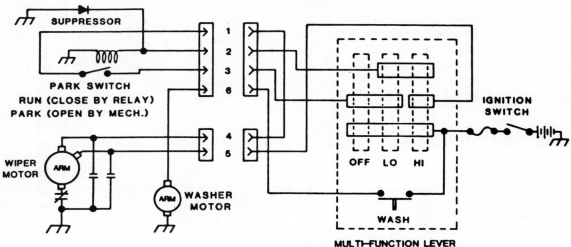

FIGURE 9-4   *Wiring diagram of a typical two-speed, three-brush wiper motor. (Courtesy of Pontiac Motor Division, GMC.)*

STANDARD TWO-SPEED WIPER-WASHER SYSTEM

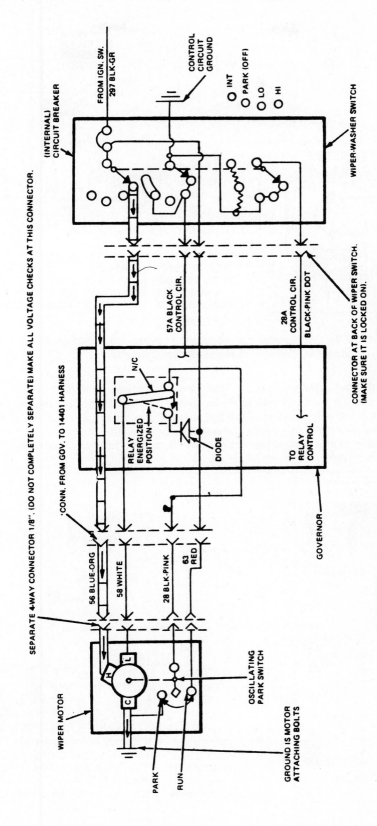

FIGURE 9-5 Wiring diagram of a typical two-speed, three-brush wiper motor showing high-speed operation. (Courtesy of Ford Motor Company.)

SEPARATE 4-WAY CONNECTOR 1/8". (DO NOT COMPLETELY SEPARATE.) MAKE ALL VOLTAGE CHECKS AT THIS CONNECTOR.

FROM IGN. SW. 297 BLK-GR

(INTERNAL) CIRCUIT BREAKER

CONTROL CIRCUIT GROUND

INT
PARK (OFF)
LO
HI

WIPER-WASHER SWITCH

57A BLACK CONTROL CIR.

28A CONTROL CIR.

BLACK-PINK DOT

CONNECTOR AT BACK OF WIPER SWITCH. (MAKE SURE IT IS LOCKED ON.)

CONN. FROM GOV. TO 14401 HARNESS

N/C

RELAY ENERGIZED POSITION

DIODE

TO RELAY CONTROL

GOVERNOR

56 BLUE-ORG
58 WHITE
28 BLK-PINK
63 RED

WIPER MOTOR

H
L
C

OSCILLATING PARK SWITCH

PARK
RUN

GROUND IS MOTOR ATTACHING BOLTS

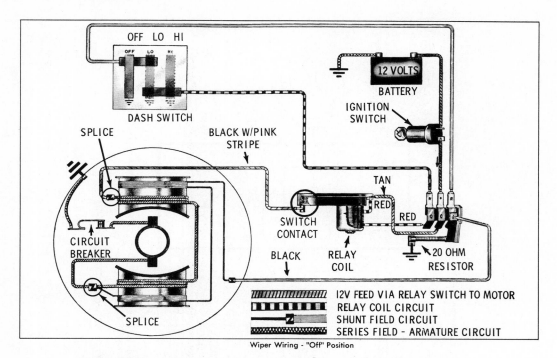

Wiper Wiring - "Off" Position

FIGURE 9-6 *GM round wiper motor circuit in the "off" position. (Courtesy of Oldsmobile Division, General Motors Corporation.)*

FIGURE 9-7 *GM round wiper motor in "low" speed. Note the current flow through the black wire at the bottom of the illustration. Current flow through this wire, starting at the splice inside the motor, causes current to flow through a shunt field coil winding that is wound in the opposite direction to the normal (series) field winding. (Courtesy of Oldsmobile Division, GMC.)*

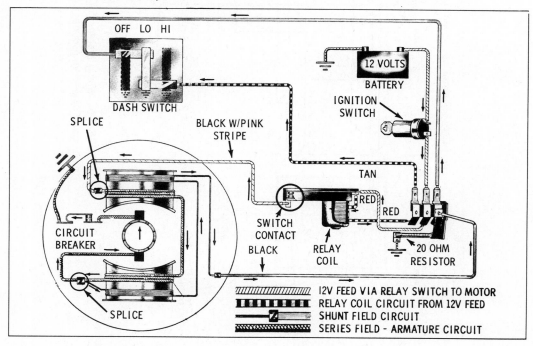

Wiper Wiring - "Lo" Speed

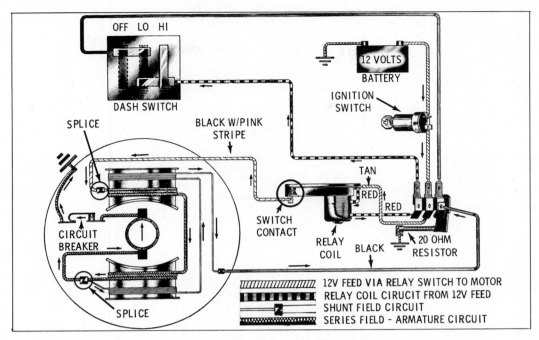

FIGURE 9-8 *GM round wiper motor in "high" speed. Current through the shunt winding flows through a 20-Ω resistor, not directly to ground through the switch. If the black shunt wire is broken (open circuit), the wiper motor would operate much faster than high speed. (Courtesy of Oldsmobile Division, GMC.)*

FIGURE 9-9 *GM round wiper motor in "park." If the relay "clicks," the relay is operating. Notice that the dash control switch must be properly grounded for proper operation of this style of wiper motor. (Courtesy of Oldsmobile Division, GMC.)*

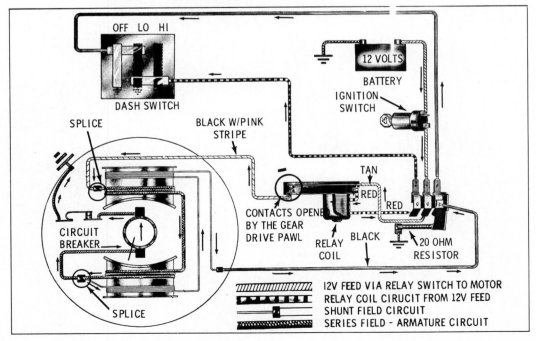

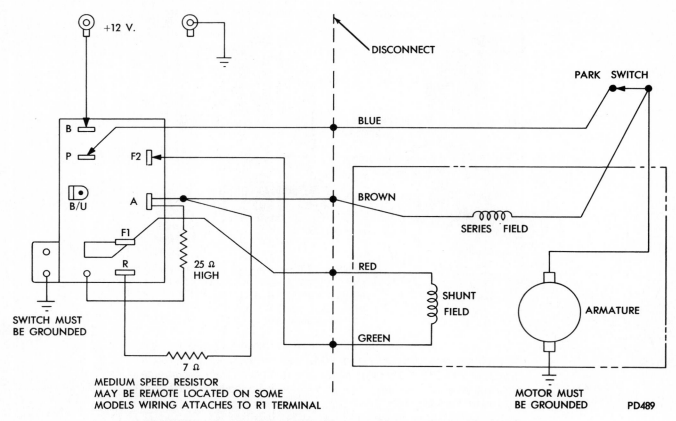

FIGURE 9-10 *Three-speed wiper motor wiring schematic. (Courtesy of Chrysler Corporation.)*

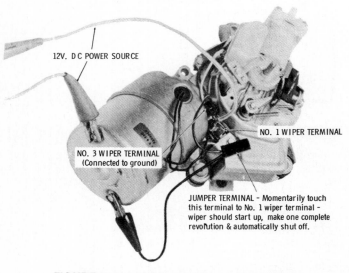

FIGURE 9-11 *Wiper motor operation check. (Courtesy of Oldsmobile Division, GMC.)*

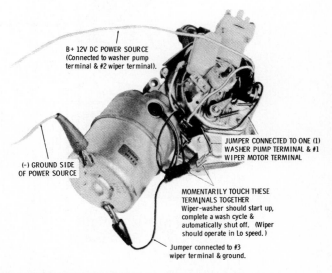

FIGURE 9-12 *Wiper washer operation check. (Courtesy of Oldsmobile Division, GMC.)*

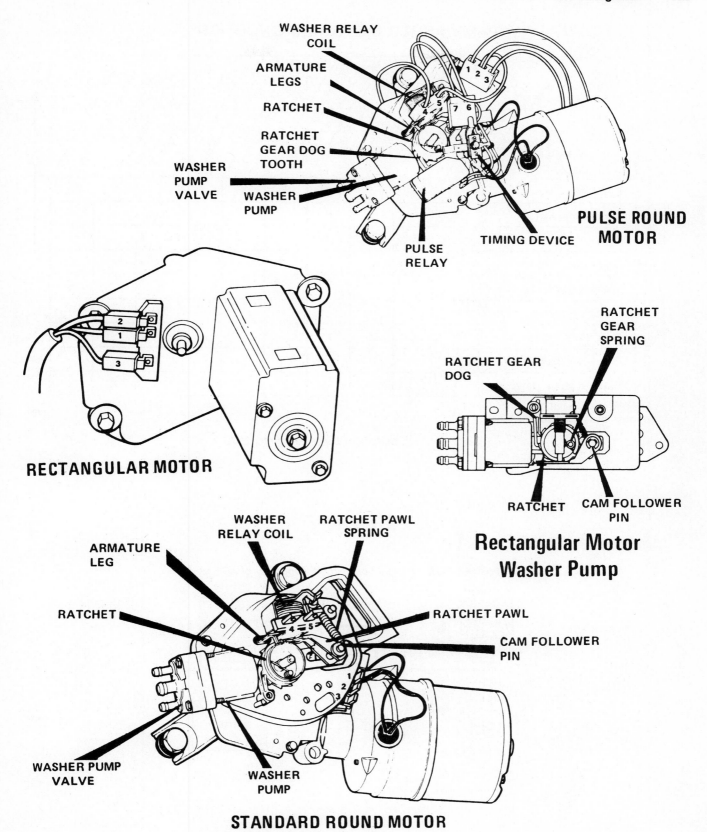

FIGURE 9-13 *Types of General Motor's wiper motors. (Courtesy of General Motors Corporation.)*

WIPER/WASHER MOTOR DIAGNOSTICS

		PROBABLE CAUSE	Wiper system inop — all modes	Wiper won't shut off	Blades cycle in & out of park	Wiper has Hi only, won't delay	Wiper inop in delay, OK in other modes	Wiper won't delay between delay wipes	Wiper has Lo speed only	Wiper intermittent operation	Washer does not operate properly in demand or program	Washer inoperative
Pulse		Defective cover/board	XX*	XX*		XX*	X	X			X	
		Motor defective	X			X			X	XX*		
		Park switch defective		X	X	X						
		Gear train damaged								X		
		Washer pump defective										X
Standard		Motor defective	X			XX*			X	XX*		
		Park switch defective		X	X	X						
		Gear train damaged								X		
		Washer pump defective									X	X

*Denotes most probable cause.

FIGURE 9-14   *(Courtesy of Delco Products, GMC.)*

FIGURE 9-15   *Typical horn circuit. (Courtesy of ASE.)*

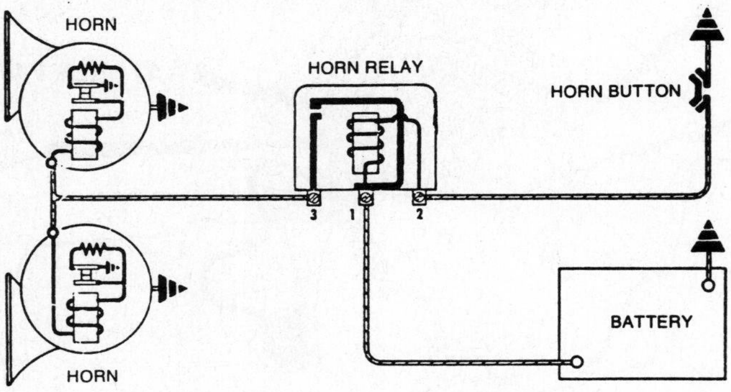

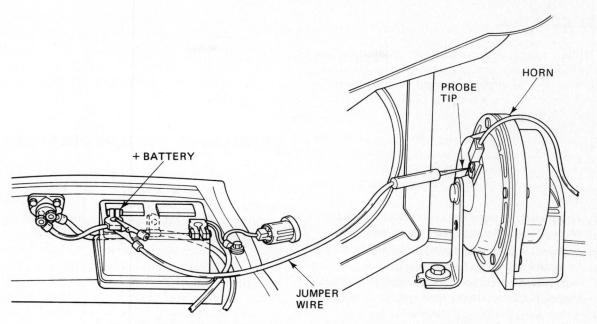

FIGURE 9-16 *To check operation, use a jumper wire between the positive (+) post of the battery and the power terminal. The horn should sound. If the horn fails to operate, the problem is either a poor ground of the horn (loose mounting) or a defective horn assembly. If the horn works, yet does not operate from the horn ring, the problem is a defective switch or defective wiring.*

FIGURE 9-17 *Horn diagnostic trouble chart (C.B. indicates a circuit breaker). (Courtesy of Ford Motor Company.)*

• Horn sounds continuously — vehicles without speed control.	• Short circuit between column disconnect and horn(s). • Short circuit in switch or column wiring.	• Service as required. • Service as required.
• Horn sounds continuously — vehicles with speed control.	• Short circuit between horn relay disconnect and horn. • Grounded circuit between horn switch and horn relay. • Open horn switch. • Defective horn relay.	• Service as required. • Service as required. • Replace horn switch. • Replace horn relay.
• Horn(s) inoperative.	• Fuse or C.B. burnt out. • Poor horn ground. • Horns out of adjustment. • Defective horn. • Open in wiring. • Defective horn switch. • Defective turn signal switch and wiring. • Defective horn relay	• Replace fuse or C.B. If fuse or C.B. goes again, check for short circuit. • Assure a good ground. • Adjust horn. • Replace horn. • Service wiring. • Replace horn switch. • Service or replace turn signal switch. and wiring. • Replace horn relay

CRUISE CONTROL

Cruise (speed) control is a combination of electrical and mechanical components designed to maintain a constant, set vehicle speed without driver pressure on the accelerator pedal. Major components of a typical cruise control system include (see Figure 9-18):

1. *Servo unit.* The servo unit attaches to the throttle linkage through a cable or chain. The servo unit controls the movement of the throttle by receiving a controlled amount of vacuum from a control unit.

2. *Transducer.* A transducer is an electrical and mechanical speed sensing and control unit.

3. *Speed set control.* A speed set control is a switch or control located on the steering column, steering wheel, dash, or console. Many cruise control units feature coast, accelerate, and resume functions.

4. *Safety release switches.* Whenever the brake pedal is depressed, the cruise control system is dis-
engaged through use of an electrical and vacuum switch, usually located on the brake pedal bracket. Both electrical and vacuum releases are used to be certain that the cruise control system is released, even in the event of failure of one of the release switches.

BASIC CRUISE CONTROL OPERATION

A typical cruise control system can be set only if the vehicle speed is above 30 mph or more. In a non-computer-operated system, the transducer contains a low-speed electrical switch that closes whenever the speed-sensing section of the transducer exceeds the minimum engagement speed. Most transducers operate by a speedometer cable driven off the transmission. The speedometer cable rotates a magnetic disk that applies a rotary force on a rubber clutch. As the rubber clutch tends to rotate, a tang on the clutch closes the low-speed switch whenever the road speed exceeds the minimum engagement speed. See Figure 9-19.

FIGURE 9-18 *Typical cruise (speed) control system components. (Courtesy of General Motors Corporation.)*

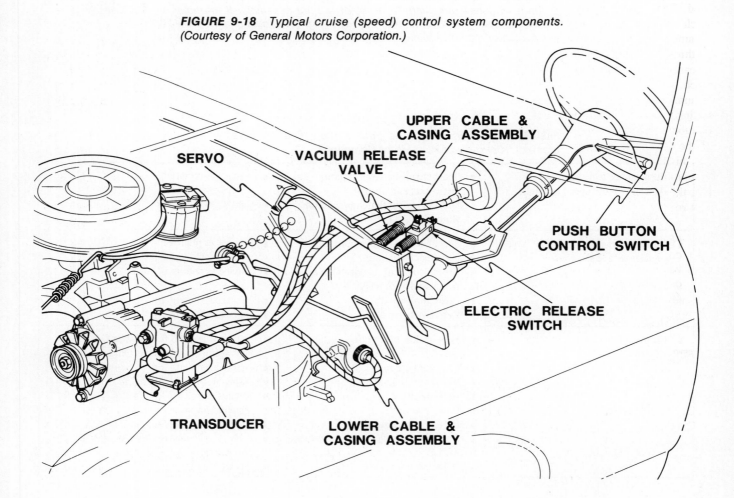

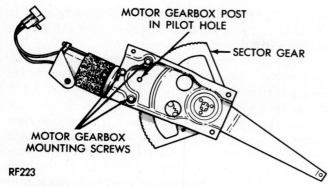

MOTOR GEARBOX POST
IN PILOT HOLE

SECTOR GEAR

MOTOR GEARBOX
MOUNTING SCREWS

RF223

FIGURE 9-24 *Typical electrical power window motor and window regulator assembly. The small gear on the motor moves the sector gear of the regulator, which moves the door glass up or down. (Courtesy of Chrysler Corporation.)*

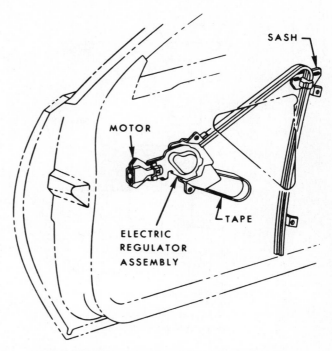

SASH

MOTOR

TAPE

ELECTRIC
REGULATOR
ASSEMBLY

FIGURE 9-25 *Typical power window regulator that uses a tape rather than gears to operate the window movement. (Courtesy of Pontiac Motor Division, GMC.)*

switch in the "on" ("run") position. This safety feature of power windows should never be defeated. Some manufacturers use a time delay for accessory power after the ignition switch is turned off. This feature permits the driver and passengers an opportunity to close all windows or operate other accessories for about 10 minutes or until a car door is opened after the ignition has been turned off.

Most power window systems use permanent-magnet (PM) electric motors. A PM motor can be run in the reverse direction simply by reversing the polarity of the two wires going to the motor. Most power window motors do not require that the motor be grounded to the body (door) of the vehicle. The ground for all the power windows is most often centralized near the driver's master control switch. The up-and-down motion of the individual window motors is controlled by double-pole, double-throw (DPDT) switches. These DPDT switches have five contacts and permit battery voltage to be applied to the power window motor and to reverse the polarity and direction of the motor. See Figure 9-23.

The power window motors rotate a mechanism called a window *regulator*. The window regulator is attached to the door glass and controls opening and closing of the glass. Door glass adjustments such as glass tilt and upper and lower stops are usually the same for both power and manual windows. See Figures 9-24 and 9-25.

TROUBLESHOOTING POWER WINDOWS

Before troubleshooting a power window problem, check for proper operation of all power windows. If one of the *control* wires that run from the independent to the master switch is cut (open), the power window may operate in just one direction. The window may go down but not up, or

vice-versa. However, if one of the *direction* wires that run from the independent switch to the motor is cut (open), the window may not operate in either direction. The direction wires and the motor must be electrically connected to permit operation and change of direction of the electric lift motor in the door. See Figures 9-23 and 9-26.

1. If *both* rear door windows fail to operate from the independent switches, check the operation of the window lock-out (if equipped) and the master control switch.
2. If one window can move in one direction only, check for continuity in the control wires (wires between the independent control switch and the master control switch).
3. If *all* windows fail to work or fail to work occasionally, check, clean, and tighten the *ground* wire(s) located either behind the driver's interior door panel or under the dash on the driver's side. A defective fuse or circuit breaker could also cause all the windows to fail to operate.
4. If one window fails to operate in either direction, the problem could be due to a defective window lift motor. The window could be stuck in the track of the door, which could cause the circuit breaker built into the motor to open the circuit to protect the wiring, switches, and motor from damage. To

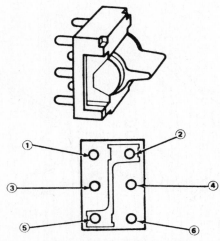

SINGLE POWER WINDOW SWITCH TEST

① POWER FEED, OR GROUND FROM DRIVERS SWITCH (ALLOWS OPERATION OF INDIVIDUAL WINDOWS FROM L.F. DOOR)

② HOT FEED FOR SINGLE SWITCH (BUS BAR TO PIN NO. 5)

③ MOTOR TERMINAL – UP

④ MOTOR TERMINAL – DOWN

⑤ HOT FEED

⑥ POWER FEED, OR GROUND FROM DRIVERS SWITCH (ALLOWS OPERATION OF INDIVIDUAL WINDOWS FROM L.F. DOOR)

FIGURE 9-26   *Typical independent door switch for the power windows. (Courtesy of Ford Motor Company.)*

check for a stuck door glass, attempt to move (even slightly) the door glass up and down, front and back, and side to side. If the window glass can move slightly in all directions, the power window motor should be able to at least move the glass. See Figure 9-27.

POWER SEATS

A typical power-operated seat includes a reversible electric motor and a transmission assembly which has three solenoids and six drive cables which turn to the six seat adjusters. A six-way power seat includes seat movement forward and backward, plus seat cushion movement up and down at the front and the rear. See Figure 9-28. The drive cables are very similar to speedometer cables because they rotate inside a cable housing and connect the power output of the seat transmission to a gear or screw jack assembly that moves the seat. A screw jack assembly is often called a *gearnut* and is used to move the front or back of the seat cushion up and down. See Figure 9-29. Between the electric motor and the transmission is usually a rubber coupling

that could permit the electric motor to continue to rotate in the event of a jammed seat. This coupling is designed to prevent motor damage.

Most power seats use a permanent-magnet (PM) motor that is reversible by simply reversing the polarity of the current sent to the motor by the seat switch. Most PM motors have a built-in circuit breaker to protect the motor from overheating. Many Ford power seat motors use three separate armatures inside one large permanent-magnet field housing. Some power seats use a series-wound electric motor with two separate field coils, one field coil for each direction of rotation. This type of power seat motor typically uses a relay to control the direction of current from the seat switch to the corresponding field coil of the seat motor. This type of power seat can be identified by the "click" heard whenever the seat switch is changed from up to down or front to back, or vise versa. The "click" is the sound of the relay switching the field coil current. Some power seats use as many as eight separate PM motors which operate all functions of the seat, including headrest height, seat length, and side balusters, in addition to the "normal" six-way power seat functions. Some power seats use a small air pump to inflate a bag or bags in the lower part of the back of the seat called the *lumbar* because it supports the lumbar section of the spine.

TROUBLESHOOTING POWER SEATS

Power seats are usually wired from the fuse panel to operate all the time without having to turn the ignition switch "on" ("run"). If a power seat does not operate or make any noise, the circuit breaker (or fuse, if equipped) should be checked first.

Step 1. Check the circuit breaker, usually located on the fuse panel using a test light. The test light should "light" on both sides of the circuit breaker even with the ignition "off." If the seat relay "clicks," the circuit breaker is functioning, but the relay or electric motor may be defective.

Step 2. Remove the screws or clips that retain the controls to the inner door panel or seat and check for voltage at the seat control.

Step 3. Also check the ground connection(s) at the transmission and clutch control solenoids (if equipped). The solenoids must be properly grounded to the vehicle body for the power seat circuit to operate.

If the power seat motor runs but does not move the seat, the most likely fault is a worn or defective rubber

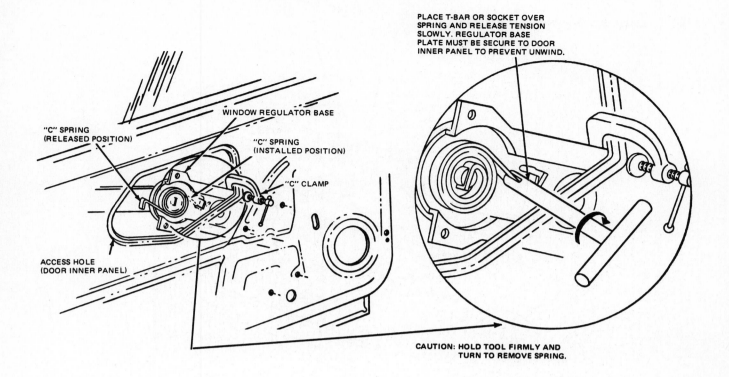

"C" SPRING (RELEASED POSITION)

WINDOW REGULATOR BASE

"C" SPRING (INSTALLED POSITION)

"C" CLAMP

ACCESS HOLE (DOOR INNER PANEL)

PLACE T-BAR OR SOCKET OVER SPRING AND RELEASE TENSION SLOWLY. REGULATOR BASE PLATE MUST BE SECURE TO DOOR INNER PANEL TO PREVENT UNWIND.

CAUTION: HOLD TOOL FIRMLY AND TURN TO REMOVE SPRING.

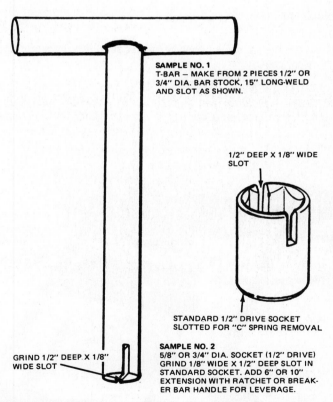

SAMPLE NO. 1
T-BAR — MAKE FROM 2 PIECES 1/2" OR 3/4" DIA. BAR STOCK, 15" LONG-WELD AND SLOT AS SHOWN.

1/2" DEEP X 1/8" WIDE SLOT

STANDARD 1/2" DRIVE SOCKET SLOTTED FOR "C" SPRING REMOVAL

SAMPLE NO. 2
5/8" OR 3/4" DIA. SOCKET (1/2" DRIVE) GRIND 1/8" WIDE X 1/2" DEEP SLOT IN STANDARD SOCKET. ADD 6" OR 10" EXTENSION WITH RATCHET OR BREAKER BAR HANDLE FOR LEVERAGE.

GRIND 1/2" DEEP X 1/8" WIDE SLOT

FIGURE 9-27 *Before removing the window regulator from a Ford car or truck, be certain that the regulator is in a fixed position, then remove the tension on the counterbalance spring before removal of the window regulator. (Courtesy of Ford Motor Company.)*

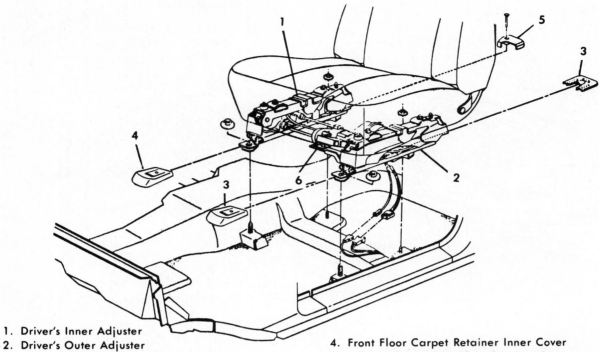

1. Driver's Inner Adjuster
2. Driver's Outer Adjuster
3. Front and Rear Floor Carpet Retainer Outer
   Cover
4. Front Floor Carpet Retainer Inner Cover
5. Adjuster Inner Rear Floor Support Cover
6. Adjuster Motor Cable Transmission Assembly

FIGURE 9-28   *Typical six-way power bucket seat. (Courtesy of Pontiac Motor Division, GMC.)*

FIGURE 9-29   *Typical six-way seat adjuster assembly with a three-motor, direct-drive system. (Courtesy of Pontiac Motor Division, GMC.)*

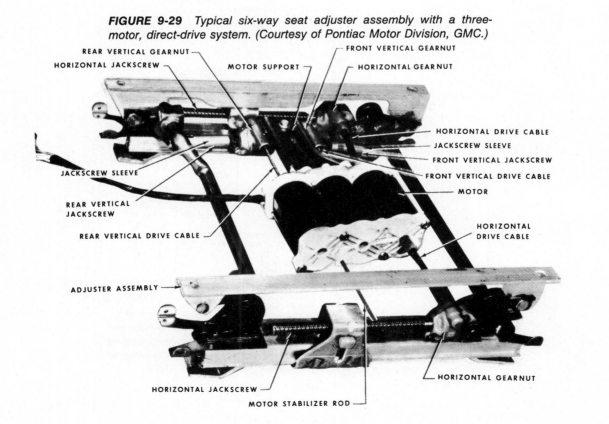

clutch sleeve between the electric seat motor and the transmission. See Figures 9-30 through 9-33.

If the seat relay clicks but the seat motor does not operate, the problem is usually due to a defective seat motor or defective wiring between the motor and the relay. If the power seat uses a motor relay, the motor has a double reverse-wound field for reversing the motor direction. This type of electric motor must be properly grounded. Perma-nent-magnet motors do not require grounding for operation. See Figures 9-34 and 9-35 for troubleshooting charts.

Power seats are often difficult to service because of restricted working room. If the entire seat cannot be removed from the vehicle because the track bolts are covered, attempt to remove the seat from the top of the power seat assembly. These bolts are almost always accessible regardless of seat position. See Figure 9-33.

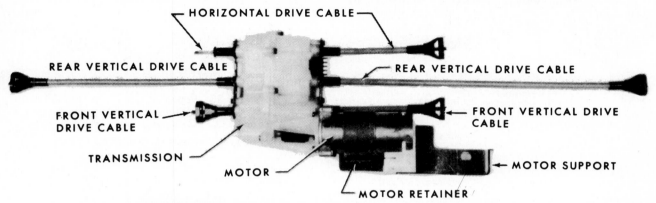

FIGURE 9-30 *Typical six-way power seat unit. One electric motor operates all six seat movements through the transmission and cable assembly. The rubber clutch mentioned in the text is located between the motor and the transmission. (Courtesy of Pontiac Motor Division, GMC.)*

FIGURE 9-31 *Six-way seat adjuster transmission component parts. The solenoid assembly (3) controls which gear is in mesh with the electric motor, which then rotates drive cables which move the seat. The direction is controlled by the master control switch, which determines the polarity of the current through the motor. The transmission assembly is usually replaced as a unit and individual internal parts may not be available. (Courtesy of Pontiac Motor Division, GMC.)*

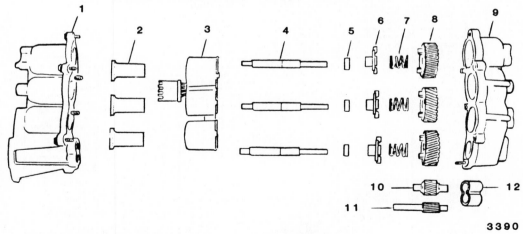

3390

1. SOLENOID HOUSING
2. SOLENOID PLUNGER
3. SOLENOID ASSEMBLY
4. SHAFT
5. DOG WASHER
6. DOG GEAR
7. DOG SPRING
8. HORIZONTAL & VERTICAL GEARS
9. GEAR HOUSING
10. IDLER GEAR
11. DRIVING GEAR
12. BUSHING

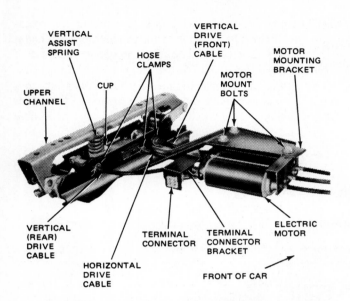

VERTICAL ASSIST SPRING
HOSE CLAMPS
VERTICAL DRIVE (FRONT) CABLE
MOTOR MOUNTING BRACKET
CUP
MOTOR MOUNT BOLTS
UPPER CHANNEL
VERTICAL (REAR) DRIVE CABLE
TERMINAL CONNECTOR
TERMINAL CONNECTOR BRACKET
ELECTRIC MOTOR
HORIZONTAL DRIVE CABLE
FRONT OF CAR

FIGURE 9-32 *Typical power bench seat track components. (Courtesy of Ford Motor Company.)*

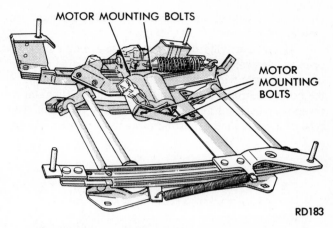

MOTOR MOUNTING BOLTS
MOTOR MOUNTING BOLTS
RD183

FIGURE 9-33 *Power seat unit showing the relationship and mounting methods to the bottom seat cushion and to the floor of the vehicle. (Courtesy of Chrysler Corporation.)*

FIGURE 9-34 *Six-way power seat troubleshooting chart. (Courtesy of Ford Motor Company.)*

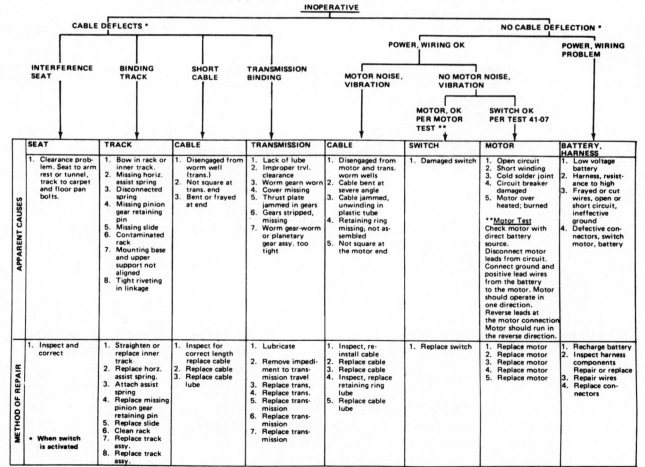

	SEAT	TRACK	CABLE	TRANSMISSION	CABLE	SWITCH	MOTOR	BATTERY, HARNESS
APPARENT CAUSES	1. Clearance problem. Seat to arm rest or tunnel, track to carpet and floor pan bolts.	1. Bow in rack or inner track. 2. Missing horiz. assist spring 3. Disconnected spring 4. Missing pinion gear retaining pin 5. Missing slide 6. Contaminated rack 7. Mounting base and upper support not aligned 8. Tight riveting in linkage	1. Disengaged from worm well (trans.) 2. Not square at trans. end 3. Bent or frayed at end	1. Lack of lube 2. Improper trvl. clearance 3. Worm gearn worn 4. Cover missing 5. Thrust plate jammed in gears 6. Gears stripped, missing 7. Worm gear-worm or planetary gear assy. too tight	1. Disengaged from motor and trans. worm wells 2. Cable bent at severe angle 3. Cable jammed, unwinding in plastic tube 4. Retaining ring missing; not assembled 5. Not square at the motor end	1. Damaged switch	1. Open circuit 2. Short winding 3. Cold solder joint 4. Circuit breaker damaged 5. Motor over heated; burned **Motor Test** Check motor with direct battery source. Disconnect motor leads from circuit. Connect ground and positive lead wires from the battery to the motor. Motor should operate in one direction. Reverse leads at the motor connection Motor should run in the reverse direction.	1. Low voltage battery 2. Harness, resistance to high 3. Frayed or cut wires, open or short circuit, ineffective ground 4. Defective connectors, switch motor, battery
METHOD OF REPAIR	1. Inspect and correct • When switch is activated	1. Straighten or replace inner track. 2. Replace horz. assist spring. 3. Attach assist spring 4. Replace missing pinion gear retaining pin 5. Replace slide 6. Clean rack 7. Replace track assy. 8. Replace track assy.	1. Inspect for correct length replace cable 2. Replace cable 3. Replace cable lube	1. Lubricate 2. Remove impediment to transmission travel 3. Replace trans. 4. Replace trans. 5. Replace transmission 6. Replace transmission 7. Replace transmission	1. Inspect, reinstall cable 2. Replace cable 3. Replace cable 4. Inspect, replace retaining ring lube 5. Replace cable lube	1. Replace switch	1. Replace motor 2. Replace motor 3. Replace motor 4. Replace motor 5. Replace motor	1. Recharge battery 2. Inspect harness components Repair or replace 3. Repair wires 4. Replace connectors

Chart headings above table:

INOPERATIVE
CABLE DEFLECTS •
NO CABLE DEFLECTION •
INTERFERENCE SEAT
BINDING TRACK
SHORT CABLE
TRANSMISSION BINDING
POWER, WIRING OK
POWER, WIRING PROBLEM
MOTOR NOISE, VIBRATION
NO MOTOR NOISE, VIBRATION
MOTOR, OK PER MOTOR TEST **
SWITCH OK PER TEST 41-07

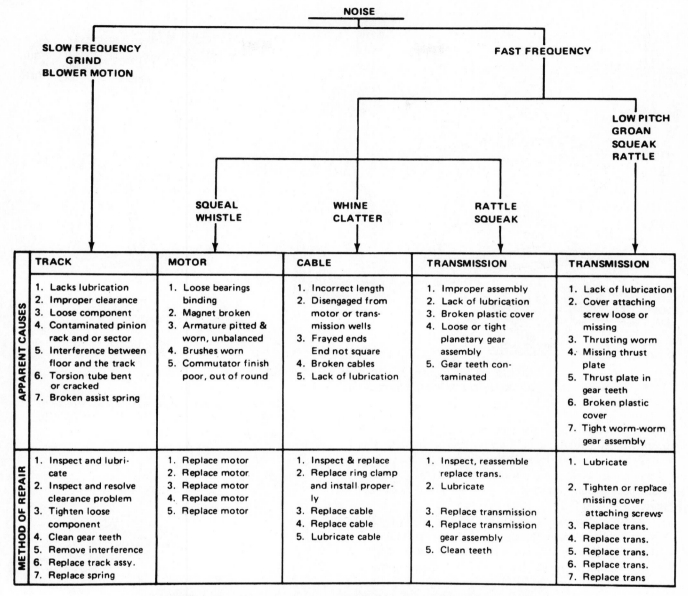

FIGURE 9-35 *Six-way seat noise diagnostic chart. (Courtesy of Ford Motor Company.)*

	TRACK	MOTOR	CABLE	TRANSMISSION	TRANSMISSION
APPARENT CAUSES	1. Lacks lubrication 2. Improper clearance 3. Loose component 4. Contaminated pinion rack and or sector 5. Interference between floor and the track 6. Torsion tube bent or cracked 7. Broken assist spring	1. Loose bearings binding 2. Magnet broken 3. Armature pitted & worn, unbalanced 4. Brushes worn 5. Commutator finish poor, out of round	1. Incorrect length 2. Disengaged from motor or transmission wells 3. Frayed ends End not square 4. Broken cables 5. Lack of lubrication	1. Improper assembly 2. Lack of lubrication 3. Broken plastic cover 4. Loose or tight planetary gear assembly 5. Gear teeth contaminated	1. Lack of lubrication 2. Cover attaching screw loose or missing 3. Thrusting worm 4. Missing thrust plate 5. Thrust plate in gear teeth 6. Broken plastic cover 7. Tight worm-worm gear assembly
METHOD OF REPAIR	1. Inspect and lubricate 2. Inspect and resolve clearance problem 3. Tighten loose component 4. Clean gear teeth 5. Remove interference 6. Replace track assy. 7. Replace spring	1. Replace motor 2. Replace motor 3. Replace motor 4. Replace motor 5. Replace motor	1. Inspect & replace 2. Replace ring clamp and install properly 3. Replace cable 4. Replace cable 5. Lubricate cable	1. Inspect, reassemble replace trans. 2. Lubricate 3. Replace transmission 4. Replace transmission gear assembly 5. Clean teeth	1. Lubricate 2. Tighten or replace missing cover attaching screws· 3. Replace trans. 4. Replace trans. 5. Replace trans. 6. Replace trans. 7. Replace trans

Noise → SLOW FREQUENCY GRIND BLOWER MOTION → TRACK

Noise → FAST FREQUENCY → SQUEAL WHISTLE → MOTOR; WHINE CLATTER → CABLE; RATTLE SQUEAK → TRANSMISSION

FAST FREQUENCY → LOW PITCH GROAN SQUEAK RATTLE → TRANSMISSION

WHAT EVERY DRIVER SHOULD KNOW ABOUT POWER SEATS

Power seats use an electric motor or motors to move the position of the seat. These electric motors turn small cables that operate mechanisms that move the seat. *Never* place rags, newspapers, or any other object under a power seat. Even ice scrappers can get caught between moving parts of the seat and often cause serious damage or jamming of the power seat.

ELECTRIC POWER DOOR LOCKS

Electric power door locks use either a solenoid or a permanent-magnet (PM) motor to lock or unlock all car door locks from a control switch or switches. Large (heavy) solenoids were typically used before the mid-1970s. These solenoids usually used two-wire connections which carried a high-ampere current through a relay controlled by a door lock switch. With a solenoid-style door lock, only one of the two wires is used at any one time. If current flows through one wire through the solenoid, the door locks; the door unlocks when current flows through the other wire to the solenoid. The solenoids must be grounded to the metal

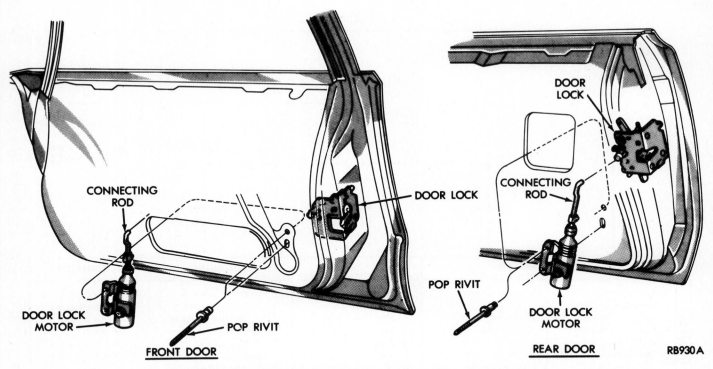

FIGURE 9-36 *Typical power door lock installation with PM motors. (Courtesy of Chrysler Corporation.)*

of the door to complete the electrical circuit. Because of constant opening and closing of a typical car door, a solenoid-style power door lock frequently vibrates loose from the mounting inside the door and fails to operate because of the poor ground connection with the metal door. See Figures 9-36 and 9-37.

Most electric door locks use a permanent-magnet (PM) reversible electric motor that operates the lock-activating rod. PM reversible motors do not require grounding because, similar to power windows, the motor control is determined by the polarity of the current through the two motor wires. Some two-door cars do *not* use a power door lock relay because the current flow for only two PM motors can be handled through the door lock switches. However, most four-door cars and vans with power locks on rear and side doors use a relay to control the current flow necessary to operate four or more power door lock motors. The door lock relay is controlled by the door lock switch and is commonly the location of the one and only *ground* connection for the entire door lock circuit. See Figure 9-38.

HEATED REAR-WINDOW DEFOGGERS

An electrically heated rear-window defogger system uses an electrical grid baked on the glass that warms the glass and clears it of fog or frost. The rear window is also called

FIGURE 9-37 *Typical power door lock assembly showing the linkage. (Courtesy of Pontiac Motor Division, General Motors Corporation.)*

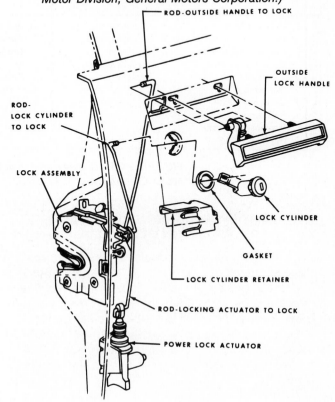

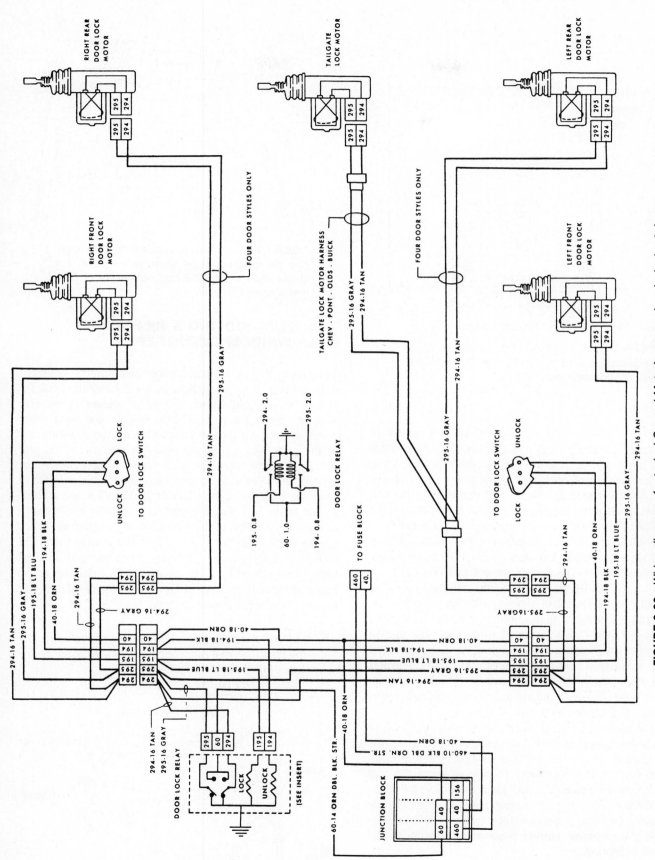

FIGURE 9-38 *Wiring diagram of a typical General Motor's power door lock circuit for a four-door. Some two-door models are similar, except that they may not use a relay. (Courtesy of General Motors Corporation.)*

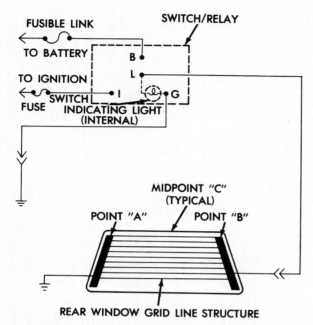

FIGURE 9-39 *Typical wiring diagram of an electrically heated rear-window defogger. (Courtesy of Chrysler Corporation.)*

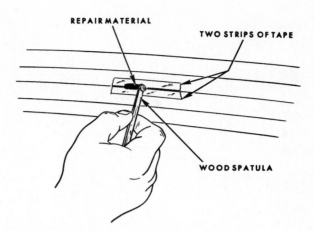

FIGURE 9-40 *A broken rear-window defogger can be repaired if the damaged area is not longer than 1½ in. (3.8 cm). (Courtesy of General Motors Corporation.)*

TROUBLESHOOTING A HEATED REAR-WINDOW DEFOGGER

Troubleshooting a nonfunctioning rear-window defogger unit involves using a test light or a voltmeter to check for voltage to the grid. If no voltage is present at the rear window, check for voltage at the switch and relay timer assembly. A poor ground connection on the opposite side of the grid from the power side can also cause the rear defogger not to operate. Because most defogger circuits use an indicator light switch and a relay timer, it is possible to have the indicator light on even if the wires are disconnected at the rear window grid. A voltmeter can be used to test the operation of the rear-window defogger grid. See Figure 9-41. With the negative (−) test terminal attached to a good body ground, carefully probe the grid conductors. There should be a decreasing voltage reading as the probe is moved from the power ("hot") side of the grid toward the ground side of the grid.

a *backlight*. The rear-window defogger system is controlled by a driver-operated switch and a timer relay. The timer relay is necessary because the window grid can draw up to 30 A, and continued operation would put a strain on the battery and the charging system. See Figure 9-39. Generally, the timer relay permits current to flow through the rear-window grid for only 10 minutes. If the window is still not clear of fog after 10 minutes, the driver can turn the defogger on again, but after the first 10 minutes any additional defogger operation is limited to 5 minutes.

Electric grid-type rear-window defoggers can be damaged easily by careless cleaning or scraping of the inside of the rear-window glass. Short broken sections of the rear-window grid can be repaired using a special epoxy-based electrically conductive material. If more than one section is damaged or if the damaged grid length is greater than approximately 1½ in. (3.8 cm), a replacement rear-window glass may be required to restore proper defogger operation. See Figure 9-40.

The electrical current through the grids depends, in part, on the temperature of the conductor grids. As the temperature decreases, the resistance of the grids decreases and the current flow increases, helping to warm the rear glass. As the temperature of the glass increases, the resistance of the conductor grids increases and the current flow decreases. Therefore, the defogger system tends to self-regulate the electrical current requirements to match the need for defogging.

FIGURE 9-41 *A voltmeter can be used to test the operation of a rear-window defogger. (Courtesy of Chrysler Corporation.)*

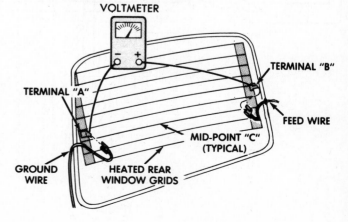

The Breath Test

It is difficult to test for the proper operation of all grids of a rear-window defogger unless the rear window happens to be covered with fog. A common "trick" that works is to turn on the rear defogger and exhale onto the outside of the rear-window glass. Similar to people cleaning eyeglasses with their breath, this procedure produces a temporary fog on the glass and the operation of all sections of the rear grids can quickly be checked for proper operation.

RADIOS

The power feed for automobile radios should be fused to an ignition switch–controlled circuit that permits radio operation only when the ignition switch is in the "on" ("run") or "accessory" positions. All radios also use electrical connections for an antenna and for one or more speakers. Most newer radios are called ETRs, meaning "electronically tuned receivers."

AM Reception. AM is an abbreviation for *amplitude modulation*, which is a method of varying the carrier signal in such a way as to vary the amplitude or strength of the signal. Frequencies range from 550 to 1600 kHz (kilohertz) and can be received a long distance from the transmitting station because the signal waves are reflected by the atmosphere (ionosphere). The AM method of transmitting is subject to noise and is, therefore, more sensitive to weak stations and lack of a properly functioning radio antenna. See Figure 9-42.

FIGURE 9-42 *AM signals can be reflected by the ionosphere and travel a greater distance than FM signals. (Courtesy of Chrysler Corporation.)*

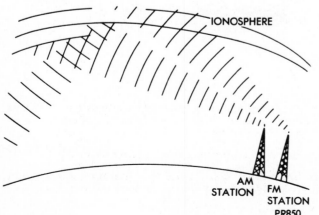

IONOSPHERE

AM STATION   FM STATION

PR850

FM Reception. FM is an abbreviation for *frequency modulation*, which is a method of varying the frequency of the carrier wave to represent the audio broadcast signal. FM is broadcast between 88 and 108 MHz and because the frequency is so high, the signal is *not* reflected by the atmosphere and the range is limited to line-of-sight distances. Because the radio antenna must be seen by the transmitting antenna, the signal can easily be blocked by a building or a hill. In cities where the radio signals are strong, the waves can bounce off tall buildings, which can provide reception to areas that are not in sight of the transmitting antenna. FM reception is usually noise-free, due to the fact that the radio receives changes in frequency rather than amplitude changes, which often contain noise.

Stereo. In stereo radio, two different sounds are broadcast and received separately to provide a fuller, more realistic sound. Originally, only FM-type transmissions could broadcast stereo sound, but since the mid-1980s, some cars are equipped to receive *AM* stereo. For best stereo separation, the car manufacturers design sound systems so that the passengers are sitting between the separate speakers.

Some factory sound systems use the lower-front door panels and the rear deck behind the rear seat, while other systems use the front dash and rear deck locations. Some manufacturers use the left speakers for the left stereo channel and the right speakers for the right channel. Other manufacturers "cross-fire" the speakers, left front and right rear speakers using the left channel and the right front and left rear speakers using the right channel.

SPEAKERS

Good-quality speakers are the key to a proper-sounding radio or sound system. Replacement speakers should be securely mounted and wired according to the correct *polarity*. All speakers used on the same radio or amplifier should have the same internal coil resistance, called *impedance*. If unequal-impedance speakers are used, sound quality may be reduced and serious damage to the radio may result.

The wire used for speakers should be as large a wire (as low a gauge number) as practical to be assured that full power is reaching the speakers. Typical "speaker wire" is about 22 gauge (0.35 mm^2), yet tests conducted by audio engineers have concluded that increasing the wire gauge to 14 (2.0 mm^2) or larger greatly increases sound quality. *All* wiring connections should be soldered after being certain that all speaker connections have the correct polarity. See Chapter 6 for soldering procedures.

Be careful when installing additional audio equipment on a General Motors vehicle or other radios that use

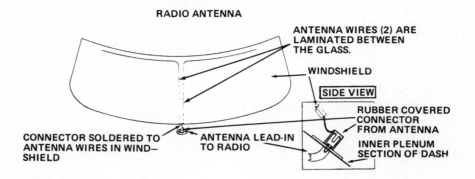

RADIO ANTENNA

ANTENNA WIRES (2) ARE LAMINATED BETWEEN THE GLASS.

WINDSHIELD

SIDE VIEW

RUBBER COVERED CONNECTOR FROM ANTENNA

INNER PLENUM SECTION OF DASH

CONNECTOR SOLDERED TO ANTENNA WIRES IN WIND—SHIELD

ANTENNA LEAD-IN TO RADIO

IMPORTANT:
For maximum efficiency, the antenna trimmer must be adjusted to exact peak (loudest signal) with radio tuned to a weak station near 1400 KC (140) and volume on full.
TO TEST ANTENNA FOR SHORT CIRCUIT
1. Remove antenna lead-in from radio.
2. Connect 12 volt test light to 12 volt source and to antenna lead-in center terminal.
  A. If test light is "OFF", antenna is not shorted.
  B. If test light is "ON", disconnect antenna from lead-in socket at base of windshield. If test light turns off, short circuit was in short wire to windshield. Tape it and position it to prevent short circuit from occuring. If test light stays on, remove lead-in and repair or replace as required.

A whip antenna can be adjusted to 25-30 inches and temporarily connected to the radio to determine if antenna or lead-in is at fault. If there is considerable improvement in reception, the connector at the base of the windshield should be inspected for corrosion and loose connections. The antenna wires can be visually inspected for breaks. A crack in the glass across one of the wires would probably break the wire. If the problem is in the antenna wires or wire to the connector and cannot be repaired, a whip antenna can be installed or the windshield replaced.

FIGURE 9-43 Typical windshield antenna.
(Courtesy of Oldsmobile Division, GMC.)

a two-wire speaker connection called a *floating ground* system. Other systems run only one power ("hot") lead to each speaker and ground the other speaker lead to the body of the vehicle.

Regardless of radio speaker connections used, *never* operate any radio without the speakers connected, or a transistor in the radio may be damaged due to the open speaker circuit.

ANTENNAS

Am radios operate best with as long an antenna as possible, but FM reception is best when the antenna height is exactly 31 in. (79 cm). Most fixed-length antennas are, therefore, exactly this height. Even the horizontal section of a windshield antenna is 31 in. (79 cm) long. See Figure 9-43.

A defective antenna will be most noticeable on AM radio reception. If poor or reduced reception is the problem, attempt first to adjust the *antenna trimmer*.

ANTENNA TRIMMERS

The antenna trimmer is an external adjustment of a radio circuit that properly matches the antenna to the radio. Whenever installing a new or replacement radio antenna, or moving the antenna to another location, the antenna trimmer adjustment should be made. Follow the procedure shown in Figure 9-44 for best AM reception.

FIGURE 9-44 Radio antenna trimmer adjustment procedure. (Courtesy of Oldsmobile Division, GMC.)

The antenna trimmer adjustment matches the antenna coil in the radio to the car antenna. Only AM radios, or the AM part of AM/FM radios, need this adjustment.

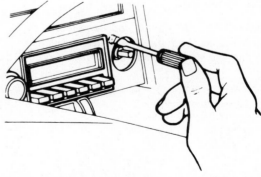

1. Tune the radio to a weak AM station near 1400 KHz. Turn the volume all the way up. You should barely hear the station.

2. Remove the right inner and outer knobs.

3. Use a small screwdriver to adjust the trimmer screw. Adjust the screw for the loudest volume.

4. Reinstall the control knobs.

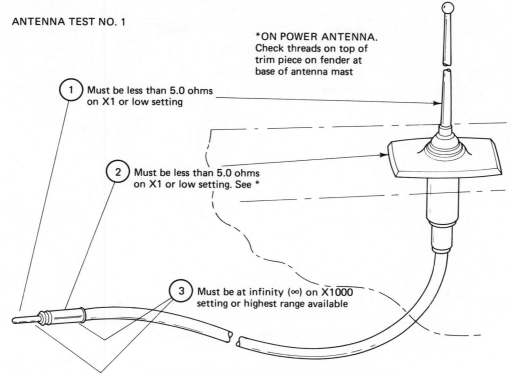

RADIO ANTENNA TEST (ALL ANTENNA TYPES)
With antenna installed on vehicle and cable unplugged from radio, perform the following resistance tests with an ohmmeter. Probes must contact antenna at points specified by arrowheads.

ANTENNA TEST NO. 1

*ON POWER ANTENNA.
Check threads on top of trim piece on fender at base of antenna mast

1. Must be less than 5.0 ohms on X1 or low setting

2. Must be less than 5.0 ohms on X1 or low setting. See *

3. Must be at infinity (∞) on X1000 setting or highest range available

FIGURE 9-45 *If all ohmmeter readings are satisfactory, the antenna is good. (Courtesy of Ford Motor Company.)*

ANTENNA TESTING

If the antenna or lead-in cable is broken (open), FM reception will be heard, but may be weak and there will be *no* AM reception. An ohmmeter should read infinity between the center antenna lead and the antenna case. For proper reception and lack of noise, the case of the antenna must be properly grounded to the vehicle body. See Figure 9-45.

POWER ANTENNAS

Most power antennas use a circuit breaker and a relay to power a reversible electric motor that moves a nylon cord attached to the antenna mast. Some cars have a dash-mounted control that can regulate antenna mast height and/or operation, while many operate automatically when the radio is turned on and off. The power antenna assembly is usually mounted between the outer and inner front fender or in the rear quarter panel. The unit contains the motor, a spool for the cord, and upper and lower limit switches. See Figures 9-46 and 9-47. The power antenna mast tests the same as a fixed mast antenna. (Infinity ohms should be noted on an ohmmeter when tested between the center an-

tenna terminal and the housing or ground.) Except for cleaning and mast replacement, most power antennas are either replaced as a unit or repaired by specialty shops.

Many power antenna problems can be prevented by making certain that the drain holes in the motor housing are not plugged with undercoating, leaves, or dirt. All power antennas should be kept clean by wiping the mast with a soft cloth and lightly oiling with a light oil.

COMMON RADIO AND TAPE PLAYER TERMS

Digital tuning. This highly accurate method of tuning uses a quartz frequency to reference station signals. This fixed frequency will not drift once tuned and permits seek-and-scan tuning.

Dynamic noise reduction (DNR). These circuits reduce noise in FM broadcasts and tapes.

Presets (memory). These radio circuits recall tuner frequency when a button is pushed, similar to the old pushbutton tuning.

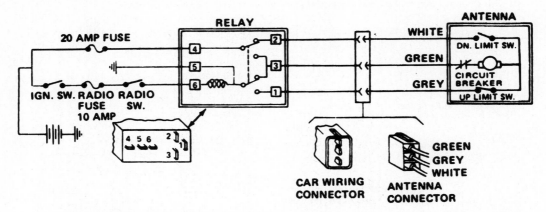

FIGURE 9-46 *Typical power antenna wiring diagram. (Courtesy of Pontiac Motor Division, GMC.)*

FIGURE 9-47 *Typical power antenna component parts. (Courtesy of Oldsmobile Division, GMC.)*

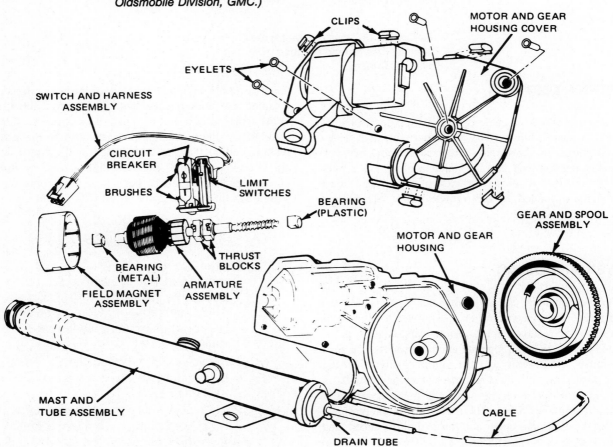

Scan tuning. These circuits in a digital radio sample all readable stations for a few seconds before tuning in the next station.

Seek tuning. These circuits in a digital radio change the tuner to the next station. Some radios can seek up or down the frequency range as determined by two separate "seek" buttons.

Auto eject. To prevent possible damage to the tape and tape player heads, the tape is ejected away from the heads whenever the ignition is switched "off." Other tape players release internal pressure on the tape when the ignition is turned off to prevent damage. Some auto-eject tape players eject the tape at the end of the tape.

Auto reverse. The tape player automatically plays the other side of the tape by reversing the tape direction at the end of each side.

Dolby noise reduction. This is a noise-suppression system, named for its inventor, that is encoded on the tape during the recording process and decoded during playback. Dolby B is the original and reduces tape "hiss" or noise by filtering out the frequencies usually associated with tape noise. Dolby C is a newer and more effective noise suppression system than Dolby B.

FM sensitivity. This is a measurement of how well an FM receiver can process weak signals usually listed in dBf; the smaller the number, the better the tuner can receive weak stations.

Frequency response. This specification indicates the range of frequencies that a component reproduces. The numbers given in the specifications should also include decibel (dB) tolerances. Human hearing is from 20 to 20,000 cycles per second (hertz); therefore, the more of this range a component can reproduce, the better.

Total harmonic distortion. Amplifiers are rated in watts of power and total harmonic distortion (THD). The power is expressed in watts per channel produced by the amplifier as an average of the highest and lowest values. The averaging method used is called root mean square (rms), and specifications usually also indicate THD as a percentage number. The THD number should be less than 1.0%, and the lower the percentage, the better the amplifier.

Signal-to-noise ratio. This ratio of background tape noise to music program is measured in decibels. The higher the number, the better the tape player.

Wow and flutter. The amount of variation in speed of the tape is expressed in percent. Wow is low-speed variation and flutter is high-speed variation. The lower the values, the better the performance.

Chassis size. Most car radios are sized according to standards established by the German Industrial Norm (DIN). These standard sizes are called: standard, mini, compact, and universal. The dimensions are given in the following order: width, height, and then depth.

POWER SUNROOFS

Most power-operated sunroofs use one reversible PM electric motor. The typical sunroof is constructed into a unit that can be removed entirely from the car after removing the interior overhead fabric material (headliner). The sunroof assembly usually consists of a metal or fiberglass pan equipped with water drains at each corner. See Figures 9-48 and 9-49.

The sunroof itself is mounted on rollers in a track assembly and moved with steel cables attached to the electric motor drive. Most adjustments of the sunroof are accessible only if the headliner is at least partially removed. Most sunroofs have adjustments for side to side, front to back, plus up and down, and all adjustments must be correct to prevent water leaks and proper operation.

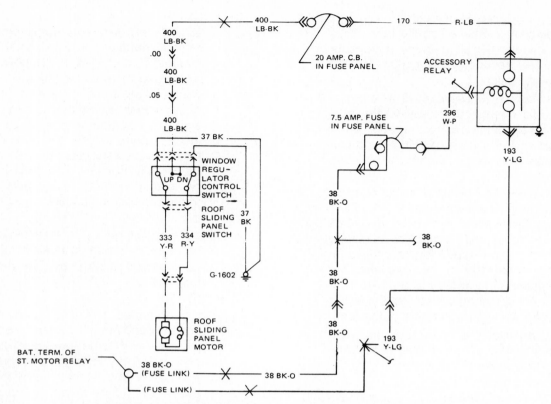

FIGURE 9-48 *Typical power sunroof wiring diagram. (Courtesy of Ford Motor Company.)*

FIGURE 9-49 *Typical wiring location for the power sunroof. The inside headliner usually has to be removed for most sunroof adjustments and part replacement. (Courtesy of Ford Motor Company.)*

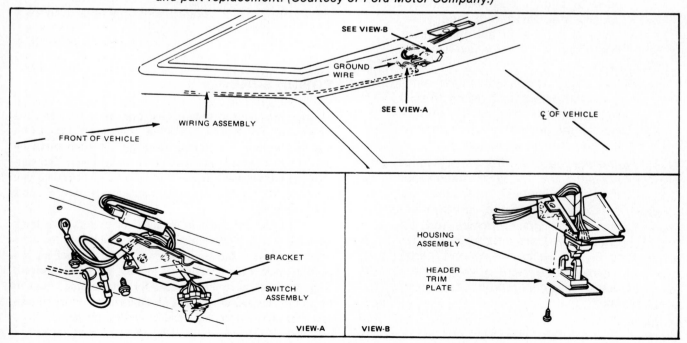

TROUBLESHOOTING POWER SUNROOFS

Most power sunroofs provide a hand crank that can be used to close the roof in the event of a failure in the power roof mechanism. Some power sunroofs require that two separate switches be pushed to close the sunroof, while only one switch is used to open the sunroof. This safety feature helps reduce the possibility of getting a finger caught in the closing sunroof panel. When troubleshooting a malfunctioning power sunroof, be certain of the correct *normal* operation. Most power sunroof switches can easily be removed and checked for proper operation using an ohmmeter. Low ohms (continuity) should exist between the power (''hot'') side of the switch and the ''open'' and ''closed'' position as the switch is actuated. If the electric motor does not operate and the switch tested good, connect a 12-V battery source to both motor connections. The motor should run in the reverse direction if the polarity of the battery leads are reversed. If the motor fails to operate correctly, it must be replaced. If the motor operates but the sunroof fails to move, the problem is probably due to a failure in the cable drive mechanism and the sunroof assembly usually has to be removed from the car for repair.

Most newer General Motors' power sunroofs (called an ''astro roof'') have a vent position that raises the rear of the vent up about 1.5 in. (3.8 cm) if the close switch is held for 2 seconds after the sunroof closes. If the sunroof fails to close properly, or does not delay before going to the vent position, the most likely cause is a defective (or disconnected) limit switch. The limit switch is located at the left (driver's side) front section of the sunroof assembly. Most electrical repairs, including the motor and limit switch, require that the headliner be removed and the assembly partially lowered to gain access to the electrical components and sunroof panel adjustments.

POWER MIRRORS

Power-operated outside mirrors usually use built-in reversible PM electric motors. The mirrors use a DPDT switch similar to power window switches. Two small motors and activators are used in each mirror for up-and-down and left-and-right mirror movement. Troubleshooting includes checking the fuse or circuit breaker and testing the switch assembly with an ohmmeter for continuity. If the switch tests good, the operation of the built-in motors can be checked by using jumper wires connecting a 12-V source directly to the motors and then reversing the polarity to attempt to reverse the direction of motor operation. If the mirrors still fail to operate, check the condition of the common ground wire connected to the mirror control switch, which is generally attached to the metal of the door or dash.

Some outside mirrors are electrically heated whenever the rear window defogger is activated. If a power mirror is found to be defective, it must be replaced as a unit since separate repair parts are seldom available.

ELECTRONIC LEVELING SYSTEMS

Some cars are equipped with rear air shocks that can be controlled electronically to adjust the ride height of the vehicle regardless of car load. A typical electronic level-control system includes the following components:

1. Air-adjustable rear shocks (or struts in some cases)
2. Small air compressor (mounted under the hood or under the car at the rear)
3. Electronic height sensor
4. Air dryer
5. Exhaust solenoid
6. Relay, wiring, and tubing

The compressor is usually a small single-piston air pump powered by a 12-V permanent-magnet (PM) electric motor. See Figure 9-50. An air dryer is usually attached to the pump to remove moisture from the air before being sent to the shocks and through the dryer (to dry the chemical dryer) during the release of air from the shocks. See Figure 9-51. The height sensor operates the compressor or exhaust solenoid, based on the height of the rear of the car. Some systems operate only when the ignition switch is ''on,'' while other systems operate anytime because the compressor is wired to a voltage source that is ''hot'' at all times.

To avoid unnecessary ride height increase or decrease due to variations in road surface, the system operates only after a time delay of about 15 seconds. Most compressors are also equipped with a timer circuit that limits compressor ''on'' time to about 3 minutes, to prevent compressor damage if the air system has a leak.

TROUBLESHOOTING ELECTRONIC LEVELING SYSTEMS

The first step of any troubleshooting procedure is to check for normal operation. Some leveling systems require that the ignition key be ''on,'' while other systems operate all the time. Begin troubleshooting by placing approximately

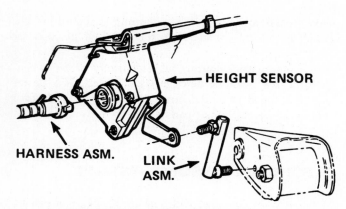

FIGURE 9-50 *Typical ride height sensor and suspension link used on an electronic ride-level system. (Courtesy of Oldsmobile Division, GMC.)*

FIGURE 9-51 *Typical ride level control pump and air dryer assembly. (Courtesy of Oldsmobile Division, GMC.)*

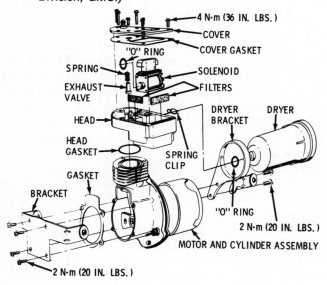

300 lbs (135 kg) on the rear of the car. After about a 15-second delay, the air compressor should operate and raise the rear of the car. If the compressor does not operate, check to see if the sensor is connected to a rear suspension member and that the electrical connections are not corroded.

Also check the condition of the compressor ground wire. It must be tight and free of rust and corrosion where it attaches to the car body. If the compressor still does not run, check to see if 12 V is available at the power lead to the compressor. If necessary, use a jumper wire directly from the positive (+) of the battery to the power lead of the compressor. If the compressor does not operate, it must be replaced.

If the ride-height compressor runs excessively, check the air compressor, the air lines, and the air shocks (or struts) with soapy water for leaks. Most air shocks or air struts are not repairable and must be replaced. Most electronic level-control systems provide some adjustments of the rear ride height by adjusting the linkage between the height sensor and the rear suspension. See Figures 9-52 and 9-53.

FIGURE 9-52 *(Courtesy of Oldsmobile Division, GMC.)*

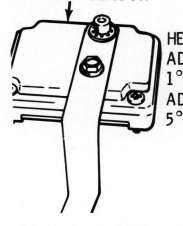

HEIGHT SENSOR

HEIGHT SENSOR ADJUSTMENT. 1° = 1/4" AT BUMPER. ADJUSTMENT OF 5° TOTAL

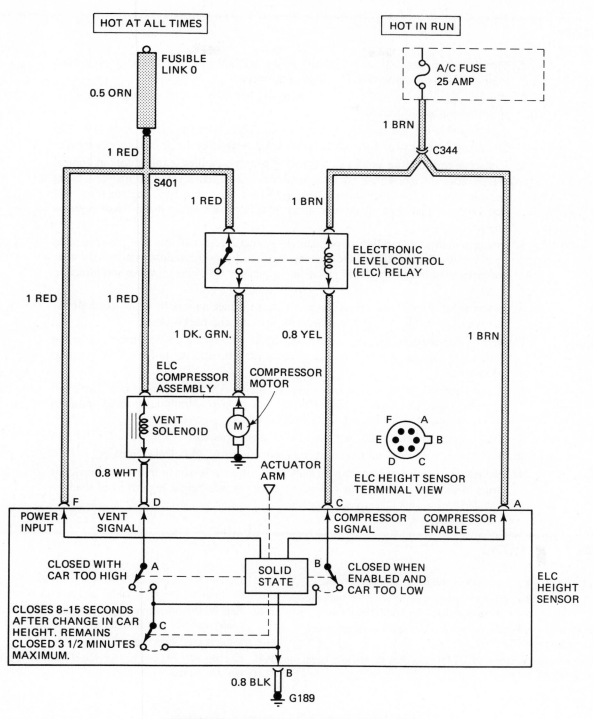

FIGURE 9-53 *Wiring diagram for an electronic level ride control system. Some systems do not operate unless the ignition is "on"; other systems are designed to operate all the time. (Courtesy of Oldsmobile Division, GMC.)*

SUMMARY

1. Blower motor operation uses a dash switch to select the current flow through blower motor resistors to provide the various motor speeds.

2. Windshield wipers usually use permanent-magnet motors with a resistor or a two-speed (three-brush) motor.

3. All horn circuits use a relay (except most Fords) to relay heavy current to the individual (one or two) horns. The horn button completes the ground for the circuit.

4. Cruise (speed) control units use a speed sensing unit called a transducer to control the throttle linkage with a vacuum-operated unit called a servo to maintain a set constant road speed. For safety, there are two release switches, one electrical and one vacuum, that shut off the operation of the cruise control if the brake pedal is applied.

5. Power windows generally use permanent-magnet electric motors that are reversible by reversing the polarity of the current. A typical power window system uses double-pole, double-throw (DPDT) switches which use both control wires and direction wires.

6. Power seats usually use reversible permanent-magnet motors to rotate gears or cables that move the seat up and down and forward and backward.

7. Heated rear-window defoggers use an electrical grid baked on the glass to clear the window of fog. Most rear defogger systems use a timer/relay to limit "on" time to 10 minutes.

8. A good antenna is absolutely necessary to receive AM radio signals. FM reception is short range but generally free of noise. For proper operation, all radio speakers should be matched for both polarity and impedance.

9. Power antennas are usually replaced rather than repaired.

10. Both power sunroofs and power mirrors use reversible PM electric motors.

11. Electronic leveling systems use a small 12-V air compressor to inflate air shocks or struts in the rear of the car to maintain the proper ride height regardless of load.

STUDY QUESTIONS

9-1. Explain how a blower motor can have different speeds.

9-2. Explain how two different speeds can be achieved with the three-brush motor commonly used for windshield wipers.

9-3. Explain the operation of the cruise control transducer and low-speed switch.

9-4. Explain the operation of DPDT switches commonly used for power windows.

9-5. Explain the difference between the "direction" wires and the "control" wires used for most power windows.

9-6. Explain the operation of power seats and how one motor could work six seat movements.

9-7. What is the "breath test"?

9-8. Explain why FM can be received if an antenna is defective, but AM cannot.

MULTIPLE-CHOICE QUESTIONS

9-1. Technician A says that a defective high-speed relay prevents high-speed blower operation, yet low-speed operates normally. Technician B says that a defective (open) blower motor resistor can prevent low-speed blower operation, yet the high-speed mode operates normally. Which technician is correct?

(a) A only.
(b) B only.
(c) both A and B.
(d) neither A nor B.

9-2. Technician A says that all automotive electric horn circuits use a relay. Technician B says that wiper motors are usually series-wound electric motors. Which technician is correct?
  (a) A only.
  (b) B only.
  (c) both A and B.
  (d) neither A nor B.

9-3. PM motors as used in power windows and power seats can be reversed by:
  (a) sending current to a reversed field coil.
  (b) using a relay.
  (c) reversing the polarity of the current to the motor.
  (d) using a relay and a two-way clutch.

9-4. Technician A says that a misadjusted brake switch could cause the cruise (speed) control to be inoperative. Technician B says that a defective low-speed switch could cause the cruise (speed) control to be inoperative. Which technician is correct?
  (a) A only.
  (b) B only.
  (c) both A and B.
  (d) neither A nor B.

9-5. Technician A says that either a defective circuit breaker or a defective ground connection can cause all power windows to fail to operate. Technician B says that if one *control* wire is disconnected, all windows will fail to operate. Which technician is correct?
  (a) A only.
  (b) B only.
  (c) both A and B.
  (d) neither A nor B.

9-6. Six-way power seats:
  (a) can use one or three motors.
  (b) must use six separate motors.
  (c) can use as many as eight motors for six-way operation.
  (d) can use permanent-magnet (PM) motors only.

9-7. When checking the operation of a rear-window defogger with a voltmeter:
  (a) the voltmeter should indicate decreasing voltage when the grid is tested across the width of the glass.
  (b) the voltmeter must be set to 110 V on the ac scale.
  (c) the voltmeter should read battery voltage anywhere along the grid.
  (d) the voltmeter must be set on a low-voltage scale.

9-8. Technician A says that a radio can receive Am signals but not FM signals if the antenna is defective. Technician B says that the speakers used with a stereo system must have matching impedance. Which technician is correct?
  (a) A only.
  (b) B only.
  (c) both A and B.
  (d) neither A nor B.

9-9. Technician A says that all speakers use a one-wire system. Technician B says that a good antenna should indicate about 500 Ω when tested with an ohmmeter between the center antenna wire and ground. Which technician is correct?
  (a) A only.
  (b) B only.
  (c) both A and B.
  (d) neither A nor B.

9-10. Technician A says that most power sunroofs and power mirrors use PM motors. Technician B says that PM motors must be properly grounded to operate correctly. Which technician is correct?
  (a) A only.
  (b) B only.
  (c) both A and B.
  (d) neither A nor B.

10

Testing Instruments

ANY ELECTRICAL TESTING MUST INCLUDE THE USE OF ELECTRICAL MEA-suring meters or test lights. This chapter includes how each type of meter works, proper hookup procedures, how to read a meter correctly, and purchasing considerations. The topics covered in this chapter include:

1. Ammeters
2. Voltmeters
3. Ohmmeters
4. Inductive meters
5. High-impedance meters
6. Meter-purchasing considerations
7. Test lights and continuity lights

TEST INSTRUMENTS

There are two basic types of test meters: *digital,* which displays numbers, and *analog,* which uses a needle to indicate readings. The basic analog meter construction and operation is described below. Digital meters are electronic but must be hooked up in the same way as for the analog type.

AMMETERS

An ammeter measures the flow of *current* through a complete circuit in units of *amperes.* The ammeter has to be installed in the circuit (in series) so that it can measure all of the current flow in that circuit, just as a water flow meter (cubic feet per minute, for example) measures the *amount* of water flow. An ammeter contains a *shunt,* a device that allows heavy current to flow through the meter without harming the meter. The greater the current flow, the stronger the electromagnet on the needle becomes and the more the needle is attracted toward the right side of the meter (higher reading). The north pole of the needle is attracted toward the south pole of the meter. See Figure 10-1. Inductive ammeters use the current generated by electromagnetic induction of the pickup around the conductor carrying the current to deflect the needle of the meter.

VOLTMETERS

A voltmeter measures the *pressure* or "potential" of electricity and measures in units of *volts.* A voltmeter is connected to a circuit in parallel. A voltmeter has a large built-in resistance so that the current flow through the meter will not affect the circuit being tested or the meter. The higher the voltage, the greater the amount of current that can get through the built-in resistances and the stronger the magnetic pull on the needle toward the right side of the dial. For a voltmeter to read different voltage readings from the same meter, changeable resistances are built in to a typical voltmeter to allow it to read various voltages on different scales. See Figure 10-2.

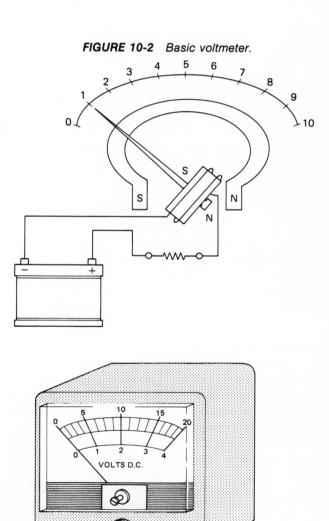

FIGURE 10-2 *Basic voltmeter.*

FIGURE 10-1 *Basic ammeter.*

TEST LEADS

INTERNAL SHUNT

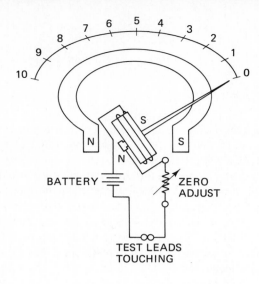

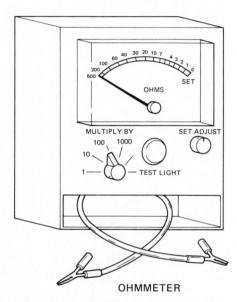

FIGURE 10-3 Basic ohmmeter. The accuracy of an ohmmeter depends on the voltage of the internal battery. The test leads must be touching to calibrate the meter (adjust ohmmeter to zero).

OHMMETERS

An ohmmeter measures the *resistance* in *ohms* of a component or circuit section when *no current* is flowing through the circuit. An ohmmeter contains a battery (or other power source). When the leads are connected together, current flows through the test leads and actually measures the difference in voltage (voltage drop) which the meter measures as resistance on its scale. An ohmmeter also contains changeable resistances so that different scales of resistance can be measured by the same meter. An ohmmeter *must* be calibrated with each use *and* with each change of scale to ensure accurate readings regardless of internal battery condition.

Zero ohms means *no* resistance between the test leads and moves the needle to the *right*. This indicates that there

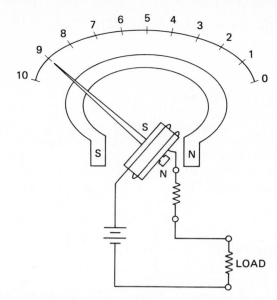

FIGURE 10-4 Ohmmeter being used to measure resistance (load).

is "continuity" or a continuous path for the current to flow in a closed circuit. Infinity (∞) ohms moves the needle all the way to the *left*. Infinity means *no* connection, as in an open circuit (extremely high resistance to current flow) through the circuit or component being tested.

> **NOTE:** The ohmmeter "reads" exactly *opposite* from all other meters. Low ohms moves the needle to the *right*. (Voltmeters and ammeters indicate low readings on the left side.)

With a closed circuit (low ohms), maximum current from the built-in battery (or other power source) causes a strong magnetic pull of the needle toward the *right*. No connection of the circuit (open circuit) would prevent any current from flowing into the needle's electromagnet. This would not cause any needle movement toward the right; therefore, the needle would remain in the far left position (ohms). See Figures 10-3 and 10-4.

HIGH-IMPEDANCE METERS

"High impedance" means high resistance in the coils of the meters. This impedance is measured in ohms. For testing computer-equipped vehicles, most automobile manufacturers recommend a multitester (a tester that can measure volts, ohms, and low amperes) that has at least 10 megohms of impedance. The prefix "meg" comes from the word "mega," which means "million" (1,000,000). Therefore, the recommended meters for use on computer controlled vehicles should have an internal resistance of at least 10 million ohms.

The purpose of the high internal resistance is to ensure that the current flow through the meter does not affect the circuit being tested. Meters are connected in parallel or in series with a circuit (depending on the meter being used, see Figure 10-5):

An ammeter is in series.

A voltmeter is in parallel.

An ohmmeter is in parallel (no current through the circuit).

FIGURE 10-5 *Summary of a test meter hookup. (Courtesy of Chevrolet Motor Division, GMC.)*

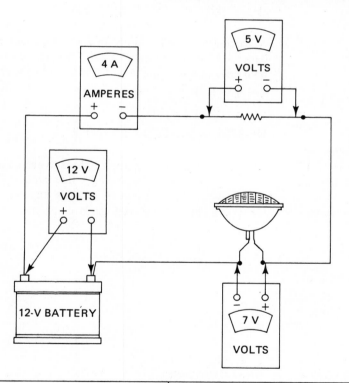
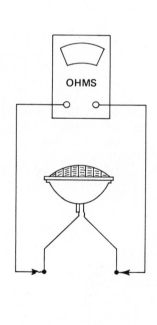

AMMETER

1. Connected in series IN a circuit according to polarity.
2. Measures current flow.
3. Used in a closed circuit.

VOLTMETER

1. Connected in parallel to a circuit or part of a circuit according to polarity.
2. Measures voltage drop:
   This is the difference between voltage at its two leads.
3. Used in a closed circuit.

OHMMETER

1. Has its own supply of power.
2. USED ONLY WHEN UNIT IS DISCONNECTED from its original circuit.
3. Measures resistance directly on meter.

ALWAYS USE A LARGE ENOUGH AMMETER AND VOLTMETER

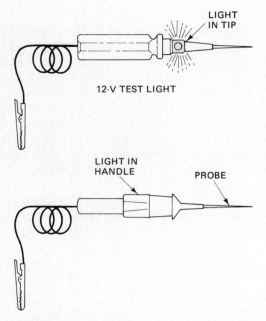

FIGURE 10-6 *Examples of probe-type test lights.*

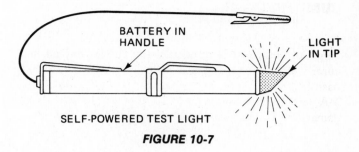

FIGURE 10-7

1. 6 to 12 V test light
2. Static timing light (because a test light can be used to set timing on point-type ignition systems)

Do not purchase a test light designed for household current (110 or 220 V). (It will not "light" with 12 V.)

TEST LIGHTS

A test light is simply a light bulb with two wires attached. See Figure 10-6. It is used to *test* for low-voltage (6 to 12 V) current. Battery voltage cannot be seen or felt and can be detected only with test equipment.

A test light can be purchased or homemade. A purchased test light could be labeled as below:

CONTINUITY TEST LIGHTS

A continuity light is similar to a test light but includes a battery. A continuity light "lights" whenever connected to both ends of a wire that has "continuity" or is not broken. See Figure 10-7.

HOMEMADE TEST LIGHTS

The easiest way to make a test light is to use a number 194 bulb, the type commonly used for side marker lights. This push-in-style bulb can easily have its exposed connector wires straightened and test leads connected. See Figure 10-8.

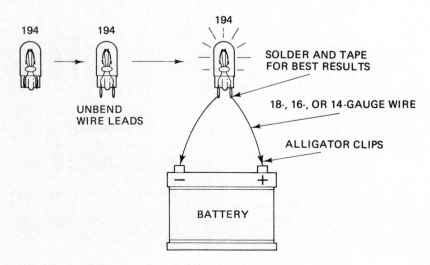

FIGURE 10-8 *Homemade test light.*

JUMPER WIRES

A jumper wire is simply a length of wire, usually with alligator clips attached to both ends. It is used to conduct current directly to a component from the battery or to bypass a suspected defective component. For safety, install a 5-A fuse in the jumper wire to protect against accidental damage.

TEST EQUIPMENT PURCHASING CONSIDERATIONS

Test meters that can measure volts and ohms are commonly called *multitesters*. Some meters can also measure very small current flow, thousandths of an ampere (milliamperes). This type of meter is called a VOM; the letters mean

$$V = volt$$
$$O = ohm$$
$$M = milliamperes$$

or if without a milliampere scale,

$$V = volt$$
$$O = ohm$$
$$M = meter$$

To avoid confusion, the recommended terms are *multitester* or *multimeter*.

Multiple-use test equipment is available in all price ranges, from very inexpensive to several thousand dollars. Regardless of price, a good meter should have the following characteristics:

1. Large, easy-to-read scale and numbers.
2. Multiple ranges for each electrical unit. In other words, a volt scale should be provided for easy measurement of very low voltage (0 to 3 V and higher voltage (1 to 18 V) by using different scales of the same meter.
3. The angle of the scale should be as wide as possible. A needle movement on a scale of 180° is preferred over only a 90° swing.
4. More expensive test meters include special circuits that prevent damage to the meter if the wrong current or voltage, or too high a current or voltage, is used. This type of test equipment is called "burnout-proof."
5. Most portable hand-held meters contain a battery for use on the ohmmeter scales. Be certain that there is an "off" position for these meters, to protect the life of the internal battery. Also, check the battery size and type to be certain that it is a common type for replacement. Some specialized batteries can be expensive.

SUMMARY

1. The proper test equipment is required for testing automotive electric and electronic systems. Standard low-cost dial-type (analog) meters, capable of measuring volts and ohms, are necessary for basic testing.
2. High-impedance meters are absolutely necessary when testing computer engine control circuits. The high internal resistance of the meter prevents the test meter from interfering and damaging the components in the circuit being tested.
3. In Chapter 11 we discuss the steps, procedures, and specifications for beginning basic tests that should be performed on any vehicle. Most of these basic tests should be performed as a part of basic service checks and used to detect defective circuits in the event of problems.

STUDY QUESTIONS

10-1. Explain the difference between a digital and an analog test meter.

10-2. Explain how an ammeter, a voltmeter, and an ohmmeter are each connected to the circuit being tested.

10-3. Explain the purpose of a shunt in an ammeter.

10-4. Explain why an analog ohmmeter "reads" in the opposite direction of an ammeter or voltmeter.

10-5. What is a high-impedance meter?

MULTIPLE-CHOICE QUESTIONS

10-1. A meter used to measure amperes is called:
   (a) an amp meter
   (b) a coulomb meter.
   (c) an ampmeter.
   (d) an ammeter.

10-2. A voltmeter should be connected to the circuit being tested:
   (a) in series.
   (b) in parallel.
   (c) only when no power is flowing.
   (d) none of the above.

10-3. An ohmmeter:
   (a) contains its own power source (battery).
   (b) actually "measures" voltage drop.
   (c) must be used in a component or circuit not carrying current.
   (d) all of the above.

10-4. A high reading on an ohmmeter (needle toward the left side of the dial) indicates:
   (a) a shorted circuit or component.
   (b) an open circuit or component (no current flows).
   (c) a high-resistance circuit or component.
   (d) that a component or circuit is defective.

10-5. A high-impedance meter is a meter that:
   (a) measures a high amount of current flow.
   (b) measures a high amount of resistance.
   (c) measures a high voltage.
   (d) has a high internal resistance.

10-6. When using a test light:
   (a) one lead must be grounded to a good vehicle ground.
   (b) both leads must be grounded to a good vehicle ground.
   (c) the device will also "light" with the high-voltage spark plug current.
   (d) none of the above.

10-7. VOM can mean:
   (a) volt-ohmmeter.
   (b) volt-ohm-milliammeter.
   (c) volts-only meter.
   (d) either (a) or (b).

10-8. A jumper wire can be used during testing:
   (a) to provide electrical power directly to an electrical component.
   (b) to jump-start another car.
   (c) to provide a ground return path for an electrical component.
   (d) (a) and (c) only.
   (e) all of the above.

11

General Electrical Systems Testing

THIS CHAPTER CONTAINS SIMPLE AND QUICKLY-PERFORMED ELECTRICAL system checks that apply to all vehicles. The testing discussed in this chapter prepares the reader for in-depth testing and troubleshooting of specific starting and charging problems covered in Chapters 12 through 17. General specifications, meter hookup, and a list of possible problem areas are included. Examples of actual readings are used to illustrate and explain the meanings of all meter readings. The topics covered in this chapter include:

1. General voltmeter test
2. Voltage-drop test
3. Battery drain test
4. Troubleshooting with a test light
5. Techniques for finding a short

GENERAL VOLTMETER TEST

Most electrical testing is done to determine the cause of an electrical problem. There is, however, a very simple three-part test that can be performed on any battery-equipped car to determine the general condition of three basic electrical "systems" of the car. The test is called a *general voltmeter test* and uses a voltmeter to determine the general condition of the:

1. The state of charge of the battery
2. The cranking circuit (starter, battery cables, and connections)
3. The charging circuit (alternator, voltage regulator)

This is the first of all electrical systems checks that should be performed to help determine the exact cause of a starting and charging problem. This three-part test does not tell you what is wrong or defective, but when compared with the range of expected readings, can help greatly in determining the exact cause of the problem.

Connect the red voltmeter lead to the positive (+) of the battery and the black lead to the negative (−) of the battery. Set the voltmeter to a scale so that readings between 6 and 18 V can be read easily. See Figure 11-1.

Step 1: *Battery voltage.* To measure the state of charge of a battery accurately, the surface charge should be removed by turning on the headlights for one full minute. Turn off the headlights and observe the voltmeter. Results: Voltmeter readings and their meanings are as follows [(at 70°F) (21°C)]:

12.6 V or higher	= 100% charged
12.4 V	= 75% charged
12.2 V	= 50% charged
12.0 V	= 25% charged
11.9 V or lower	= discharged

FIGURE 11-1 *General voltmeter test hookup.*

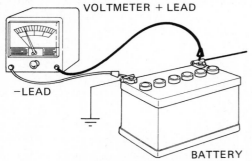

HOW CAN A 12-V BATTERY BE ONE-QUARTER CHARGED IF IT HAS 12 V?

A fully charged automotive lead-acid battery has 2.1 V per cell. A 12-V battery has six cells and could read 12.6 V (2.1 × 6). A reading higher than 12.6 V is possible and represents a "surface charge" on the battery. If a 12-V battery has only 10.5 V (1.75 V per cell) or less, it cannot perform satisfactorily in a car. A battery could be considered "dead" at 10.5 V, fully charged at 12.6 V, and therefore, only one-quarter charged at 12.0 V.

Step 2: *Cranking voltage.* Remove and ground the coil wire from the distributor or remove the power lead from the GM HEI distributor (white clip) to prevent the engine from starting. Crank the engine with the ignition key for 15 seconds. The voltmeter should read *above* 9.6 V. If it reads at or below 9.6 V, there is a possible problem with:

(a) Defective (or dirty) battery cables and connections.
(b) Defective (or discharged) battery (under load).
(c) Defective starter, solenoid, or relay.

Step 3: *Charging Voltage.* Reconnect the coil wire and start the engine. With the engine running at approximately 2000 rpm (fast idle) it should read 13.5 to 15 V or a *maximum* of 2 V higher than basic battery voltage. See Figure 11-2. If the reading is over 15 V, there could be a possible defective voltage regulator or bad connections. To test the charging circuit accurately, the battery must be loaded. With the engine running at approximately 2000 rpm, turn on all lights and accessories. If the reading is

FIGURE 11-2

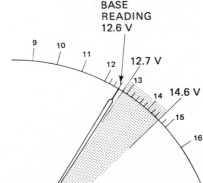

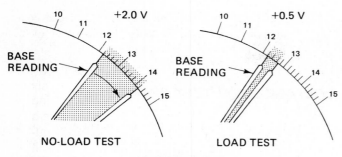

FIGURE 11-3

under 13.5 V [or under ½ V over the basic battery voltage (see Figure 11-3)], there could be a possible:

(a) Loose alternator belt

(b) Dirty or defective electrical connections

(c) Defective voltage regulator

(d) Defective alternator

EXAMPLE SITUATIONS

EXAMPLE 11-1

A car was tested using the foregoing procedure with the following results:

basic battery voltage	= 12.4 V
cranking voltage	= 10.8 V
charging voltage	= 12.8 V

Conclusion: The battery is 75% charged. The cranking voltage is excellent. Remember, the voltage should be *above* 9.6 V, and 10.8 V *is* above 9.6 V. This indicates that the battery under load, the starter motor, and the cables are all functioning correctly. The charging voltage, however, is too low according to both guidelines (less than 13.5 V *and* less than 0.5 V higher than basic battery voltage). There may be a problem with the alternator and/or voltage regulator. A loose alternator belt could also cause a low charging voltage.

EXAMPLE 11-2

A car was tested using the foregoing procedure with the following results:

basic battery voltage	= 12.8 V
cranking voltage	= 9.2 V
charging voltage	= 14.2 V

Conclusion: Even though the battery and charging circuit are functioning correctly, the cranking circuit has excessive current drain from the battery, which causes battery voltage to decrease.

Therefore, the following items should be checked to determine the exact cause of the problem:

1. Check and/or test all battery cables and connections (see "Voltage-Drop testing" below).
2. Check the battery using the load test (see Chapter 13).
3. Check the starter and solenoid or relay (see Chapter 15).

TESTING IS BETTER THAN GUESSING

As illustrated in Examples 11-1 and 11-2, a general voltmeter test can be used to check any vehicle *quickly* to determine if the starting and charging systems are functioning correctly. Electrical components and circuits (unlike automotive components such as brakes) *cannot* be judged to be good or bad simply by looking at them without testing.

WHEN TO PERFORM THE GENERAL VOLTMETER TEST

Because the general voltmeter test is a quick and easy test, it is recommended that this test be performed every time the vehicle is serviced so that possible problems can be corrected *before* they become a serious and expensive repair. For example, if the charging voltage is too low, it may be due to a slightly loose belt, which could slip at higher engine speeds and cause reduced charging of the battery. Therefore, the general voltmeter test should be performed:

1. During routine service to make certain that the starting and charging systems are functioning correctly
2. As the first test when troubleshooting the battery, cranking circuit, or charging circuit

VOLTAGE-DROP TESTING

Voltage-drop testing is a method of testing the condition of wires, cables, and connections. Since the amount of voltage drop in a circuit is related directly to the resistance, it would seem that the resistance of a cable or wire should be tested directly with an ohmmeter, rather than indirectly using a voltmeter.

Automotive cable and wire are constructed of multiple-strand conductors. If only one small strand of a 30-strand wire were connected between the battery and the starter, an ohmmeter would indicate very low resistance. However, the high current flow required for starter motor operation would quickly overload the one small strand of wire. An ohmmeter uses low voltage and very low current flow to measure the resistance, while the circuit being tested must be able to carry very high starting current. The most accurate method to measure resistance of wires, cables, or connections is to measure the loss of voltage from one end of a wire to the other with current flowing through the circuit. The results of a voltage-drop test include:

1. A zero or very low voltage drop indicates *no* or low resistance.

2. A higher-than- 0.2 V drop across any cable or connection is considered a high-voltage drop and therefore indicates high resistance.

NOTE: Some battery cables are extra long, and therefore the allowable voltage drop should be adjusted for the extra-long cables. For example, an 8-ft-long battery cable could have 0.4-V drop and still perform satisfactorily.

3. The circuit to be tested *must* be operating so that current is flowing through the circuit. Always remember that there must be voltage before there can be a voltage drop. (Use the ignition key switch or a remote starter connected to the starter solenoid or relay to operate the starter during testing.)

4. A voltmeter with a very low scale (0 to 3 V) is recommended. Attempting to read very low volts (less than 1 V) on a high voltmeter scale is not practical.

5. The maximum acceptable voltage drop for each connection is 0.1 V.

6. The maximum acceptable *total* voltage drop of all cables and connections in the starting circuit should not exceed 0.7 V.

WHEN TO PERFORM A VOLTAGE-DROP TEST

The most common circuit tested using the voltage-drop method is the cranking circuit (starter motor, starter solenoid, and battery cables and connections). The voltage-drop test should be performed on all cables and connections of the cranking circuit (as outlined below) whenever these conditions exist:

1. *Slow cranking of the starter.* This could be due to high resistance in the cables or connections. (It is much easier and less expensive to replace a battery cable than a starter motor, which may not need replacement.)

2. If the cranking section of the general voltmeter test indicates a cranking voltage at or *below 9.6 V.* Again, the voltage-drop test can confirm or condemn high-resistance cables as the cause of a low cranking voltage test.

3. Any other time that high-resistance cables or connections are suspected to be the cause of intermediate or slow engine cranking.

VOLTAGE-DROP TEST PROCEDURE

1. Set the voltmeter on the lowest scale so that a reading of less than 1 V can easily be observed.

2. The red lead of the voltmeter is always connected to the *most* positive end of the cable or component. The *most* positive is always the end closest to the positive (+) post of the battery with the current flowing from the positive (+) post to the negative (−) post. The black lead of the voltmeter is always connected to the most negative (−) end of the cable or component. The end of the cable that is *most* negative (−) is the end closest to the negative (−) post of the battery. See Figure 11-4.

HINT: When testing battery cables, it can be confusing which color voltmeter lead attaches to which end of the cable before cranking the engine. Just remember:

(a) There is no *more positive* place on the entire car than the positive (+) post of the battery.

FIGURE 11-4 Voltage-drop hookups for checking battery cable and resistance. Connect the voltmeter on the low scale as shown and crank the engine. The voltage drop should not exceed 0.1 V. If over 0.1 V, remove and clean, or replace, the cable ends. (Courtesy of Ford Motor Company.)

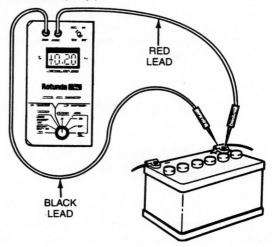

VOLTAGE DROP IN CABLE CONNECTOR

(b) There is no *more negative* place on the entire car than the negative (−) post of the battery.

(c) If voltmeter leads are not long enough to reach *both* ends of the cable being tested, test the voltage at each end separately, then subtract the low reading from the high reading to determine the difference. The difference should not be greater than 0.2 V.

3. Disconnect the ignition coil wire out of the center of the distributor cap and ground the coil wire to prevent possible damage to the coil while cranking the engine.

4. For each wire or connection tested, the voltmeter leads must be connected correctly, as indicated below.

5. For each connection or cable tested, the engine must be *cranked* just long enough to read the voltmeter. The engine can be cranked using a remote starter connected to the starter solenoid or relay or by turning the ignition key.

CAUTION: Excessive and prolonged engine cranking can overheat the starter, bleed down hydraulic valve lifters, and rapidly discharge the battery. Noisy hydraulic lifters, caused by extended engine cranking, will usually quit after several miles of driving.

NO-TOOL TESTING

High resistance in cables or connections can easily be determined by touching or holding the battery cables and connections and then cranking the engine with the starter. There is absolutely no danger involved in touching the battery cables or connections because 12 V is too low a voltage to feel. If cables and connections are in satisfactory condition, no sensation should be felt. However, if the connections are warm or hot to the touch, there is *high resistance* and the connections should be cleaned or replaced. A hot battery cable during cranking means that the cable should be replaced.

GM-TYPE CRANKING CIRCUITS

The GM-type cranking circuit is a general classification used on many foreign and some other American-made vehicles. To test the positive (+) battery cable from the battery to the starter, connect the voltmeter as illustrated in Figure 11-5 (connection 1). Crank the engine just long enough to read the voltmeter. If zero volts, there is no resistance in the cable. The maximum allowable voltage drop is 0.2 V (two-tenths of a volt). To test the negative (−) battery cable, see Figure 11-5 (connection 2). To test the solenoid, connect the voltmeter leads as shown in Figure 11-5 (connection 3).

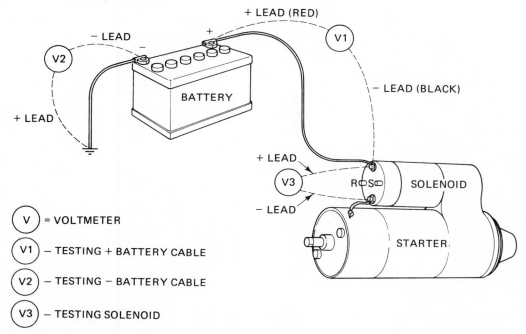

FIGURE 11-5 *Voltmeter hookups for voltage-drop testing a GM-type cranking circuit.*

V = VOLTMETER

V1 — TESTING + BATTERY CABLE

V2 — TESTING − BATTERY CABLE

V3 — TESTING SOLENOID

CAUTION: Before cranking the engine, the voltmeter will indicate *full* battery voltage because it is connected across an open switch. The voltmeter should be set on a higher scale before cranking to prevent possible damage to the voltmeter.

PROCEDURE FOR TESTING A SOLENOID OR RELAY

To protect the voltmeter from possible damage due to full battery voltage being applied to the meter set on a low-voltage setting, it is important to follow the procedure listed below while testing all solenoids and relays by the voltage-drop method.

1. Set the voltmeter on a high scale (to register battery voltage).
2. Connect the voltmeter leads [red to the most positive (+) and black to the most negative (−)].
3. Crank the engine.
4. While cranking the engine, switch the voltmeter to the low scale.
5. Observe the voltmeter reading.
6. Switch the voltmeter back to the high voltmeter scale.

7. Stop cranking.
8. If the voltage drop is greater than 0.2 V, there is excessive resistance in the solenoid or relay, and it must be repaired or replaced.

FORD-TYPE CRANKING CIRCUITS

1. Test the battery cable between the battery and the starter relay. Hook up the voltmeter as shown in Figure 11-6 (connection 1).
2. Test the battery cable between the starter relay and the starter motor as shown in Figure 11-6 (connection 2).
3. Test the negative battery cable with the voltmeter connected as shown in Figure 11-6 (connection 3).

CHRYSLER-TYPE CRANKING CIRCUITS

Chrysler starter circuits can be tested for the resistance of the main current-carrying cable and the negative (−) battery cable as illustrated in Figure 11-7 (connections 1 and 2).

FIGURE 11-6 Voltmeter hookups for voltage-drop testing a Ford-type cranking circuit.

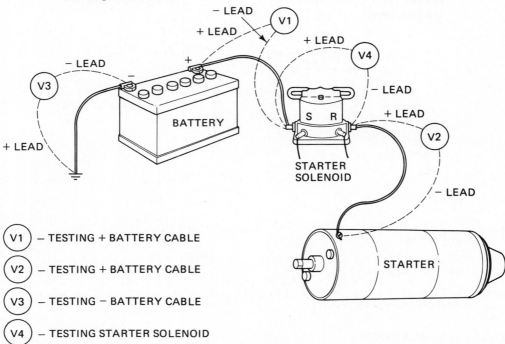

V1 — TESTING + BATTERY CABLE

V2 — TESTING + BATTERY CABLE

V3 — TESTING − BATTERY CABLE

V4 — TESTING STARTER SOLENOID

ACCESSORY	DRAIN TIME
HEADLIGHTS	1 TO 1½ HOURS
INTERIOR LIGHTS	2½ HOURS
1 SEAT BELT RETRACTOR	7 DAYS
2 SEAT BELT RETRACTORS	3½ DAYS
PARKING LIGHTS	4-6 HOURS
TRUNK LIGHT	2½ DAYS
TRUNK CLOSING MOTOR	18 HOURS

FIGURE 11-10 *Several examples of electrical units and the time it would take to drain a fully charged battery to a level that may not start the engine. (Courtesy of General Motors Corporation.)*

1. A defective clock could drain the battery.
2. A trunk or under-the-hood light remaining on can be caused by a defective mercury switch.
3. A glove box light that remains on could drain the battery.
4. Courtesy lights remaining on due to a defective door switch or headlight switch knob turned fully counterclockwise would leave the interior lights on all the time.

WHAT TO DO IF A BATTERY DRAIN STILL EXISTS AFTER ALL THE FUSES ARE DISCONNECTED

If all the fuses have been disconnected and the drain still exists, the source of the drain has to be between the battery and the fuse box. The most common sources of drain under the hood include:

1. The alternator. Disconnect the alternator wires and retest. If the test light is now "off," the problem is a defective diode(s) in the alternator. (See Chapter 16 for details.)
2. The starter solenoid (relay) or wiring near these components. These are also a common source of battery drain, due to high current flows and heat, which can damage the wire or insulation.

TROUBLESHOOTING WITH A TEST LIGHT

The most useful test device in automotives is the 12-V test light. Since 12-V electricity cannot be seen or felt, a meter or a light bulb must be connected to the circuit to be assured that current is available at the test points. See Figure 11-11. Always remember two things:

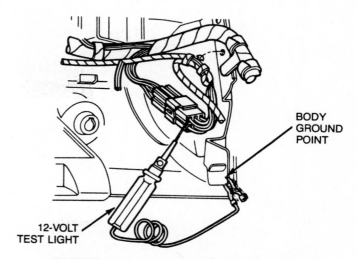

BODY GROUND POINT

12-VOLT TEST LIGHT

FIGURE 11-11 *Use a test light to check for 12 V at various points in the circuit. (Courtesy of Ford Motor Company.)*

1. The test light is grounded to metal or the negative (−) of the battery during testing. If the test light "lights," there *is* voltage at that location. It does *not* confirm that the circuit being tested has its own *good* ground return path. Many electrical problems are caused by a poor ground connection.
2. Always check the test light before troubleshooting. Connect the test light to the battery posts to check that the test light is working correctly.

If a test light confirms that current is available at a taillight harness and socket, for example, the taillight should work if there is a good ground return path and a good bulb. A corroded light bulb socket or a defective bulb could be the cause of a bulb not working. Bulbs can be tested with an ohmmeter.

HOW TO LOCATE A SHORT CIRCUIT

A short circuit usually "blows" a fuse, and a replacement fuse often also blows in the attempt to locate the source of the short circuit. There are several different methods that can be used to locate the short.

1. Disconnect one component at a time and then replace the fuse. If the new fuse "blows," continue the process over and over until the location of the short is determined. This method uses many fuses and is *not* a preferred method for finding a short circuit.

2. Another method is to connect an automotive circuit breaker to the contacts of the fuse holder with alligator clips. See Figure 11-12. The circuit breaker will alter-

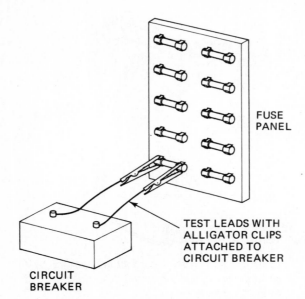

FIGURE 11-12 *Replace a fuse with a circuit breaker in the fuse panel in an attempt to locate the source of a short circuit.*

nately open and close the circuit, protecting the wiring from possible overheating damage while providing current flow through the circuit. All of the components included in the defective circuit should be disconnected one at a time until the circuit breaker stops clicking. The last unit disconnected is the unit causing the short circuit. If the circuit breaker continues to click with all circuit components unplugged, the problem is in the wiring *from* the fuse panel *to* any one of the units in the circuit. Visual inspection of all the wiring or further disconnecting will be necessary to locate the problem.

WHAT IS A SMOKE TEST?

A smoke test is sometimes used in the repair of electronic equipment such as televisions, radios, and stereos. It involves plugging the unit in and turning it on. The component that starts to smoke is obviously defective and must be replaced. The smoke test is *not* a recommended method of automotive electrical diagnosis!

---

3. The third method uses an ohmmeter connected to the fuse holder and ground. This is the recommended method of finding a short circuit, which is an electrical connection to another wire or to ground before the current flows through some or all of the resistance in the circuit. An ohmmeter will indicate low ohms when connected to a short circuit. An ohmmeter should never be connected to an operating circuit. The correct procedure for locating a short using an ohmmeter includes the following steps:

(a) Connect one lead of an ohmmeter (set to a low scale) to a good clean metal ground and the other lead to the *circuit side* of the fuse holder. (*Caution:* Connecting the lead to the power side of the fuse holder will cause current flow through and damage to the ohmmeter.)

(b) The ohmmeter will read zero or almost zero ohms if the circuit is shorted.

(c) Disconnect one component in the circuit at a time and watch the ohmmeter. If the ohmmeter goes to high ohms or infinity, the component just unplugged was the source of the short circuit.

SHORT FINDING USING A GAUSS GAUGE

A Gauss gauge is a meter sensitive to the magnetic field surrounding a wire conducting current. The gauss is a unit of magnetic induction named for Karl Friedrich Gauss (1777–1855), a German mathematician.

If a short circuit blows a fuse, a special pulsing circuit breaker (similar to a flasher unit) can be installed in the circuit in place of the fuse. Current will flow through the circuit until the circuit breaker opens the circuit. As soon as the circuit breaker opens the circuit, it closes again. This "on"-and-"off" current flow creates a pulsing magnetic field around the wire carrying the current. By using the small hand-held Gauss gauge, this pulsing magnetic field is indicated on the gauge as needle movement to the left, then to the right of center. This pulsing magnetic field will register on the Gauss gauge even through the metal

FIGURE 11-13 *A Gauss gauge can be used to determine the location of a short circuit.*

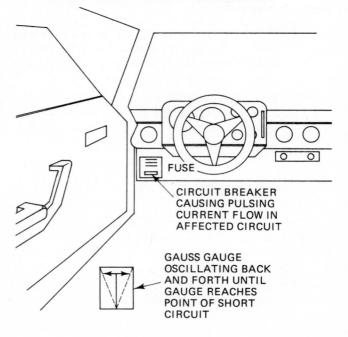

body of the car. A needle-type compass can also be used to observe the pulsing magnetic field. To locate the position of the actual short (or short to ground), move the gauge along the *outside* of the car until the needle *stops* oscillating. Where the needle movement stops along the circuit is the location of the short. See Figure 11-13. Gauss gauge testers are available at most automotive full-service parts stores.

FEEDBACK

When current that lacks a good ground goes backward in the power side of the circuit in search of a return path (ground) to the battery, this reverse flow is called "feedback" or "reverse-bias" current flow. Feedback can cause other lights or gauges to work which should not be working.

AN EXAMPLE OF FEEDBACK

A customer complained that when the headlights were "on," the left-turn signal indicator light in the dash remained on. The cause was found to be a poor ground connection for the left-front parking light socket. The front parking light bulb is a dual filament: one filament for the parking light (dim) and one filament for the turn signal operation (bright). A corroded socket did not provide a

good enough ground to conduct all of the current required to light the dim filament of the bulb.

The dual-filament bulb "shares" the same ground connection and is electrically connected. When all of the current could not flow through the bulb's ground in the socket, it caused a feedback or reversed its flow through the other filament, "looking" for ground. The turn signal filament is electrically connected to the dash indicator light; therefore, the reversed current on its path toward ground could light the turn signal indicator light. Cleaning or replacing the socket usually solves the problem if the ground wire for the socket is making a secure chassis ground connection.

THE BATTERY SIDE IS THE CORROSION AND STARTER SIDE

Most poor connections are due to rust or corrosion of the sockets, wire connectors, or ground wires. Since batteries produce corrosive fumes, many front lighting problems can be traced to the battery side of the car. Also, because of the need to keep battery cables as short as possible to prevent resistance losses, the battery side is usually on the same side of the engine as the starter motor.

SUMMARY

1. The general voltmeter test is one of the most important electrical tests. The general voltmeter test should be performed as a part of any regular automotive service to be certain of the proper operation of the starting and charging system. The general voltmeter test is also the first test that should be performed on any vehicle with starting or charging problems, to help determine the exact cause of the problem.
2. Voltage-drop testing is used to determine if there is high resistance in the circuit being tested. High resistance also means heat. Therefore, if a battery cable connection is hot to the touch, excessive resistance is indicated. By checking each cable and connection, a voltage-drop test can pinpoint the exact location of the high resistance.
3. The battery drain test can be used to locate the cause of a battery being discharged.
4. Basic test light, jumper wire, and short-finding techniques are used every day by automotive technicians. Many automotive electrical problems can be located using these simple, yet accurate testing procedures. In this chapter we have prepared the technician for the next six chapters on batteries, starters, and alternators.

STUDY QUESTIONS

11-1. Explain the meter hookup, procedure, and results of the general voltmeter test.

11-2. Explain the meter hookup and procedure for voltage-drop testing of the starter circuit.

11-3. Describe the procedure used to find the source of a battery drain using a test light, a voltmeter, and an ohmmeter.

11-4. Describe four methods that can be used to locate a short circuit.

MULTIPLE-CHOICE QUESTIONS

11-1. A voltage drop is commonly used to test:
   (a) the starter ampere draw.
   (b) the condition of the starter brushes.
   (c) the condition of the cables and connections.
   (d) the spark plug wires.

11-2. If cranking voltage is below 9.6 V:
   (a) normal—everything is okay.
   (b) perform a voltage-drop test to check cables and connections.
   (c) the alternator is possibly defective.
   (d) none of the above.

11-3. The correct specification for a voltage-drop test is:
   (a) above 9.6 V.
   (b) more than 0.2 V.
   (c) 0.2 V maximum.
   (d) below basic battery voltage.

11-4. The correct specification for a battery-drain test using a voltmeter is:
   (a) below battery voltage.
   (b) 13.5 to 15.0 V.
   (c) 0.2 V maximum.
   (d) above 9.6 V.

11-5. If the ohmmeter reads 1000 Ω during a battery drain test, the technician should:
   (a) perform the general voltmeter test to determine the problem.
   (b) perform a voltage-drop test.
   (c) perform a battery drain test using a test light.
   (d) look elsewhere for the problem because the test indicates no battery drain.

11-6. If the charging voltage is above 15.0 V, there is:
   (a) a possible alternator problem.
   (b) a possible voltage regulator problem.
   (c) a possible starter problem.
   (d) a possible spark plug problem.

11-7. After removing the surface charge, a 12-V battery indicating 12.6 V means that the battery is:
   (a) 100% charged.
   (b) 60% charged.
   (c) 25% charged.
   (d) overcharged.

11-8. If a voltmeter indicates 14.4 V while the engine is run at 2000 rpm, this means that:
   (a) the battery is being overcharged.
   (b) the alternator and voltage regulator are okay.
   (c) the starter and alternator are okay.
   (d) the battery is defective.

11-9. Voltage-drop testing should be used to determine the condition of:
   (a) the starter.
   (b) the battery.
   (c) the cables and connections.
   (d) the alternator and voltage regulator.

11-10. To find a short circuit:
   (a) one ohmmeter lead can be connected to the circuit side of the fuse holder and the other lead to a good ground.
   (b) one ohmmeter lead can be connected to the power side of the fuse holder and the other lead to the positive (+) post of the battery.
   (c) a Gauss gauge can be used.
   (d) both (a) and (c) are correct.

12

Batteries

THE BATTERY IS THE HEART OF THE AUTOMOTIVE ELECTRICAL SYSTEM. This chapter describes the construction and operation of an automotive battery. The topics covered in this chapter include:

1. Battery construction
2. How a battery works
3. Specific gravity
4. Maintenance-free batteries
5. Battery ratings
6. Causes of battery failure
7. Characteristics of a weak battery

PURPOSE OF A BATTERY

The primary purpose of an automotive battery is to provide a source of electrical power for starting and for electrical demands that exceed generator output. The battery also acts as a *stabilizer* to the voltage for the entire electrical system. The battery is a voltage stabilizer because it acts as a reservoir where large amounts of current (amperes) can be removed quickly during starting and replaced gradually by the alternator during charging. Because a battery should be able to smooth out short-duration high or low voltages, the battery *must* be in good (serviceable) condition before testing the charging and the cranking system. For example, if a battery is discharged, the cranking circuit (starter motor) could test as being defective because the battery voltage may drop below specifications. (See the description of the general voltmeter test in Chapter 11.)

The charging circuit could also be tested as being defective because of a weak or discharged battery. It is important to test the vehicle battery before further testing of the cranking or charging system.

BATTERY CONSTRUCTION

Most automotive battery cases (container/cover) are constructed of polypropylene, a thin [approximately 0.08 in. (2.0 mm) thick], strong, and lightweight plastic. Containers for industrial batteries and some truck batteries are constructed of a hard, thick rubber material.

Inside the case there are six cells (for a 12-V battery). Each cell has positive and negative plates. Built into the bottom of many batteries are ribs that support the lead-alloy plates and provide a space for sediment to settle. This space prevents spent active material from causing a short circuit between the plates at the bottom of the battery and is called the *sediment chamber*. Some maintenance-free batteries do not have a sediment chamber, but enclose the plates in an envelope-type separator which prevents material from settling to the bottom of the battery case. See Figure 12-1.

Maintenance-free is a term used to describe batteries that use little water during normal service because of the alloy material used to construct the battery plate grids. Maintenance-free batteries are also called "low water loss" batteries.

GRIDS

Each positive and negative plate in a battery is constructed on a framework or grid made primarily of lead. Lead is a soft material and must be strengthened for use in an automotive battery grid. Adding antimony or calcium to the pure lead adds strength to the lead grids. See Figure 12-2. Battery grids hold the active material and provide the electrical pathways for the current created in the plate.

MAINTENANCE-FREE VERSUS STANDARD BATTERY GRIDS

A normal battery uses up to 5% antimony in the construction of the plate grids to add strength. However, the greater the amount of antimony, the greater the amount of gassing (hydrogen gas and oxygen gas released), and therefore the more water the battery will use. Maintenance-free batteries use calcium instead of antimony because 0.2% calcium has the same strength as 6% antimony. A typical lead-calcium grid use only 0.09–0.12% calcium.

Low-maintenance batteries use a low percentage of antimony about 2%–3% or use antimony only in the

FIGURE 12-2 *Lead-alloy grid. The active battery materials are "pasted" onto these grids.*

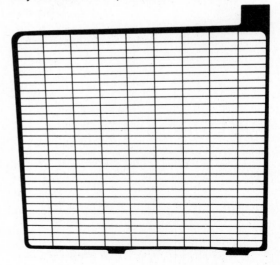

FIGURE 12-1 *Typical battery case.*

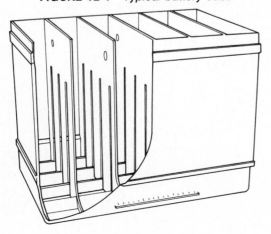

positive plates and use calcium for the negative plates. The percent of alloy of the plate *grids* is the major difference between standard and maintenance-free batteries. The chemical reactions that occur inside each battery are identical regardless of the type of material used to construct the grid plates.

RADIAL-GRID DESIGN

Some batteries use a grid design with only vertical and horizontal strips. The battery plate creates electrical energy from chemical energy, and this current must flow from where it is generated (for example, location *B* in Figure 12-3) to where it is connected to the outside battery post indicated by point *A*. The current must move over and up along the grid strips to reach point *A*.

With a radial grid design ("radial" means branching out from a common center), the current generated near point *B* in Figure 12-4 can travel directly to point *A*. Therefore, a grid with a radial design has lower resistance and can provide more current more rapidly than can conventional non-radial-grid design batteries. The radial spokes act as a superhighway system for the current to travel from all areas of the grid to the battery post.

POSITIVE PLATES

The positive plates have *lead dioxide (peroxide)* placed onto the grid framework. This process is called "pasting." This active material can react with the sulfuric acid of the battery and is dark brown in color.

FIGURE 12-3

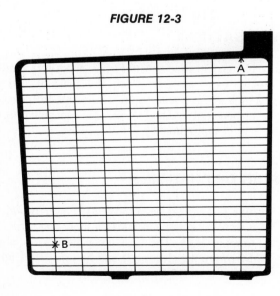

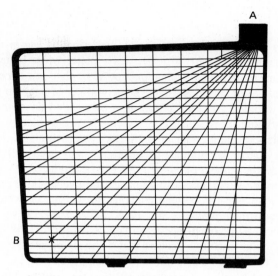

FIGURE 12-4   *Radial design battery grids permit lower resistance for current generated at point B to reach point A.*

NEGATIVE PLATES

The negative plates are "pasted" with a pure porous lead called sponge lead and are gray in color.

SEPARATORS

The positive and the negative plates must be installed alternately next to each other without touching. Nonconducting separators are used, which allow room for the reaction of the acid with both plate materials, yet insulate the plates to prevent shorts. These separators are porous (many small holes) and have ribs facing the positive plate. Some batteries also use a glass fiber between the positive plate and the separator to help prevent the loss of the active material from the grid plate. Separators can be made from resin-coated paper, porous rubber, fiberglass, or expanded plastic. Many batteries use envelope-type separators that encase the entire plate and help prevent any material that may shed from the plates from causing a short circuit between plates at the bottom of the battery.

SMALL YET POWERFUL BATTERIES

It used to be that the larger the battery, the more power the battery was able to produce. Designs that include radial grids, polypropylene cases, and fiberglass envelopes around the plates have all resulted in smaller, lighter-weight, more powerful batteries.

CELLS

Cells are constructed of positive and negative plates with insulating *separators* between each plate. Most batteries use one more negative plate than positive plate in each cell. Many newer batteries use the same number of positive ($+$) and negative ($-$) plates. A cell is also called an *element*. See Figure 12-5. Each cell is actually a 2-V battery, regardless of the number of positive or negative plates used. The greater the number of plates used in each cell, the greater the amount of *current* that can be produced. Typical batteries contain four positive plates and five negative plates per cell. A 12-V battery contains six cells connected in series, which produces the 12 V ($6 \times 2 = 12$ V) and contains 54 plates (9 plates per cell $\times$ 6 cells). If the same 12-V battery had five positive plates and six negative plates, for a total of 11 plates per cell (5 plus 6), or 66 plates (11 plates $\times$ 6 cells), it would have the same voltage, but the amount of current that the battery could produce would be increased. The capacity of a battery is determined by the amount of active plate materials in the battery and the area of the plate material exposed to the liquid, called electrolyte, in the battery.

FIGURE 12-5 *Battery cell or element containing positive (+) and negative (−) plates separated by insulating separators. Each cell is capable of producing 2.1 V when installed in an acid solution electrolyte.*

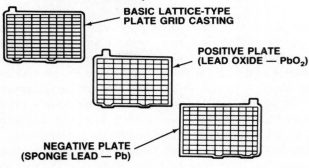

BASIC LATTICE-TYPE
PLATE GRID CASTING

POSITIVE PLATE
(LEAD OXIDE — PbO_2)

NEGATIVE PLATE
(SPONGE LEAD — Pb)

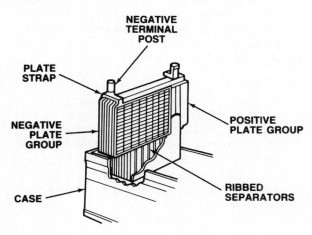

NEGATIVE
TERMINAL
POST

PLATE
STRAP

NEGATIVE
PLATE
GROUP

POSITIVE
PLATE GROUP

CASE

RIBBED
SEPARATORS

PARTITIONS

Each cell is separated from the other cells by partitions, which are made of the same material as that used for the outside case of the battery. Electrical connections between cells are provided by lead connectors which loop over the top of the partition and connect the plates of the cells together. Many batteries connect the cells directly through the partition connectors, which provides the shortest path for the current and the lowest resistance. Older-style truck and industrial batteries commonly used connectors that extended through the top of the case and over and then down through the case to connect the cells.

ELECTROLYTE

The electrolyte used in automotive batteries is a solution (liquid combination) of 36% sulfuric acid and 64% water. This electrolyte is used for both lead-antimony and lead-calcium (maintenance-free) batteries. The chemical symbol for this sulfuric acid solution is H_2SO_4

H = symbol for hydrogen (the subscript 2 means that there are two atoms of hydrogen)

S = symbol for sulfur

O = symbol for oxygen (the subscript 4 indicates that there are four atoms of oxygen)

This electrolyte is sold premixed in the proper proportion and is factory installed or added to the battery when sold. Additional electrolyte must never be added to any battery after the original electrolyte fill. It is normal for some water (H_2O) to escape during charging due to the "gassing" that is produced by the chemical reactions. Only pure distilled water should be added to a battery. If distilled water is not available, clean drinking water can be used.

HOW A BATTERY WORKS

A fully charged lead-acid battery has a positive plate of lead dioxide (peroxide) and a negative plate of lead surrounded by a sulfuric acid solution (electrolyte). The difference in potential (voltage) between lead peroxide and lead in acid is approximately 2.1 V.

DURING DISCHARGING

The positive plate PbO_2 (lead dioxide) combines with the SO_4 from the electrolyte and releases its O_2 into the electrolyte, forming H_2O. The negative plate also combines with the SO_4 from the electrolyte and becomes lead sulfate ($PbSO_4$). See Figure 12-6.

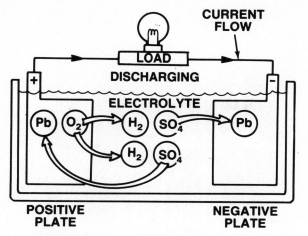

FIGURE 12-6 *The chemical reaction for a lead-acid battery that is fully charged being discharged by the attached load. (Courtesy of Chrysler Corporation.)*

WHEN FULLY DISCHARGED

When the battery is fully discharged, both the positive and the negative plates are $PbSO_4$ (lead sulfate) and the electrolyte has become water (H_2O). It is usually not possible for a battery to become discharged 100%. However, as the battery is being discharged, the plates and electrolyte approach the completely dead situation. There is also the danger from freezing when a battery is discharged because the electrolyte is mostly water.

DURING CHARGING

During charging, the sulfate (acid) leaves both the positive and the negative plates and returns to the electrolyte, where it becomes normal-strength sulfuric acid solution. The pos-

FIGURE 12-7 *The chemical reaction for a lead-acid battery that is fully discharged being charged by the attached generator. (Courtesy of Chrysler Corporation.)*

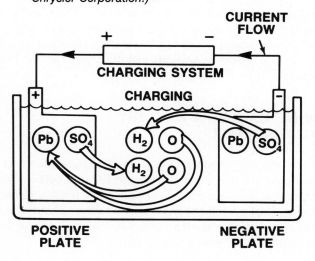

itive plate returns to lead dioxide (PbO_2) and the negative plate is again pure lead (Pb). See Figure 12-7.

SPECIFIC GRAVITY

The amount of sulfate in the electrolyte is determined by its specific gravity. Specific gravity is the ratio of the weight of a given volume of a liquid divided by the weight of an equal volume of water. In other words, the more dense the material (liquid), the higher its specific gravity. Pure water is the basis for this measurement and is given a specific gravity of 1.000 at 80°F. Pure sulfuric acid has a specific gravity of 1.835; the *correct* concentration of water and sulfuric acid (called *electrolyte*—64% water, 36% acid) is 1.260 to 1.280 at 80°F. The higher the battery's specific gravity, the more fully it is charged. See Figure 12-8.

FIGURE 12-8 *As the battery becomes discharged, the specific gravity of the battery acid decreases.*

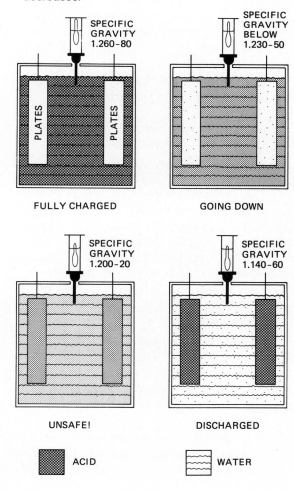

SPECIFIC GRAVITY VERSUS STATE OF CHARGE AND BATTERY VOLTAGE

Values of specific gravity, state of charge, and battery voltage at 80°F (27°C) are given in the following table.

Specific Gravity	State of Charge	Battery Voltage
1.265	Fully charged	12.6 V or higher
1.225	75% charged	12.4 V
1.190	50% charged	12.2 V
1.155	25% charged	12.0 V
Below 1.120	Discharged	11.9 V or lower

WHAT IS THE ".84 FACTOR?"

There is a relationship between the specific gravity of a battery cell and its voltage. For the normal operating range of a battery, add 0.84 to the specific gravity of the cell and the resulting number should equal the voltage of the cell. For example:

$$1.265 = \text{the specific gravity of the cell}$$
$$+\ .840 = \text{added factor}$$
$$2.105 = \text{volts of the cell}$$

If all cells of a 12-V battery were the same, then the battery should produce:

$$2.105 = \text{volts per cell}$$
$$\times\ \ \ 6 = \text{number of cells}$$
$$12.630 = \text{battery voltage}$$

The same factor can be used to estimate the specific gravity of the cell if the cell voltage is known. For example, if a battery has a voltage of exactly 12.0 V, then the average cell voltage is exactly 2.0 V. To determine the specific gravity of the cell, subtract .84 from the cell voltage:

$$2.000 = \text{volts per cell}$$
$$-\ .840 = \text{factor}$$
$$1.160 = \text{calculated specific gravity.}$$

CAUSES AND TYPES OF BATTERY FAILURE

Most batteries have a useful service life of 2 to 5 years. However, proper care can help increase the life of a battery and abuse can shorten the life. The major cause of pre-

FIGURE 12-9 *Battery that was accidentally left on the battery charger overnight. The plates warped and the top blew off.*

mature battery failure is overcharging. The automotive charging circuit, consisting of an alternator (generator), voltage regulator, and connecting wires, must be operating correctly to prevent damage to the battery. Charging voltages higher than 15.5 V can damage a battery by warping the plates due to the heat of overcharging. See Figure 12-9.

Overcharging also causes the active material to disintegrate and fall out of the supporting grid framework. Vibration or bumping can also cause internal damage similar to overcharging. It is important, therefore, to ensure that all automotive batteries are securely clamped down in the vehicle. The shorting of cell plates can occur without notice. If one of the six cells of a 12-V battery is shorted, the resulting voltage of the battery is only 10 V (12 − 2 = 10). With only 10 V available, the starter *usually* will not be able to start the engine.

BATTERY HOLD-DOWNS

All batteries must be attached securely to the vehicle to prevent battery damage. Normal vehicle vibrations can cause the active materials inside the battery to shed. Battery hold-down clamps or brackets help reduce vibration, which can greatly reduce the capacity and life of any battery. See Figure 12-10.

WHAT CAUSES A BATTERY TO WEAR OUT

Every automotive battery has a limited service life of approximately 2 to 5 years. During the life of a battery, the active material sheds from the surface of the positive plates. This gradually limits the power of the battery. This cycling can also cause the negative plates to become soft, which will also cause eventual battery failure.

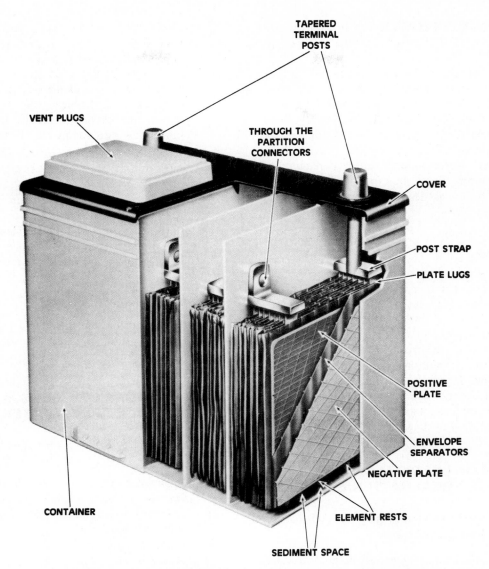

FIGURE 12-10 *Typical automotive battery. Notice the use of envelope-style separators to help prevent vibration damage. (Courtesy of Battery Council International.)*

BATTERY RATINGS

Batteries are rated according to the amount of current that a battery can produce during specific conditions.

Cold Cranking Performance. Every automotive battery must be able to supply electrical power to crank the engine in cold weather and still provide battery voltage high enough to operate the ignition system for starting. The cold cranking power of a battery is the number of amperes that can be supplied by a battery at 0°F (−18°C) for 30 seconds and still maintain a voltage of 1.2 V per cell or higher. This means that the battery voltage would be 7.2 V for a 12-V battery and 3.6 V for a 6-V battery. The cold cranking performance rating is abbreviated CCA for "cold cranking amperes." The CCA rating of an automotive battery should equal or exceed the number of cubic inches of engine displacement. For example, a 300 cu.-in. (4.9-liter) engine should be equipped with a battery rated at 300 CCA or higher. See vehicle manufactures' specifications for recommended battery capacity.

Reserve Capacity. The reserve capacity rating for batteries is *the number of minutes* the battery can produce 25 A and still have a battery voltage of 1.75 V per cell (10.5 V for a 12-V battery). This rating is actually a measurement of the time a car can be driven in the event of a charging system failure.

WHAT DETERMINES BATTERY CAPACITY

The capacity of any battery is determined by the amount of active material in the battery. A battery with a large number of thin plates can produce high current for a short period. If a few thick plates are used, the battery could produce low current for a long period. A trolling motor battery used for fishing needs to supply a low current for a long period of time. An automotive battery is required to produce a high current for a short period for cranking. Therefore, every battery is designed for a specific application.

DEEP CYCLING

Deep cycling means that a battery is almost fully discharged and then completely recharged. Golf cart batteries are an example of lead-acid batteries that must be designed to be deep-cycled. A golf cart must be able to travel two 18-hole rounds of golf and then be fully recharged overnight. Since charging is hard on batteries because the internal heat generated can cause plate warpage, these specially designed batteries use thicker plate grids, which resist warpage. Normal automotive batteries are not designed for repeated deep cycling.

ADVANTAGES OF LEAD-CALCIUM (MAINTENANCE-FREE) BATTERIES

Batteries that use lead-calcium-alloy grids for their battery plates have the following advantages:

1. Lead-calcium batteries generally do not use water during normal usage. Water use is due to the "gassing" that occurs during charging and rapid discharge. Gassing is the electrical separation of water into its component parts: hydrogen and oxygen. Hydrogen is released at the negative plate and is a very explosive gas. Oxygen is released at the positive plate, and even though it does not burn, it helps anything else burn more rapidly. Since the water leaves the battery electrolyte as separate gases, the water must be replaced in normal batteries to maintain the proper acid strength and electrolyte level. Maintenance-free batteries also use envelope-type separators, allowing the plate material to rest lower in the battery case. This permits a higher level of electrolyte above the plates.

2. Lead-calcium batteries reduce the corroding of battery cables, terminals, and battery trays. Lead-

calcium cells do not "gas" until the cell voltage exceeds 2.15 V (12.9 battery voltage), whereas non-maintenance-free (lead-antimony) batteries start to gas at 2.0 V per cell (12.0 battery voltage). Gassing causes the battery to release gases and sulfuric acid fumes, which can cause corrosion to the battery terminals, battery tray, and other areas surrounding the battery.

3. A lead-calcium battery has a longer shelf life than most other types of batteries (three times longer than a lead-antimony battery). "Shelf life" means how long a battery can remain in storage (on a shelf) at a store before being sold and still be able to provide new-battery performance. Maintenance-free batteries are shipped with electrolyte installed in the battery at the factory. This eliminates the hazards of acid handling and the need for expensive equipment and clothing at the store that sells the battery.

4. Because of the higher internal resistance of lead-calcium plates, a maintenance-free battery resists overcharging.

DISADVANTAGES OF LEAD-CALCIUM (MAINTENANCE-FREE) BATTERIES

Maintenance-free (lead-calcium) batteries have the following disadvantages:

1. Lead-calcium batteries are difficult to jump-start because of the same high internal resistance that is listed above as an advantage against overcharging. To jump-start a vehicle equipped with a maintenance-free battery, the jumper cables should be left connected for at least one full minute (longer, if possible) to allow time for the good battery to overcome the high internal resistance of the dead battery and charge the battery before attempting to start the stalled vehicle. See Chapter 13 for jumper cable hookup and procedure.

2. Lead-calcium batteries cannot tolerate deep cycling as well as do lead-antimony batteries. The life of a maintenance-free battery is severely shortened (three times shorter than lead-antimony batteries) if subjected to repeated cycles of complete discharge and full recharge.

3. Many maintenance-free batteries have sealed filler caps which prevents access to each cell for testing or adding water. Maintenance-free batteries *do* use some water during their service life. The *rate* of water use is very low and should not require the addition of water in *normal* service. Many lead-calcium batteries can be opened for service to al-

low adding water, if necessary. Even though many maintenance-free batteries have sealed filler caps, the battery is vented to allow gases to escape during operation.

HYBRID BATTERIES

A hybrid battery is a battery that uses two different alloys in the plate grids. The positive plates are usually a low-antimony (2% to 3%) alloy, while the negative plates are a calcium alloy. This hybrid (combining two types of construction) produces a battery with lower water usage and corrosion than that of a lead-antimony battery, with improved deep-cycling capacity.

HINTS FOR COLD-WEATHER STARTING

Every battery loses power at lower temperatures, because a battery does not "store" electricity. It produces electrical power by chemical reactions. These chemical reactions occur much more slowly when the temperature is low. In fact, a fully charged 12-V battery has only 40% of its capacity at 0°F (−18°C), due to the cold temperature reducing the speed of the chemical reactions in the battery.

If the engine does not start after 15 seconds of cranking, the ignition switch should be turned "off" to allow the starter motor to cool. The battery also becomes warmer because the chemical reaction needed to produce the current necessary to turn the starter motor creates heat. This heat warms the battery. A warm battery has greater current-producing capacity than that of a cold battery, and therefore may be able to produce the needed power to start a cold engine on the second attempt. If a battery will not crank a cold engine at all on the first try, turn on the headlights for a short time. The chemical reaction required to produce the current for the headlights often creates enough heat to warm the battery, which *may* then produce enough current required to start the engine.

SUMMARY

1. Batteries convert chemical energy into electrical energy.
2. A warm battery can convert more energy faster than can a cold battery.
3. A battery at 0°F (−18°C) can only produce 40% of its capacity at 80°F (27°C).
4. Maintenance-free batteries use lead-calcium alloy battery grids instead of the lead-antimony of standard batteries. The lead-calcium prevents "gassing" and therefore reduces battery water use and battery corrosion.
5. All batteries must be held down with rigid brackets to prevent vibration damage.
6. Battery ratings list the amount of electrical work the battery can perform under certain conditions.
7. The higher the CCA, the longer the battery will live, because battery life depends on proper maintenance, charging rates, and the amount of original active battery material.
8. The starting and charging systems must be in proper operating condition for the battery to function correctly.

STUDY QUESTIONS

12-1. Explain the difference between maintenance-free and standard battery grids.

12-2. Describe the operation of a battery during discharging and charging.

12-3. Explain why specific gravity can indicate the state of charge and the condition of a battery.

12-4. List the advantages and disadvantages of a maintenance-free (lead-calcium) battery.

12-5. List and describe two battery ratings.

MULTIPLE-CHOICE QUESTIONS

12-1. The positive (+) plate of an automotive battery is:
(a) PbO_2.
(b) Pb.
(c) H_2SO_4.
(d) $PbSO_4$.

12-2. The negative (−) plate of an automotive battery is:
(a) PbO_2.
(b) Pb.
(c) H_2SO_4.
(d) $PbSO_4$.

12-3. The electrolyte of an automotive battery is:
(a) PbO_2.
(b) Pb.
(c) H_2SO_4.
(d) $PbSO_4$.

12-4. A maintenance-free battery:
(a) does not contain water.
(b) uses lead-antimony plates.
(c) uses lead-calcium plates.
(d) does not contain electrolyte.

12-5. Each cell of an automotive battery:
(a) contains one more negative plate than positive plate.
(b) contains one more positive plate than negative plate.
(c) produces 2.1 V when fully charged.
(d) both (a) and (c).

12-6. A fully charged 12-V automotive battery should indicate:
(a) 12.6 V or higher.
(b) a specific gravity of 1.265 or higher.

(c) 12.0 V.
(d) both (a) and (b).

12-7. "Deep cycling" means:
(a) overcharging the battery.
(b) the battery is fully discharged and then recharged.
(c) the battery is overfilled with water.
(d) the battery is overfilled with acid.

12-8. During discharge, the positive and negative plates become:
(a) sulfuric acid.
(b) lead peroxide.
(c) porous lead.
(d) lead sulfate.

12-9. "Reserve capacity" for batteries means:
(a) the number of *hours* the battery can supply 25 A and stay above 10.5 V.
(b) the number of *minutes* the battery can supply 25A and stay above 10.5 V.
(c) the number of *minutes* the battery can supply 50 A and stay above 9.6 V.
(d) the number of *minutes* the battery can supply 150 A and stay above 9.6 V.

12-10. As the battery temperature decreases, the specific gravity:
(a) increases.
(b) decreases.
(c) stays the same.
(d) none of the above.

13

Battery Testing and Service Procedures

THE BATTERY IS THE HEART OF THE AUTOMOTIVE ELECTRICAL SYSTEM. THE battery must be in good usuable condition for the rest of the electrical system to function correctly. In this chapter we describe the battery service and testing methods necessary to assure proper battery operation. The topics covered in this chapter include:

1. Visual inspections
2. Hydrometer testing
3. Voltmeter testing
4. Carbon-pile load testing
5. Battery load testing using a voltmeter
6. Battery-charging procedures
7. Jump-starting procedures

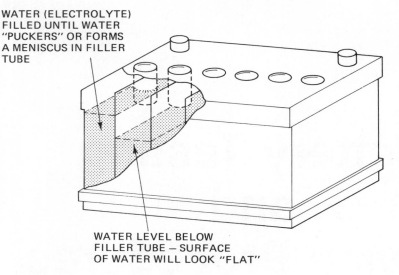

WATER (ELECTROLYTE)
FILLED UNTIL WATER
"PUCKERS" OR FORMS
A MENISCUS IN FILLER
TUBE

WATER LEVEL BELOW
FILLER TUBE – SURFACE
OF WATER WILL LOOK "FLAT"

FIGURE 13-1

ELECTROLYTE-LEVEL INSPECTION

The battery electrolyte level should be checked regularly if possible. Even though maintenance-free batteries do not normally require water, the electrolyte level should be checked and filled (if possible) as needed to ensure long battery life and proper operation. The battery should be filled with distilled water, or if it is not available, clean drinking water can be used. The water level should be above the plates and below the filler tube. This correct level is usually 1 to 1 ½ in. from the top of the battery. If the electrolyte level reaches the filler tube, the liquid "puckers" and looks curved (concave) when the liquid level is correct. This "puckering" of a liquid in a tube is called the *meniscus*. Therefore, water should be added to a battery until the water just puckers (forms a meniscus). See Figure 13-1.

NORMAL WATER LOSS

For conventional lead-antimony batteries, normal acceptable water use is 1 to 2 oz (30 to 60 ml) of water per cell per 1000 miles (1600 km). If water use exceeds this amount, the charging system should be checked and repaired as necessary to correct the overcharging of the battery.

Maintenance-free (lead-calcium) batteries are resistant to overcharging and do not require that additional water be added to the electrolyte during *normal* service. However, if the charging voltage is too high, even a lead-calcium battery will "gas" and may require water. Many maintenance-free batteries can be opened to gain access to the cells so that the battery water can be checked and water added. See Figure 13-2.

FIGURE 13-2 *Many maintenance-free batteries can be opened to check the electrolyte level without damaging the battery. Notice the surface dirt, which could cause the battery to self-discharge.*

WHAT SHOULD BE CHECKED IF A BATTERY NEEDS WATER

1. Whenever any battery requires water, the charging system should be checked for possible overcharging (charging voltage should be 15 V or less at 80°F (27°C).

2. Many maintenance-free batteries use a hybrid battery design that utilizes lead-antimony positive (+) plates with lead-calcium negative (−) plates. This style of battery combines the advantages and disadvantages of both lead-calcium and lead-antimony construction. Some water use could be considered normal with this type of construction, especially as the battery becomes older.

3. Check the heat shields that many automobile manufacturers install around the battery. These plastic shields or covers are designed to keep engine heat from the battery. Excessive heat can cause water use due to evaporation and shorter battery life by causing grid (plate) warpage and shedding.

4. The Freedom battery built by Delco and installed in most General Motors vehicles as standard equipment since the late 1970s cannot be opened to allow the adding of water. If the charge indicator is white or clear, the electrolyte level is too low to float the green ball in the indicator. This means that the battery must *not* be charged and the battery *must* be replaced.

5. Do not fill a battery in freezing weather without charging the battery afterward to allow the added water to mix thoroughly with the acid. This prevents the battery from freezing.

6. Do not overfill the battery. If battery water is overfilled, normal charging action will allow the sulfuric acid solution to overflow the filler tubes and cause corrosive damage to the battery cables, the battery tray, and surrounding components under the hood.

CAUTION: Do not allow electrolyte to get in the eyes or on the skin, clothing, or painted surfaces. Electrolyte will darken and and eventually eat through the painted surface of a car and will start to attack the steel or plastic of the body.

BATTERY VOLTAGE TEST

Testing the battery voltage with a voltmeter is a simple method to determine the state of charge of any battery. The voltage of a battery does not necessarily indicate whether or not the battery can perform satisfactorily, but it does indicate to the technician more about the battery's condition than does a simple visual inspection. A battery that "looks good" may not be good. This test is commonly called an "open-circuit battery voltage test" because with an open circuit, no current flows and no load is applied to the battery.

1. If the battery has just been charged or the car has recently been driven, it is necessary to remove the "surface charge" from the battery before testing. A surface charge is a higher-than-normal voltage charge which is just on the surface of the battery plates. The surface charge is quickly removed whenever the battery is loaded and therefore does not accurately represent the true state of charge of the battery.

2. To remove the surface charge, turn the headlights on high beam (brights) for 1 minute, then turn the headlights off and wait 2 minutes.

3. With the engine and all electrical accessories off, and the doors shut (to turn off the interior lights), connect a voltmeter to the battery posts on the correct scale to read 12 to 13 V. Connect the red positive (+) lead to the positive (+) post and the black negative (−) lead to the negative (−) post. See Figure 13-3.

NOTE: If the meter reads negative (−), the battery has been reverse charged (has reversed polarity) and should be replaced; or the meter has been connected incorrectly.

4. Read the voltmeter and compare the results with the state-of-charge chart shown. The voltages shown are for a battery at or near room temperature (70 to 80°F) (21 to 27°C).

Battery Voltage	State of Charge
12.6 V or higher	100% charged
12.4 V	75% charged
12.2 V	50% charged
12.0 V	25% charged
11.9 V or below	Discharged

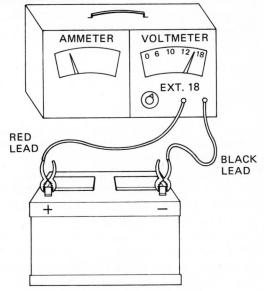

FIGURE 13-3 *Voltmeter hookup for testing the battery voltage.*

DRY-CELL BATTERIES CAN ALSO BE TESTED USING A VOLTMETER

Dry-cell batteries, which are commonly used for test equipment and flashlights, can be checked easily with a voltmeter. A new and therefore fully charged dry-cell battery usually has a higher voltage than that indicated on the battery. For example, a new D cell, which is a 1.5-V battery, will test approximately 1.7 V when new. As the battery is used, its voltage decreases. Depending on the use, whenever the voltage decreases to 80% of its indicated value, the battery should be replaced. Rechargeable dry-cell batteries are fully charged at a lower voltage than that for a nonrechargeable battery. The following chart indicates basic battery voltages for the most popular dry-cell sizes.

Battery	Voltage New	Discard Voltage
D (1.5 V)	1.7 V	1.2 V
AA (1.5 V)	1.7 V	1.2 V
AAA (1.5 V)	1.7 V	1.2 V
C (1.5 V)	1.7 V	1.2 V
9-V	9.5 V	7.2 V

Rechargeable dry-cell batteries are usually of lower voltage (1.5-V sizes are 1.25 V) and may not be recommended for some applications. For the most accurate results when testing the condition of any battery, test the voltage of the battery under *load.* Therefore, to test the condition of a dry-cell battery, measure the voltage of the battery while the battery is in the unit and working. Replace or recharge any battery that does not produce 80% or more of its original voltage as indicated in the chart above.

HYDROMETER TESTING

All lead-antimony and many maintenance-free batteries can be tested using a special test instrument called a *hydrometer*, which measures the concentration of the acid solution in the battery. The greater the concentration of acid in the electrolyte, the higher the state of charge of the battery.

A hydrometer is a hollow-glass tube with a suction bulb used to draw electrolyte solution into the tube. Inside the tube is a calibrated float which is suspended at various depths in the electrolyte, depending on the density of the liquid. The higher the specific gravity (the more dense the solution), the higher the float reading. See Figure 13-4.

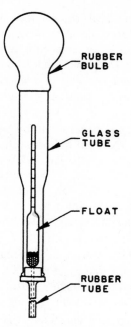

FIGURE 13-4 *Glass tube battery hydrometer for measuring the specific gravity of the battery electrolyte.*

Radiator coolant is also tested with an hydrometer, but is calibrated for use only with antifreeze solutions. Only use an hydrometer designed for testing battery electrolyte. To test a battery using an hydrometer, use the following procedures:

1. Open the hood or deck lid to gain access to the battery. Put a fender cover or other protection over the painted areas to guard against accidental dripping or spilling of battery acid, which could discolor or damage a painted surface.

2. Remove the filler caps. These caps may be individual (older style) or two larger caps covering three cells each, or a long, narrow plastic strip that covers all six cells. Even though many maintenance-free batteries cannot have their electrolyte checked, some maintenance-free batteries have removable filler caps. If the filler caps cannot be removed, check the state of charge of the battery by using a voltmeter.

3. Squeeze the bulb of the hydrometer and insert the pickup tube into the electrolyte of a cell. If the electrolyte level is too low to fill a battery hydrometer properly, distilled water should be added. However, since water is lighter than battery electrolyte, the added water simply remains on the surface until charging the battery or normal vehicle operating vibrations thoroughly mix the water with the electrolyte. Do not test a battery

with a hydrometer if water has recently been added, or the readings will be incorrect.

4. Slowly release the pressure on the rubber bulb. This forces the electrolyte into the glass barrel of the hydrometer. The hydrometer should be held vertically (straight up and down) and have enough electrolyte in the glass tube to allow the float scale to be suspended freely in the barrel.

5. Read the float scale at the liquid line.

6. The temperature of the electrolyte is important because the specific gravity increases (becomes more dense) when the temperature decreases. Battery hydrometers are calibrated [indicate correct readings only for electrolyte that is 80°F (27°C)]. For temperatures (of electrolyte) below 80°F (27°C), 0.004 point should be subtracted from the reading for every 10 Fahrenheit degrees below 80. Most battery hydrometers have built-in thermometers that measure the temperature of the electrolyte while it is in the barrel of the hydrometer. For example, if the reading on the float indicates 1.240 and the temperature of the electrolyte is 60°F (16°C), then 0.008 point must be subtracted (80 − 60 = 20 difference × 0.004 for each 10° increment = 0.008 point): 1.240 − 0.008 = 1.232, which is the final reading. See Figure 13-5.

7. Repeat the procedure above for all battery cells. If the *corrected* specific gravity readings for all of the individual cells are 1.225 or higher, the state of charge is satisfactory.

8. If one or more cells has a corrected specific gravity of less than 1.225, charge the battery for 20 minutes at 35 A and retest.

9. If there is a 50-point or greater variation between cells of the same battery, the battery must be replaced. [Example of a 50-point variation: With readings of 1.225, 1.230, 1.215, 1.200, 1.180, and 1.170, the difference between the highest reading (1.230) and the lowest reading (1.170) is 60 points (1.230 − 1.170 = 0.060 = 60 "points").]

Many maintenance-free batteries use a *charge indicator* that is actually a hydrometer. The charge indicator uses a green ball that reflects light and shows green when the battery (actually only one cell) has a 65% or greater state of charge. See Figures 13-6 and 13-7.

FIGURE 13-5 *Temperature-compensation thermometer. The specific gravity must be adjusted for temperatures of the electrolyte above or below 80°F (27°C) for accurate hydrometer results.*

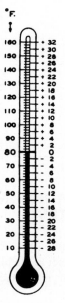

FIGURE 13-6 *Delco Freedom battery with a built-in hydrometer. (Courtesy of Oldsmobile Division, GMC.)*

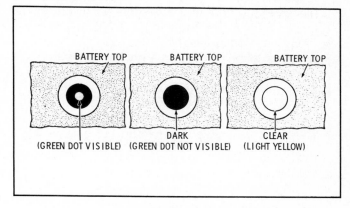

BATTERY TOP   BATTERY TOP   BATTERY TOP

(GREEN DOT VISIBLE)   DARK (GREEN DOT NOT VISIBLE)   CLEAR (LIGHT YELLOW)

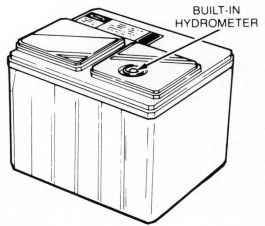

BUILT-IN HYDROMETER

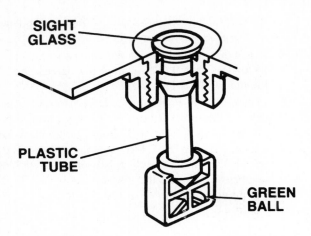

SIGHT GLASS

PLASTIC TUBE

GREEN BALL

FIGURE 13-7 Cutaway view of a built-in hydrometer. The green ball floats if the specific gravity is high. (Courtesy of Oldsmobile Division, GMC.)

WHAT DECIMAL POINT?

Specific gravity measurements for battery electrolyte are expressed in decimal form with numbers slightly greater than 1 and expressed to the thousandths place, which is three decimal places to the right of the decimal point. For example, a hydrometer reading of 1.225 *should* be "read" as "one point, two, two, five." Some automotive technicians ignore decimal points and commonly read a hydrometer reading of 1.225 as "twelve twenty-five" (1225). Just remember that regardless of how you hear specific gravity readings pronounced, battery electrolyte is only slightly more dense than water, not over 1000 times as dense.

BATTERY TESTING USING A LOAD TESTER

The best method that can be used to test a battery is to place an electrical load on the battery. By measuring the current and voltage output of the battery, the battery condition and capacity can be determined. A battery load tester is usually called a *carbon pile* because inside the tester are carbon piles or plates which provide the necessary resistance paths for the battery load. The carbon plates provide increasing numbers of resistances in parallel with the battery terminals. As explained in Chapter 3, the greater the number of paths the current can flow through, the greater the possible current flow. The high current flow from the

battery through the carbon plates of a carbon-pile tester place an electrical load on the battery. The "load knob" control on a carbon-pile tester connects additional carbon resistors in parallel as the knob is turned. A battery load tester is also equipped with an ammeter to measure the current leaving the battery (load) and a voltmeter to indicate the battery voltage during the test. To get accurate test results, the battery *must* be at least 75% charged. If the battery does not have a specific gravity of 1.225 or higher or an open-circuit voltage of 12.4 V or higher, the battery must be charged until the state of charge is high enough for testing.

To perform a load test correctly, turn the load knob to one-half of the CCA (cold cranking amperes) rating for the battery. Continue the load on the battery and read the voltmeter after 15 seconds, *then* remove the load. The battery voltage should be 9.6 V or higher. If the battery voltage is below 9.6 V, recharge the battery and repeat the test. If the battery voltage is below 9.6 V, replace the battery.

If the cold cranking amperes are unknown, load the battery to three times the ampere-hour rating of the battery. For example, if a battery has an ampere-hour rating of 50, the correct load to apply is 150 A (50 × 3 = 150). If neither the cold cranking rating nor the ampere-hour rating is known, the correct load can be estimated by loading the battery to one-half of the cubic-inch displacement of the engine, but never less than a 150-A load.

The minimum acceptable voltage is lowered with lowered temperatures. This lowered acceptable voltage during load testing is consistent with the reduced capacity of a battery to produce electricity at lower temperatures plus the greater cranking power required to start a cold engine.

Battery Temperature	Minimum Acceptable Voltage
70°F (21°C)	9.6 V
60°F (16°C)	9.5 V
50°F (10°C)	9.4 V
40°F (5°C)	9.3 V
30°F (−1°C)	9.1 V
20°F (−7°C)	8.9 V
10°F (−12°C)	8.7 V
0°F (−18°C)	8.5 V

TEST EQUIPMENT HOOK-UP FOR BATTERY LOAD TEST

Battery load testers (carbon-pile testers) use heavy battery cable leads which connect to the battery posts. The positive (+) tester lead (usually red) attached to the positive (+)

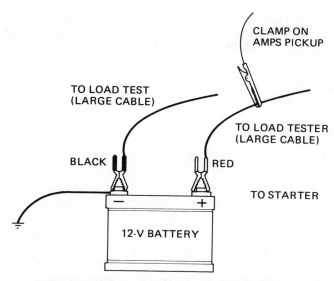

FIGURE 13-8 *Hookup for the battery load test using a Sun Electric VAT-40.*

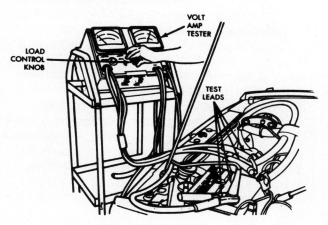

FIGURE 13-9 *Typical load tester showing the load control knob. (Courtesy of Chrysler Corporation.)*

battery post or if the inductive type, the cable. The negative (−) tester lead (usually black) attaches to the negative (−) post or cable of the battery. If an inductive-type ammeter is used, the inductive lead for the ammeter must be attached around the positive meter lead (not the vehicle's positive cable!). See Figures 13-8 and 13-9.

> **NOTE:** Most inductive pickups on ammeters have an arrow shown on the pickup. If the pickup is attached with the arrow facing the wrong direction, the needle of the ammeter will simply indicate charge rather than discharge and no harm will be done to the meter.

See the instructions for the test equipment being used for the proper hookup and testing procedure. After deter-

mining the proper battery load (no less than 150 A), use the load control to increase the load to the desired amount. The battery voltage will drop as the load is applied. The voltage of a new fully charged battery will drop only 1 or 2 V (11.6 - 10.6 V) when the load is applied and will continue to register this same high voltage for the entire 15-second test period. A weak battery tends to drop its voltage quickly (over 2 V) and continues to drop the voltage the longer the load is applied. If the battery is above 9.6 V at the end of the 15-second test while the load is still being applied, the battery can be used for service.

BATTERY LOAD TESTING USING A VOLTMETER

Similar to load testing a dry-cell battery by testing the battery voltage while the battery is "working," an automotive battery can and should be tested under load. The "load" that can be used is the starter motor. If the battery voltage drops below a certain voltage (9.6 V), the battery *could* be defective because it is not able to produce the necessary current to operate the starter motor. However, if the voltage does read below 9.6 V during engine cranking, the problem *could* be the starter, solenoid, or cables in the cranking circuit and *not* the battery. A typical starter motor requires from less than 100 A up to 250 A to crank the engine. *If* the starter is operating correctly, using the starter to load the battery is generally an acceptable method of battery load testing because the battery installed in the vehicle should be able to handle the starting current of that vehicle. The unknown factor in many cases is the condition of the starter and the rest of the cranking circuit.

Following is the procedure for load testing a battery using only a voltmeter and the car's starter motor.

1. Connect a voltmeter that is set to the proper scale to read voltages between 6 and 12 V. Connect the positive (+) lead to the positive (+) post of the battery and the negative (−) lead (usually the black) of the voltmeter to the negative (−) post of the battery. The car's battery cables are to remain connected to the battery throughout the entire test.

2. To load test the battery, the starter must be used to crank the engine. To prevent the engine from starting, the ignition system must be disconnected to prevent the engine from starting.

3. Using the ignition switch or remote starter, crank the engine for 15 seconds. Read the voltmeter during cranking at the end of the 15-second period. The voltage should be 9.6 V or higher.

4. If the battery voltage is at or below 9.6 V during cranking, the battery *may* not be able to supply

the necessary current required for engine cranking *or* the cranking circuit or engine condition is requiring excessive current (amperes) from the battery.

5. If the voltage is above 9.6 V during cranking, the battery, as well as the cranking-circuit components, are operating correctly. The higher this cranking voltage is, the better. For example, a reading of 11.1 V is better than 10.3 V.

BATTERY CHARGING

To recharge a battery fully, the same amount of electricity that was drained from the battery must be returned to the battery during charging. For example, if the headlights (20 A) were left on for 5 hours, they would drain 100 ampere-hours (ah) (20 × 5 = 100). The same amount of electricity has to be returned to the battery plus an extra 20% because battery recharging is only 80% efficient. Therefore, the battery charger should replace the 100 ah of drain plus 20% (20 ah) to recharge this battery fully. A charging rate of 20 A could fully recharge the battery in 6 hours (20 A × 6 h = 120 ah). Most battery manufacturers recommend a charging rate of 1% of the CCA rating (usually, 3 to 6 A).

The chart from the Battery Council International (BCI) shown in Figure 13-10 recommends charging rates and times as determined by the reserve capacity (in minutes) rating of the battery.

A lead-calcium battery or a battery that has been discharged for a period of time may not accept a charge up to several hours after starting the charger. It may be necessary to increase the charge rate to approximately 35 A for 30 minutes, and then reduce to a slow-charge rate for charging maintenance-free batteries. A battery is fully charged whenever the specific gravity is not increased by additional charging.

SHOULD A BATTERY BE KEPT OFF CONCRETE FLOORS?

Many technicians are told that if a battery is placed on a concrete floor, the battery will quickly become discharged. Some people believe that a battery should not be charged while sitting on a concrete floor.

Since self-discharge can occur whenever dirt and moisture are present between the battery posts, it is best to store batteries in a cool and *dry* area. The plastic battery case insulates the battery plates from discharging through to the concrete. Therefore, batteries *can* be stored or charged on concrete, but to help reduce the *possibility* of self-discharge due to the dampness of the concrete, it is *best* to keep batteries away from concrete. Placing batteries on a board helps, but on a workbench would be better because of the greater distance from the cool, damp concrete. Top post batteries are more likely to suffer self-discharge due to the accumulation of dirt and moisture on the battery than side post batteries.

BATTERY CHARGING GUIDE
(6-Volt and 12-Volt Batteries)

Figure 13-10

Caution - Do not use for Low Water Loss Batteries

Recommend Rate and Time for Fully Discharged Condition

Rated Battery Capacity (Reserve Minutes)	Slow Charge	Fast Charge
80 Minutes or Less	10 hrs. @ 5 Amperes 5 hrs. @ 10 Amperes	2.5 hrs. @ 20 Amperes 1.5 hrs. @ 30 Amperes
Above 80 to 125 Minutes	15 hrs. @ 5 Amperes 7.5 hrs. @ 10 Amperes	3.75 hrs. @ 20 Amperes 1.5 hrs. @ 50 Amperes
Above 125 to 170 Minutes	20 hrs. @ 5 Amperes 10 hrs. @ 10 Amperes	5 hrs. @ 20 Amperes 2 hrs. @ 50 Amperes
Above 170 to 250 Minutes	30 hrs. @ 5 Amperes 15 hrs. @ 10 Amperes	7.5 hrs. @ 20 Amperes 3 hrs. @ 50 Amperes
Above 250 Minutes	24 hrs. @ 10 Amperes	6 hrs. @ 40 Amperes 4 hrs. @ 60 Amperes

BATTERY SERVICING/CHARGING PRECAUTIONS AND RECOMMENDATIONS

1. Reduce the charging rate or turn off the charger if the battery case feels hot [125°F (52°C)]. Heat can seriously damage the battery by causing excessive gassing and plate warpage.

2. Be certain that the area is well ventilated around the charging battery to prevent accumulation of hydrogen and oxygen gases.

3. Any heat source or open flames (including smoking) should be kept away from any battery. The battery can explode if the gases from the battery are ignited.

4. Battery vent (filler) caps should remain *on* the battery during charging. All batteries are vented and these vent openings are designed to allow for the gases to disperse when released by charging. Removing the caps exposes these gases to any spark or flame and could cause an explosion due to the concentrated area of gases.

5. Do not attempt to charge a frozen battery. For best results, the battery should be at room temperature.

6. Make the battery charger connections to the battery with the charger turned "off." Also be certain to turn the charger "off" before disconnecting the charger cables from the battery, to avoid possible sparks.

7. Batteries should be stored in a cool [50°F (10°C)], dry place. Storage above 80°F (27°C) increases self-discharge.

8. Use charging adapters when connecting side-post batteries to battery charger cables. Using bolts can result in arcing and damage to the lead battery terminals.

9. Whenever servicing any electrical component on a vehicle, such as the alternator, starter, or relays, always disconnect the *negative* battery cable. This will *open* the entire electrical system and no current will flow. If there is current flow, a spark could cause a battery explosion or damage to any electronic equipment in the car.

NOTE: After reconnecting the battery after service remember to reset the car clock. Electronically tuned radios will also need to be reset because the loss of battery power erases the radio's memory for preset stations.

10. Safety glasses and proper acid-resistant clothing should be worn whenever working with battery electrolyte. Flush with water any area of skin or eyes that comes into contact with battery acid. Also avoid breathing fumes from a battery being charged.

JUMP STARTING

To jump-start another car with a dead battery, connect good-quality copper jumper cables as indicated in Figure 13-11. The last connection made should always be on the engine block or an engine bracket as far from the battery as possible. It is normal to create a spark when the jumper cables finally complete the jumping circuit, and this spark

FIGURE 13-11
THIS HOOK-UP FOR NEGATIVE GROUND VEHICLES

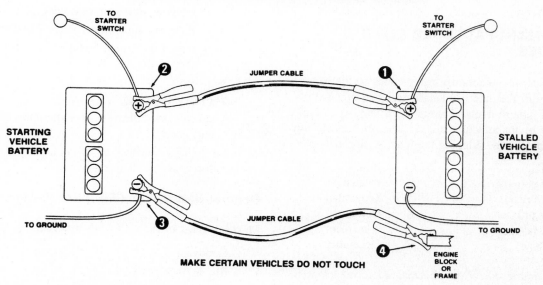

could cause an explosion of the gases around the battery. Many newer cars have special ground connections built away from the battery just for the purpose of jump starting. Check the owner's manual or service manual for the exact location.

WHY VEHICLES SHOULD NOT TOUCH WHILE JUMP STARTING

The negative battery cable (on negative-grounded vehicles) is attached to the engine block. The engine is mounted on rubber motor mounts and is therefore insulated from the frame. The exhaust system and the transmission are attached to the body or frame with rubber hangers and mounts to eliminate vibrations. The suspension is also insulated from the frame and body of the vehicle with rubber bushings and rubber insulator blocks at both ends of the coil springs (if so equipped). Therefore, the only electrical path for the current is through the body or frame. The small engine-to-body ground wires are designed only to carry the ground return path current for the lights and accessories (approximately 25 to 35 A) and are not large enough to handle the current required for starting. Not only could the starting current damage the bumper finish if the bumpers of the vehicles touched, but the excessive current flow through these small ground wires could be damaged by overheating. Some current could travel through the frame, drive shaft, and transmission, due to dirt or slight metal-to-metal contacts in the suspension system. Thus current could cause U-joint or transmission damage due to the electrical arcing through the bearings. The voltage drop caused by the small body ground wires would also reduce the current available for starting.

BATTERY MANUFACTURER BUILD DATE CODES

All major battery manufacturers stamp codes on the battery case that identify the date of manufacture and other information. Most battery manufacturers use a number to indicate the year of manufacture and a letter to indicate the month of manufacture, skipping the letter I. For example:

A	= January	G	= July
B	= February	H	= August
C	= March	J	= September
D	= April	K	= October
E	= May	L	= November
F	= June	M	= December

The shipping date from the manufacturing plant is usually indicated by a *sticker* on the end of the battery.

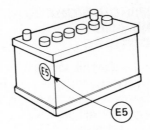

FIGURE 13-12

Almost every battery manufacturer uses just one letter and one number to indicate the month and year. For example, a shipping sticker with an "E5" indicates that the battery was shipped in May, 1985.

BATTERY BUILD DATE CODES AND MANUFACTURERS

Johnson Control (formerly Globe Battery)

Maker of most of the following batteries:

Sears	TBC (Tire Battery Corp.)
Standard Oil	Farm and Fleet
Interstate	Battery Associates
NAPA	Walmart
Motorcraft	

Build Code: Y J 9 B J 6 is as follows:

Y = plant location

J = special code

9 = manufacturer's process code

B = February (the letter I is used here)

J = day of the month (begins with 1 through 9 and then A through W is used for 10–31)

6 = year

(February 19, 1986)

A seventh identification is sometimes used for the shift code.

An eighth identification is sometimes used for the line code.

General Battery Co. (GBC) (old Prestolite)

Maker of most of the following batteries:

Goodyear	Firestone
Big A Auto Parts	Prestolite
some NAPA	Titan

Build Code: B 29 6 3 C R 3 is as follows:

B = shift
29 = day of the month
6 = year
3 = month
C = type
R = manufacturing code
3 = plant location
(March 29, 1986)

NOTE: Some older batteries may still use the Prestolite Code (until about October, 1985):

3 L 16 V 02 is as follows:
3 = year
L = month
16 = day of the month
V = plate construction
02 = plant location
(November 16, 1983)

Exide

Maker of most of the following batteries:

Western Auto   Chrysler (Mopar)

Build Code: Lo 23 B 5 B 1 is as follows:

Lo = plant location
23 = day of the month
B = month
5 = year
B = paste line
1 = shift
(February 23, 1985)

DELCO

Plant locations given for DELCO:

I = Indiana
G = Georgia

K = Kansas
N = New Jersey
F = France
Z = Canada
C = California

Build Code: 5 B I   17 is as follows:

5 = year
B = month
I = plant location
17 = day of the month
(February 17, 1985 in Indiana)

GNB, Inc. (formerly Gould National Battery)

Maker of most of the following batteries:

Champion Batteries
Power Breed
Action Pack

Build Code: 6 A 01 H is as follows:

6 = year
A = month
01 = day of the month
H = manufacturer code
(January 1, 1986)

East Penn Manufacturing Co.

Maker of the following batteries:

DEKA   Lynx
Federal   Start Rite

Build Code: 6 234 6 is as follows:

6 = year
234 = day of the year
6 = manufacturer code
(August 21, 1986)

NOTE: The day of the year is called the Julian date.

SUMMARY

1. Battery testing must include as many different test procedures as possible. Many battery tests cannot determine a weak discharged battery from a weak defective battery. All batteries should be tested with a voltmeter and a hydrometer (if possible) before load testing to determine if the battery can supply an adequate amount of current for its size.

2. Battery testing is extremely important. All starter and charging circuit testing must be performed with a known-good battery, or inaccurate test results will incorrectly indicate a defective starter or alternator.

3. The major cause of premature battery failure is overcharging. Another major cause is vibration caused by the lack of proper battery hold-downs, resulting in shorted plates. If one or more (but not all) cells start to "gas" (bubble) when under a heavy electrical load (approximately 200 A), such as during an attempted engine start, the cause is shorted cells. The cells that are shorted will "gas" sooner than normal cells. The battery is the first and most important item that needs to be tested in the starting circuit.

4. To be assured of accurate test results, all other starting circuit tests depend on having a known-good charged battery.

STUDY QUESTIONS

13-1. Explain how a battery voltage test can determine the state of charge of any battery.

13-2. Explain why battery hydrometer readings must be temperature compensated.

13-3. Describe the battery load testing procedure.

13-4. Explain the safety precautions necessary when working with batteries.

13-5. Describe the proper jump-starting procedures.

MULTIPLE-CHOICE QUESTIONS

13-1. Which of the following is (are) indicated by a battery that uses excessive water?
 (a) the charging voltage may be too high.
 (b) the battery plates could be sulfated.
 (c) both (a) and (b).
 (d) none of the above.

13-2. A fully charged 12-V battery has:
 (a) 12.0 V.
 (b) 12.2 V.
 (c) 12.4 V.
 (d) 12.6 V. or higher.

13-3. The maximum allowable difference between the highest and lowest hydrometer (specific gravity) reading is:
 (a) 0.010.
 (b) 0.020.
 (c) 0.050.
 (d) 0.500.

13-4. A battery load test:
 (a) tests the condition of the starter.
 (b) tests the condition of the battery cables.
 (c) tests the condition of the battery.
 (d) tests the state of charge of the starter.

13-5. A battery high-rate discharge (load, capacity) test is being performed on a 12-V battery. Technician A says that a good battery should have a voltage reading *below* 9.6 V while under load. Technician B says that the battery should be discharged (loaded) to twice its cold cranking rating. Which technician is correct?

 (a) A only.
 (b) B only.
 (c) both A and B.
 (d) neither A nor B.

13-6. When charging a maintenance-free (lead-calcium) battery:
 (a) the initial charging rate should be about 35 A for 30 minutes.
 (b) the battery may not accept a charge for several hours.
 (c) the battery temperature should not exceed 125°F.
 (d) all of the above.

13-7. Whenever jump starting:
 (a) the last connection should be the positive (+) post of the dead battery.
 (b) the last connection should be on the engine block of the dead vehicle.
 (c) the alternator must be disconnected on both vehicles.
 (d) the vehicles should touch, if possible, to be assured of a good ground connection.

13-8. The charge indicator on some batteries:
 (a) "lights" if the battery voltage is 12.6 V or higher.
 (b) is a built-in hydrometer on one cell only.
 (c) can indicate if the electrolyte level is too low.
 (d) both (a) and (c).
 (e) both (b) and (c).

13-9. When performing a capacity test (load test) on a battery, what is the proper load to apply?
 (a) 15 A for 45 seconds.

(b) three times the ampere-hour rating for 15 seconds.

(c) four times the ampere-hour rating for 30 seconds.

(d) 10 times the ampere-hour rating for 15 seconds.

13-10. Technician A says that a battery hydrometer test can be used to find a defective battery. Technician B says that a battery load test can be used to determine a discharged or defective battery. Which technician is correct?

(a) A only.

(b) B only.

(c) both A and B.

(d) neither A nor B.

14

Starter Motors and the Cranking Circuit

THE CRANKING CIRCUIT IS AN IMPORTANT COMBINATION OF MECHANICAL and electrical parts that have to work together correctly or the engine will not start. The cranking circuit must be able to crank the engine (up to 250 engine rpm) in all types of weather. This chapter includes descriptions and operating principles for all of the components in the automotive cranking circuit. The topics covered in this chapter include:

1. Starter motor construction
2. Starter motor operation
3. Starter drives
4. Solenoid and relay operation

FIGURE 14-12 *High-torque armatures can often be identified by the size and shape of the armature loops. The armature on the right has square armature loops and is a high-torque unit.*

SERIES MOTOR

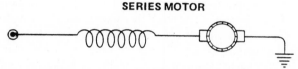

FIGURE 14-13 *Wiring diagram of a series motor. (Courtesy of General Motors Corporation.)*

TYPES OF STARTER MOTORS

Starter motors must provide high power at low starter motor speeds to crank an automotive engine at all temperatures and at the required cranking speed for the engine to start (60 to 250 engine rpm). Electric motors are classified according to the internal electrical motor connections. The method used determines the power-producing characteristics of the electric motor. Many starter motors are series wound, which means that the current flows first through the field coils, then in series through the armature, and finally to a ground through the ground brushes. See Figure 14-13.

SERIES MOTORS

A series motor develops its maximum torque at the initial start (zero rpm) and develops less torque as the speed increases. A series motor is commonly used for an automotive starter motor because of its high starting power characteristics. A series starter motor develops less torque at high rpm because a current is produced in the starter itself which acts against the current from the battery. Because this current works against battery voltage, it is called counter electromotive force or CEMF. This counter EMF is produced by electromagnetic induction (see Chapter 4) in the armature conductors which are cutting across the

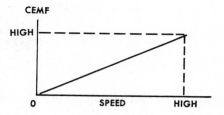

FIGURE 14-14 *As the speed of a series motor increases, the current necessary decreases because the CEMF increases. (Courtesy of General Motors Corporation.)*

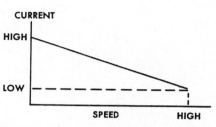

FIGURE 14-15 *A series motor develops its greatest torque at low rpm. (Courtesy of General Motors Corporation.)*

magnetic lines of force formed by the field coils. This induced voltage operates against the applied voltage supplied by the battery, which reduces the strength of the magnetic field in the starter. See Figure 14-14.

Since the power (torque) of the starter depends on the strength of the magnetic fields, the torque of the starter decreases as the starter speed increases. See Figure 14-15. It is also characteristic of series-wound motors to keep increasing in speed under light loads. This could lead to the destruction of the starter motor unless controlled or prevented.

SHUNT MOTORS

Shunt-type electric motors have the field coils in parallel (or shunt) across the armature as shown in Figure 14-16. A shunt motor does not decrease in torque at higher motor rpm because the CEMF produced in the armature does not decrease the field coil strength. A shunt motor, however,

SHUNT MOTOR

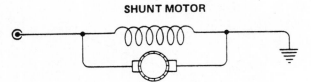

FIGURE 14-16 Wiring diagram of a shunt-type electric motor. (Courtesy of General Motors Corporation.)

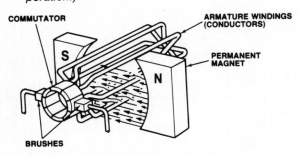

PERMANENT MAGNET MOTOR

FIGURE 14-17 (Courtesy of Chrysler Corporation.)

does not produce as high a starting torque as that of a series-wound motor, and is not used for starters. Many small electric motors used in automotive blower motors, windshield wipers, power windows, and power seats use permanent magnets rather than electromagnets. Because these permanent magnets maintain a constant field strength, the same as a shunt-type motor, they have similar operating characteristics. Permanent-magnet starter motors were developed by General Motors for automotive use in the mid-1980s. The permanent magnets used are an alloy of neodymium, iron, and boron. This magnet is almost 10 times more powerful than permanent magnets used previously. See Figure 14-17.

COMPOUND MOTORS

A compound motor has the operating characteristics of a series motor *and* a shunt-type motor because some of the field coils are connected to the armature in series, and some (usually only one) field coil(s) are connected directly to the battery in parallel (shunt) with the armature. See Figure 14-18.

Compound-wound starter motors are commonly used in Ford, Chrysler, and some GM starters. The shunt-wound field coil is called a "shunt coil" and is used to limit the maximum speed of the starter. Since the shunt coil is energized as soon as the battery current is sent to the starter, it is used to engage the starter drive on Ford positive-engagement starters.

COMPOUND MOTOR

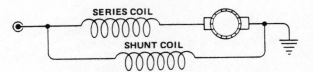

FIGURE 14-18 Wiring diagram of a compound wound electric motor. (Courtesy of General Motors Corporation.)

GEAR-REDUCTION STARTERS

Gear-reduction starters are used by many automotive manufacturers. The purpose of the gear reduction (typically 2 to 4:1) is to increase starter motor speed and provide the torque multiplication necessary to crank an engine. See Figure 14-19. As a series-wound motor increases in rotational speed, the starter produces less power, and less current is drawn from the battery because the armature generates greater CEMF as the starter speed increases. However, a starter motor's maximum torque occurs at zero rpm and decreases with increasing rpm. A smaller starter using a gear-reduction design can produce the necessary cranking power with reduced starter amperage requirements. Lower current requirements mean that smaller battery cables can be used. General Motors' permanent-magnet starters use a planetary gear set to provide the necessary torque for starting.

STARTER DRIVES

A starter drive includes a small pinion gear that meshes and rotates the larger gear on the engine for starting. The pinion gear must engage with the engine gear slightly *before* the starter motor rotates, to prevent serious damage to either the starter gear or the engine, but must be disengaged after the engine starts. The ends of the starter pinion gear

FIGURE 14-19

FIGURE 14-20 *Starter drive unit. Notice the tapered end of the pinion gear and the internal bushing.*

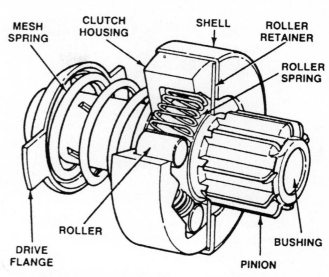

FIGURE 14-21 *Cutaway view of a starter drive unit, showing the overrunning clutch and related parts. (Courtesy of Ford Motor Company.)*

are tapered to help the teeth mesh easier without damaging the flywheel ring-gear teeth. See Figure 14-20. The ratio of the number of teeth on the engine gear and on the starter pinion is 15 to 20:1. A typical small starter pinion gear has nine teeth which turn an engine gear with 166 teeth. This provides an 18:1 gear reduction. This means that the starter motor is rotating approximately 18 times faster than the engine. Normal cranking speed for the engine is 200 rpm. This means that the starter motor speed is 18 times faster, or 3600 starter rpm (200 × 18 = 3600). If the engine started and was accelerated to 2000 rpm (normal cold engine speed), the starter would be destroyed by the high speed (36,000 rpm) if the starter were not disengaged from the engine.

Older-model starters (before the early 1960s) often used a Bendix drive mechanism, which used "inertia" to engage the starter pinion with the engine flywheel gear. "Inertia" means that a stationary object tends to remain stationary, because of its weight, unless forced to move. The small starter pinion gear is attached to a shaft with threads, and the weight (inertia) of this gear causes it to be spun along the threaded shaft and mesh with the flywheel whenever the starter motor spins. If the engine speed is greater than the starter speed, the pinion gear is forced back along the threaded shaft and out of mesh with the flywheel gear. The Bendix drive mechanism has generally not been used since the early 1960s.

All starter drive mechanisms use a type of one-way clutch that allows the starter to rotate the engine, but turns freely if the engine speed is greater than the starter motor speed. This clutch is called an *overrunning clutch* and protects the starter motor from damage if the ignition switch is held in the start position after the engine starts. The overrunning clutch, which is built in as a part of the starter drive unit, uses steel balls or rollers installed in tapered

notches. See Figure 14-21. This taper forces the balls or rollers tightly into the notch when rotating in the direction necessary to start the engine. Whenever the engine rotates faster than the starter pinion, the balls or rollers are forced out of the narrow tapered notch, allowing the pinion gear to turn freely (overruns).

The spring between the drive tang or pulley and the overrunning clutch and pinion is called a "mesh" spring and helps cushion and control the engagement of the starter drive pinion into mesh with the engine flywheel gear. This spring is also called a *compression* spring because the starter solenoid or starter yoke compresses the spring and the spring tension causes the starter pinion to engage the engine flywheel.

SYMPTOMS OF A DEFECTIVE STARTER DRIVE

A starter drive is generally a very dependable unit and does not require replacement unless defective or worn. The major wear occurs in the overrunning clutch section of the starter drive unit. The steel balls or rollers wear and often do not wedge tightly into the tapered notches necessary for engine cranking. See Figure 14-22.

A worn starter drive can cause the starter motor to operate freely and not rotate the engine. Therefore, the starter makes a "whine" noise. The whine indicates that the starter motor is operating and that the starter drive is not rotating the engine flywheel. The entire starter drive is replaced as a unit. The overrunning clutch section of the starter drive cannot be serviced or repaired separately because it is a sealed unit. Starter drives are more likely to

FIGURE 14-22 *Cutaway view of rollers, springs, and notches of a typical overrunning clutch.*

fail intermittently at first and then gradually more frequently, until replacement becomes necessary to start the engine. Intermittent starter drive failure (starter whine) is often most noticeable during cold weather. See Chapter 15 for starter drive replacement procedures.

STARTER DRIVE OPERATION

The starter drive (pinion gear) must be moved into mesh with the engine ring gear before the starter motor starts to spin. Most automotive starters use a solenoid or the magnetic "pull" of the shunt coil in the starter to engage the starter pinion.

POSITIVE-ENGAGEMENT STARTERS

Positive-engagement starters, used on many Ford engines, utilize the shunt coil winding of the starter to engage the starter drive. The high starting current is controlled by an ignition-switch-operated starter solenoid, usually mounted near the positive (+) post of the battery. When this control circuit is closed, current flows through a hollow coil (called a drive coil) that attracts a movable pole shoe. The movable metal pole shoe is attached through a lever (called the plunger lever) and engages the starter drive. See Figure 14-23.

As soon as the starter drive has engaged the engine flywheel, a tang on the movable pole shoe "opens" a set of contact points. The contact points provide the ground return path for the drive coil operation. After these grounding contacts are opened, all of the starter current can flow through the remaining three field coils and through the brushes to the armature, causing the starter to operate. The movable pole shoe is held down (which keeps the starter drive engaged) by a smaller coil on the inside of the main drive coil.

FIGURE 14-23 *Positive-engagement starter. (Courtesy of Ford Motor Company.)*

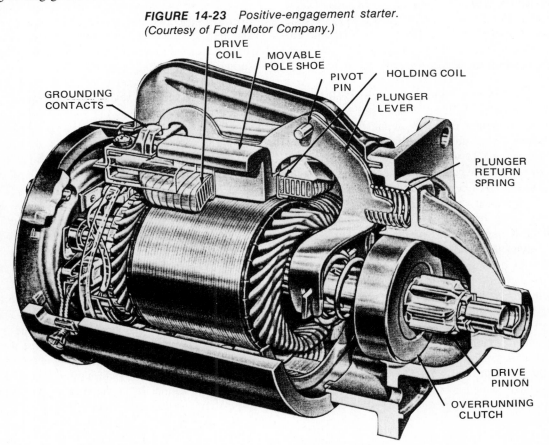

GROUNDING CONTACTS

DRIVE COIL

MOVABLE POLE SHOE

PIVOT PIN

HOLDING COIL

PLUNGER LEVER

PLUNGER RETURN SPRING

DRIVE PINION

OVERRUNNING CLUTCH

This coil is called the *holding coil* and is strong enough to hold the starter drive engaged while permitting the maximum current possible to operate the starter. If the grounding contact points are severely pitted, the starter may not operate the starter drive or the starter motor because of the resulting poor ground for the drive coil. If the contact points are bent or damaged enough to prevent them from opening, the starter would "clunk" the starter drive into engagement, but would not allow the starter motor to operate. See Chapter 15 for service and testing procedures.

SOLENOID-OPERATED STARTERS

A starter solenoid is an electromagnetic switch containing two separate, but connected electromagnetic windings. This switch is used to engage the starter drive and control the current from the battery to the starter motor. See Figure 14-24.

The two internal windings contain approximately the same number of turns but are made from different gauge wire. Both windings together produce a strong magnetic field which pulls a metal plunger into the solenoid. The plunger is attached to the starter drive through a shift fork lever. When the ignition switch is turned to the *start* position, the motion of the plunger into the solenoid causes the starter drive to move into mesh with the flywheel ring gear. The heavier gauge winding (called the *pull-in winding*) is needed to draw the plunger into the solenoid. The

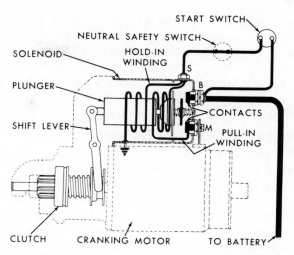

FIGURE 14-24 *Wiring diagram of a starter solenoid. (Courtesy of General Motors Corporation.)*

lighter gauge winding (called the *hold-in winding*) produces enough magnetic force to keep the plunger in position. The main purpose of using two separate windings is to permit as much current as possible to operate the starter and yet provide the strong magnetic field required to move the starter drive into engagement. The instant the plunger is drawn into the solenoid enough to engage the starter drive, the plunger makes contact with a metal disk that connects the battery terminal post of the solenoid to the motor terminal. See Figure 14-25. This permits full battery

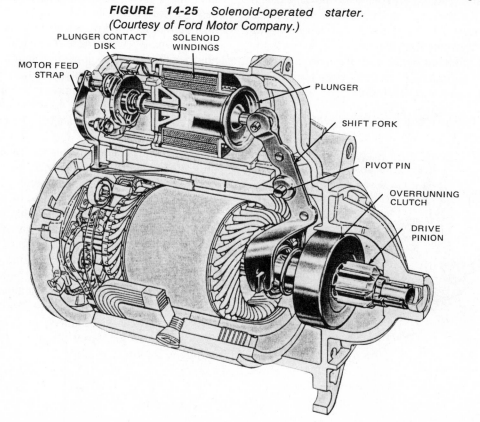

FIGURE 14-25 *Solenoid-operated starter. (Courtesy of Ford Motor Company.)*

current to flow through the solenoid to operate the starter motor. The contact disk also electrically disconnects the pull-in winding. The solenoid *has* to work to supply current to the starter. Therefore, if the starter motor operates at all, the solenoid is working even though it may have high external resistance which could cause slow starter motor operation.

RELAY VERSUS SOLENOID

The terms *relay* and *solenoid* are frequently confused since both function to *relay* heavy current (closing the circuit) between the battery and starter motor. Both are electromagnetic switches, but relays use contact points attached to a *movable arm* to open and close a circuit, whereas a solenoid uses a *movable iron core* to open and close the circuit. If the solenoid is mounted on top of the starter, the movable core also operates the starter drive. See Figure 14-26. Ford and some American Motors cars utilize a starter solenoid mounted to the inner fender. It is called a starter solenoid because it uses a movable core to open and close the starter circuit. See Figure 14-27. If a starter solenoid is mounted near the battery, it must be properly grounded by the mounting bolts because the ground return paths for the pull-in and hold-in windings are attached to the metal mounting bracket.

The solenoid should also be mounted vertically (straight up and down) to ensure proper operation and movement of the internal iron core. A solenoid can transfer higher current than a relay. Some solenoids (movable cores) are called a *relay* if they carry heavy current such as that used in many automotive diesel engines in the glow plug circuit. If an electromagnetic switch does outside work, the unit is always identified as a solenoid. Refer to the manufacturer's service information for the proper term to use when purchasing replacement parts.

FIGURE 14-26 *Typical starter-mounted solenoid.*

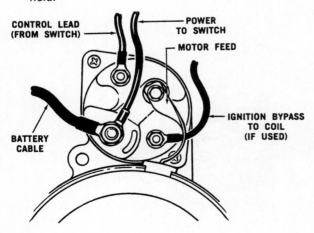

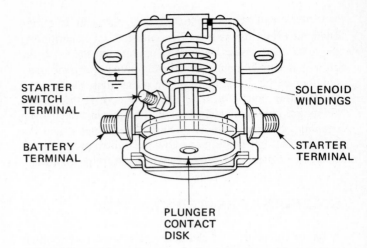

FIGURE 14-27 *Typical remote starter solenoid. Note that the mounting flange* must be *grounded properly for the solenoid windings to operate. (Courtesy of Ford Motor Company.)*

SOLENOID CLICKING

A weak battery (discharged) or high resistance in the battery cables or connections can cause the starter solenoid to "click" or "chatter" and not operate the starter. Often the noise is repeated rapidly, like a machine gun. This is caused by the current energizing the windings inside the solenoid, causing a movable iron core to be drawn into a magnetic field and a metal plate to make the electrical connection between the battery and the starter. However, due to the high current required by the starter, the battery voltage decreases rapidly. When battery voltage decreases, the solenoid windings can no longer maintain the magnetic field strength necessary to hold the metal plate in contact with the battery and starter terminals inside the solenoid. Therefore, the circuit is "opened" as soon as the circuit is closed, releasing the load on the battery. The battery voltage once again is high enough to operate the pull-in and hold-in windings of the solenoid. This causes the solenoid to click on and off rapidly and not operate the starter.

The life of any starter motor can be increased by keeping the engine in a good state of tune to permit rapid engine starts. It is not the miles that a car is driven that wear out a starter, it is the amount of time the starter is used that usually determines the life of brushes, bushings, and starter drives. To protect the starter, do not crank an engine for longer than 15 seconds without stopping to allow the starter to cool. Heat is damaging to any starter.

See Chapter 11 for voltage-drop testing of the battery-to-starter cables, connections, and solenoids/relays. See Chapter 15 for starter diagnosis, troubleshooting, and testing procedures.

SUMMARY

1. Starter motors consist of three major exterior parts: drive-end housing, brush-end housing, and field-coil housing.

2. Full battery current flows through the starter solenoid to the field coils. The current through the field coils creates a strong magnetic field. The current then flows through the hot brushes and through the armature loops. It is the magnetic forces between the field coils and the armature that forces the armature to turn. The method used to electrically connect the field coils and armature determines its operating characteristics and name: series, shunt, and compound.

3. A series motor creates a counter-electromotive current in the loops of the armature in the opposite direction of applied battery voltage. As a starter motor speed increases, the higher the CEMF and the less current it draws and the lower its power. Therefore, a series starter motor produces its maximum power at 0 rpm.

4. Low battery voltage or high-resistance battery cables can cause a ''no crank'' situation or a solenoid clicking sound. Solenoid-activated starters use the solenoid to engage the starter drive *before* the starter motor is rotated. Some starters use the magnetic force of the shunt coil to engage the starter.

STUDY QUESTIONS

14-1. Describe the operation of the cranking circuit when the ignition switch is turned to the ''start'' position.

14-2. What is a series-wound motor?

14-3. Describe the operation of a starter drive.

14-4. What is a positive-engagement starter?

14-5. Describe the operation of the pull-in and hold-in windings of a solenoid.

MULTIPLE-CHOICE QUESTIONS

14-1. The starter control circuit:
   (a) prevents starter operation unless the gear selector is in park or neutral.
   (b) prevents starter operation on some manual transmission vehicles unless the clutch is depressed.
   (c) controls the operation of the starter solenoid.
   (d) all of the above.

14-2. Starter brushes:
   (a) are constructed of a combination of copper and carbon.
   (b) consist of one hot brush and one ground brush.
   (c) consist of two hot brushes and one ground brush.
   (d) both (a) and (c).

14-3. Starter motors operate on the principle that:
   (a) the field coils rotate in the opposite direction from the armature.
   (b) opposite magnetic poles repel.
   (c) like magnetic poles repel.
   (d) the armature rotates from a strong magnetic field toward a weaker magnetic field.

14-4. Series motors:
   (a) produce electrical power.
   (b) produce maximum power at 0 rpm.
   (c) produce maximum power at 5000 rpm.
   (d) use a shunt coil.

14-5. Automotive starters are:
   (a) series wound.
   (b) compound wound.
   (c) shunt wound.
   (d) both (a) and (b).

14-6. Technician A says that a defective solenoid can cause a starter whine. Technician B says that a defective starter drive can cause a starter whine. Which technician is correct?
   (a) A only.
   (b) B only.
   (c) both A and B.
   (d) neither A nor B.

14-7. Starters that do not have a solenoid mounted on the starter:

(a) are called positive-engagement starters.

(b) are called gear-reduction starters.

(c) use a drive coil in the field housing to move a metal pole shoe.

(d) both (a) and (c).

14-8. The instant the ignition switch is turned to the "start" position:

(a) both pull-in and hold-in windings are energized.

(b) the hold-in winding is energized.

(c) the pull-in winding is energized.

(d) the starter motor starts to rotate before energizing the starter pinion.

14-9. A solenoid:

(a) is an electromagnetic switch.

(b) uses a movable-arm electrical contact.

(c) uses a movable core.

(d) is the same as a Bendix.

(e) both (a) and (c).

14-10. Technician A says that a discharged battery (lower than normal battery voltage) is harmful to the starter motor. Technician B says that a discharged battery or dirty (corroded) battery cables can cause solenoid clicking. Which technician is correct?

(a) A only.

(b) B only.

(c) both A and B.

(d) neither A nor B.

15

Starting System Testing and Service

THE FIRST INDICATION OF MANY AUTOMOTIVE FAILURES IS, "THE CAR WILL not start." The electrical starting system contains many possible parts, pieces, and connections. It is, therefore, important to understand basic troubleshooting and testing of the cranking circuit to be certain of finding the exact problem. The topics covered in this chapter include:

1. Basic starting system testing
2. Starter ampere testing
3. Starter disassembly and component testing:
   (a) Hold-in and pull-in windings of solenoids
   (b) Growler testing starter armatures
   (c) Field coil testing
4. Reassembly and bench testing of starter motors

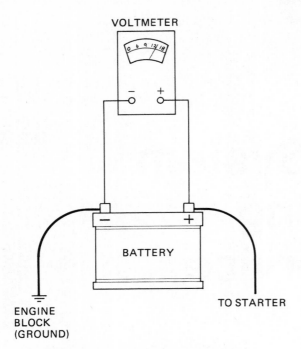

VOLTMETER

ENGINE
BLOCK
(GROUND)

BATTERY

TO STARTER

FIGURE 15-1 *Typical voltmeter hookup.*

BASIC STARTER SYSTEM TESTING

The starter system requires a known-good battery to be able to supply the required current necessary to start an engine, especially during cold weather. Even a new battery can only supply 40% of its rated capacity at 0°F (−18°C). Cold weather also increases the power required to start an engine because the crankcase oil is thicker, especially if the oil is of the wrong viscosity (thickness) or has not been changed regularly. Therefore, if the starting system fails to crank the engine, the first thing that should be tested is the general operation of the electrical system, including the battery. See Figure 15-1.

MAKE CERTAIN THE BATTERY IS GOOD

It is important that the battery being used during starting system testing be at least 75% charged (specific gravity of 1.225 or higher or 12.4 V or higher). See Chapter 13 for details. If the battery is unsatisfactory, a known-good replacement battery must be used before further testing is performed on the starting system.

VOLTMETER STARTER TEST

After the battery has been *tested* and confirmed to be usable, the general condition of the starting circuit can be determined with the same voltmeter connections used above

(Figure 15-1). If the starting circuit requires more amperes than normal, the battery voltage will decrease. If the voltmeter reads below 9.6 V during cranking, the problem is excessive resistance in the battery cables, connections, solenoid/relay, or starter motor.

> **NOTE:** Internal engine problems could also cause the starter to draw too much current. The slower the starter operates, the less CEMF it produces and the more current it draws.

If the voltage is *above* 9.6 V, the starting system is satisfactory according to this test. If the starter is still cranking the engine too slowly, remove the starter and test all individual components.

If the voltage is at or below 9.6 V, further testing is needed to eliminate simple items such as battery cables or dirty connections. See Chapter 11 for detailed procedures for voltage-drop testing, which determines if any battery cable or connection has excessive resistance. High resistance in the cables or connections would "drop" or lower the voltage available to the starter. Low battery voltage will not only prevent proper operation of the starter, but could damage the starter.

STARTER SOLENOID TESTING

Starter solenoids conduct the high cranking system currents from the battery to the starter. If the solenoid is attached to the starter, it is also used to engage the starter drive. When the ignition switch is rotated to the "start" position, 12 V from the ignition switch is sent to the S (start) terminal of the solenoid. The current flows through the pull-in winding and hold-in winding, creating a strong magnetic pull and moving a metal plunger. This moving plunger, inside the solenoid, causes a metal disk to contact two internal metal terminals which are attached to the battery on one side and the starter motor on the other side. Typical solenoid-related problems and their causes include:

1. *Slow cranking speed.* Slow cranking speed can arise from a number of conditions, including a low or discharged battery or high-resistance battery connections or cables. The internal contacts and metal disk become pitted and corroded with use, and this is a common source of high resistance. Use the voltage-drop test described in Chapter 11 to determine if the solenoid should be overhauled or replaced.

2. *Solenoid "chatter" or "clicking."* Solenoid chatter or clicking is often *not* caused by the solenoid, but is caused by a low or discharged battery or high resistance in the battery cables and connections. See Chapter 11 for voltage-drop testing procedures.

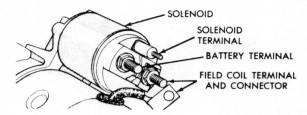

— SOLENOID

— SOLENOID TERMINAL

— BATTERY TERMINAL

— FIELD COIL TERMINAL AND CONNECTOR

FIGURE 15-2 *Typical starter solenoid. The solenoid terminal is often called the S terminal. Current sent to the solenoid terminal operates the solenoid and the starter motor. (Courtesy of Chrysler Corporation.)*

3. *"Nothing happens" on the start position.* If nothing happens in the "start" key position, an open circuit is preventing current from activating the hold-in and pull-in windings of the solenoid. (If the solenoid windings were operating, a click would be heard from the solenoid.) All solenoid windings are operated by current sent from the ignition switch to the S (start) terminal of the solenoid. Current flows through the windings and completes the circuit to the ground through the starter motor brushes (if the solenoid is mounted on the starter) or through the metal mounting brackets. See Figure 15-2. A loose or corroded connection at the S terminal of the solenoid is a common source of this problem. If current is not available at the S terminal, the problem is an open circuit between the ignition switch and the S terminal. To confirm the operation of the solenoid and starter, connect a remote starter or jumper wire between the positive (+) post of the battery and the S terminal of the solenoid. If neither the starter nor the solenoid operate, the solenoid must be tested separately.

SIMPLE STARTER TESTS

Some cars are equipped with a voltmeter, but they are usually designed to indicate voltage only with the ignition in the "on" position and not in the "start" position. However, a voltmeter reading could be obtained by turning the ignition "on" and using a remote starter switch to crank the engine.

If a voltmeter is not available, the brightness of the lights of the car can help indicate battery voltage. An automotive light bulb becomes brighter with higher voltage and dimmer with lower voltage. See Figure 15-3.

It is human nature to think of the most serious item that could be defective instead of the simplest cause for a problem. For example, dirty battery connections can cause the starter solenoid/relay to click or chatter when attempting to start an engine. Instead of replacing the solenoid or

relay and finding that the same problem still exists, check the simple things first, such as the condition of the battery cables and connections. (See Chapter 11 for voltage-drop testing.)

Other simple items that could be overlooked when troubleshooting a "no crank" or "slow cranking" situation include:

1. *Gear selector in "drive" instead of "neutral" or "park":*
   (a) Misadjusted gear selector.
   (b) Misadjusted or defective neutral safety switch.
   (c) Not depressing the clutch all the way on manual transmission cars equipped with a clutch safety switch.

2. *Battery connections:*
   (a) Dirty positive (+) and/or negative (−) cables at the battery.
   (b) Loose negative cable connections on the engine.
   (c) Loose positive cable connections on the starter or the solenoid/relay.

3. *Ignition switch not operating the solenoid/relay:*
   (a) Misadjusted ignition switch not allowing the current to be sent to the starter solenoid or relay that closes the electrical circuit for the starter motor current.
   (b) Loose connection on the solenoid or relay at the wire sending current from the ignition switch. This connection is usually labeled with a small letter "s" and can loosen enough (especially in cold weather) to intermittently not operate the starter at all.

4. *Overadvanced ignition timing.* Overadvanced ignition timing can cause slow and often jerky cranking because the spark plugs are igniting the air/fuel mixture in the cylinders too soon in the upward movement of the piston. This causes the expanding burning air/fuel mixture to exert a force down on the piston as it travels upward on the compression stroke. Check the ignition timing (see Chapter 20) or disconnect the ignition system and operate the starter. If the starter cranks the engine at normal speed, the cause is due to overadvanced ignition timing. If cranking speed is still too slow, the problem is due to engine mechanical problems or cranking circuit problems.

Since it is difficult to observe the brightness of the headlights during the cranking (except at night), the car door can be opened and the dome (interior) light can be observed. It is normal for the brightness of the light to dim slightly because battery voltage will drop approximately 2 V (for example, from 12.6 V to 10.6 V) during normal engine cranking. However, excessive dimming of the light

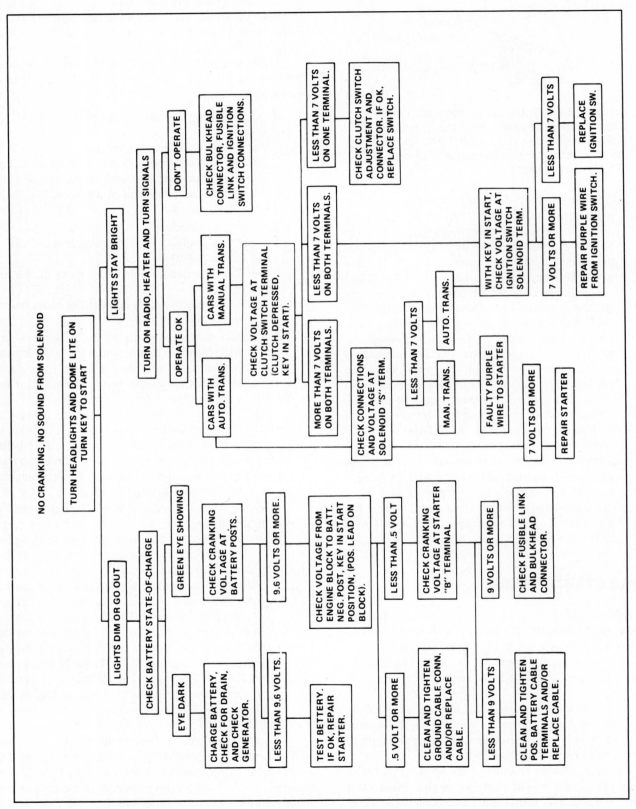

FIGURE 15-3 *(Courtesy of Oldsmobile Division, GMC.)*

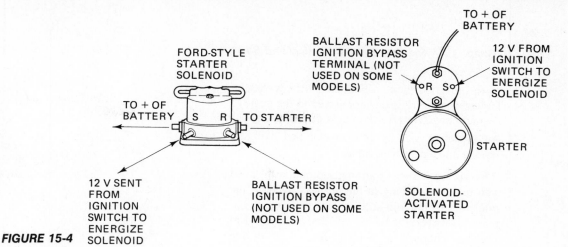

FORD-STYLE STARTER SOLENOID

TO + OF BATTERY

12 V SENT FROM IGNITION SWITCH TO ENERGIZE SOLENOID

TO STARTER

BALLAST RESISTOR IGNITION BYPASS (NOT USED ON SOME MODELS)

BALLAST RESISTOR IGNITION BYPASS TERMINAL (NOT USED ON SOME MODELS)

TO + OF BATTERY

12 V FROM IGNITION SWITCH TO ENERGIZE SOLENOID

STARTER

SOLENOID-ACTIVATED STARTER

FIGURE 15-4

indicates possible battery or starter problems because the battery voltage is lower than normal. If no change occurs in the brightness of the light during cranking, there is an open circuit in the starting system. No current flows in an open circuit; therefore, the starter will not operate at all, and the voltage of the battery would remain unchanged. Some examples of open-circuit problems in the starter circuit include:

If the starter will not operate and the lights remain bright

1. Defective or misadjusted neutral safety switch.
2. Loose connection on the starter solenoid or relay at the S terminal labeled. See Figure 15-4.
3. Defective or misadjusted ignition switch. The ignition switch conducts battery voltage to the S terminal of the starter solenoid or relay in the ''start'' or ''crank'' position.
4. Defective starter solenoid or relay.
5. Open field coil armature, or defective brushes in the starter motor.
6. Excessively loose or corroded battery cable connections.

If the lights dim and the starter does not operate normally

1. Discharged or defective battery.
2. Loose or corroded battery connections.
3. Field coils in the starter motor shorted to ground. This requires replacement of the starter or having the field coils rewound and reinsulated.
4. Armature shorted to ground. This requires starter motor or armature replacement.

STARTER TESTING USING AN AMMETER

A high-reading ammeter (up to 500 A) should be used to test the amount of current required to crank an engine to start. Automobile manufacturers specify maximum acceptable current draw for each starter and engine application.

NOTE: Most specifications for starter motors indicate maximum current draw with the starter being tested on the bench, not installed in the vehicle.

If the exact specifications are unknown, the following general *maximum* allowable starter current specifications can be used for testing the starter in the vehicle:

Four- or six-cylinder engines	150 A maximum
V-8 and V-6 engines (except GM V-8s)	200 A maximum
GM V-8s	250 A maximum

TEST EQUIPMENT HOOK-UP FOR STARTER AMPERE TEST

Most automotive testers have both an ammeter and a voltmeter with the built-in carbon-pile tester that is necessary for battery and alternator testing. Modern test equipment uses inductive-type ammeters which do not require that the battery cables be disconnected. See Figure 15-5 for the usual hookup for an inductive-style ammeter required for starter ampere testing.

FIGURE 15-5 (Courtesy of Sun Electric Corporation.)

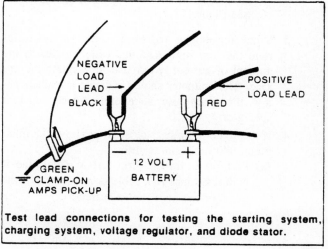

NEGATIVE LOAD LEAD → BLACK

POSITIVE LOAD LEAD RED

GREEN CLAMP-ON AMPS PICK-UP

12 VOLT BATTERY

Test lead connections for testing the starting system, charging system, voltage regulator, and diode stator.

1. Most inductive pick-up ammeters simply require large-diameter test leads connected to the positive (+) and negative (−) terminals of the battery.

2. An inductive pick-up is then clamped over the *vehicle's* negative (−) cable.

NOTE: If the inductive ammeter probe is installed with the arrow facing in the wrong direction, the ammeter reading will still indicate the correct amperes. However, the meter will read backward (indicating "charge" rather than "discharge" during starter motor operation).

PROCEDURE FOR STARTER AMPERE TESTING

1. Connect the ammeter in series with either battery cable or clamp the inductive pickup around a battery cable according to the ammeter manufacturer's instructions.

2. Disconnect the ignition system so that the engine will not start when the starter is operated by the ignition switch.

   (a) Disconnect the coil wire from the center of the distributor cap and connect it to the engine block.

NOTE: If the coil wire is *not* grounded, the high voltage produced in the coil could cause coil damage. If necessary, use a jumper wire between the coil wire and a good engine ground (a clean unpainted metal surface) as shown in Figure 15-6.

   (b) Disconnect the battery lead to the distributor if the car is a GM vehicle built after 1974 equipped with HEI (High Energy Ignition). See Figure 15-7.

3. Crank the engine and observe the ammeter. Ignore the first high reading, which indicates the amount of current required to first rotate the engine. The ammeter should indicate a reading below the maximum acceptable starter current specifications.

4. If the starter current is close to or above the maximum allowable starter draw:
   (a) Perform a voltage-drop test to determine if high resistance in a cable or connection is the cause of the high starter current.

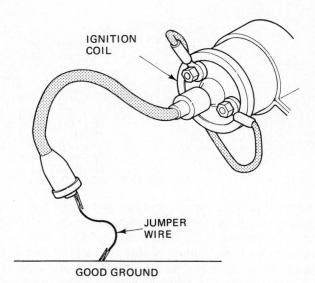

FIGURE 15-6 *During engine cranking, make certain that the coil wire is grounded to prevent the engine from starting and to protect the coil.*

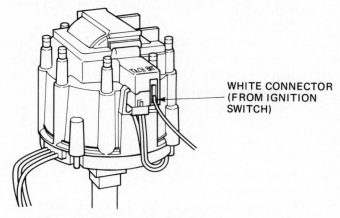

FIGURE 15-7 *GM vehicles equipped with HEI ignition can be disconnected by unplugging the white connector wire to keep the engine from starting.*

NOTE: If a voltmeter is not available, touch the battery cables and connections with your hands during engine cranking. Excessive resistance creates heat. Replace cables that are hot to the touch.

   (b) Be certain that the battery is capable of supplying the necessary cranking current by testing the battery separately. A weak battery will cause a higher-than-normal amperage draw by the starter because the voltage decreases excessively under load with a weak battery.

   (c) Remove the starter for disassembly, testing, and service.

WHY DOES A WEAK BATTERY CAUSE AN INCREASE IN THE STARTER CURRENT?

A starter motor requires a certain amount of *power* to start an engine. What is power? Power expressed in electrical terms is amperes times volts (power = I × E). The power required to start an engine remains the same even if the battery voltage decreases. For example:

Good Battery	Weak Battery
11.0 V during cranking	9.8 V during cranking
190 A	213 A
power = 11.0 × 190 = 2090 W	power = 9.8 × 213 = 2090 W

Notice that the power required is the same for the starter motor to crank the engine (2090 W). However, the good battery can maintain 11.0 V while supplying the necessary current for starter operation (190 A). A weak battery decreases in voltage while supplying the high current required for cranking. The *power* required by the starter to crank an engine is constant. If the battery voltage decreases, the amount of current must *increase* to compensate for the drop in voltage. Notice that the required current is increased from 190 A (good battery) to 213 A for the weak battery. Therefore, to get accurate test results, a known good battery at least 75% charged should be used during starter testing.

STARTER REMOVAL

If the starter has been determined to be defective, it must be removed from the vehicle. The following procedure should be followed to remove a starter safely.

1. Disconnect the negative (−) battery cable. This "opens" the entire electrical system and prevents possible damage or personal injury in the event of an accidental connection between a hot battery connection and a chassis ground.

NOTE: Make certain that the ignition switch is "off" to prevent possible damage to the computer (if equipped). Damage could be caused by a high-voltage spike or arc occurring when disconnecting a battery cable with the ignition switch "on."

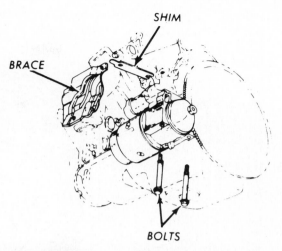

FIGURE 15-8 *Typical starter installation using shims between the starter mounting pad and the engine block. (Courtesy of Lester Catalog Company.)*

2. Unbolt the starter from the engine. Mark and disconnect all the electrical wires. Do not allow the starter to hang from the wires.

NOTE: Many starters have thin strips of metal (called shims) between the starter housing and the engine block. See Figure 15-8. The shims must be reinstalled in the exact position as originally installed. This is true of all extra support brackets and heat shields that may be installed on the starter.

3. Clean the exterior housing of the starter, but do not submerge the starter in solvent (cleaning solution). Cleaning solution can wash out the lubricants in the starter drive mechanism and damage the insulation on the armature and field coil windings.

SERVICING SOLENOID-ACTIVATED STARTERS

General Motors, Chrysler, many Ford, and many import starters use a solenoid to engage the starter drive with the flywheel ring gear. The solenoid is bolted to the starter and should be tested and inspected during starter motor inspection. See Figure 15-9. See Chapter 11 for voltage-drop testing of starter cables and solenoids.

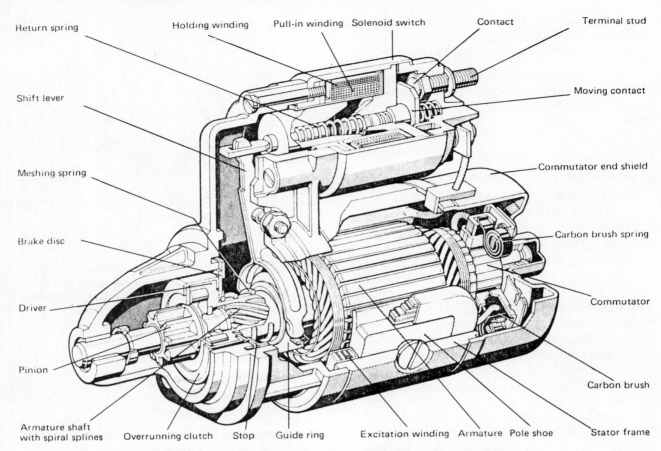

Return spring · Holding winding · Pull-in winding · Solenoid switch · Contact · Terminal stud · Shift lever · Moving contact · Meshing spring · Commutator end shield · Brake disc · Carbon brush spring · Driver · Commutator · Pinion · Carbon brush · Armature shaft with spiral splines · Overrunning clutch · Stop · Guide ring · Excitation winding · Armature · Pole shoe · Stator frame

FIGURE 15-9 *Typical solenoid-activated starter. (Courtesy of Robert Bosch Corporation.)*

TESTING SOLENOID WINDINGS

Whenever the starter solenoid is off the car for service or during troubleshooting, the pull-in and hold-in windings of the solenoid should be tested. Continuity can be checked with an ohmmeter between the S and M terminals for the pull-in coil, and between the S and the solenoid metal housing for the hold-in winding. The ohmmeter should indicate low ohms (0.2 to 0.6 Ω). See Figure 15-10. An ohmmeter test, however, is not as accurate an indication of the correct operation of the coils as actually testing the coils during operation as indicated below.

TO CHECK PULL-IN WINDINGS

The pull-in windings are made from heavy-gauge wire which creates a strong magnetic "pull" on the metal plunger moving the starter drive into mesh with the flywheel. However, this strong magnetic force does require current from the battery that must be conserved for use by the starter motor. Therefore, the pull-in winding is de-

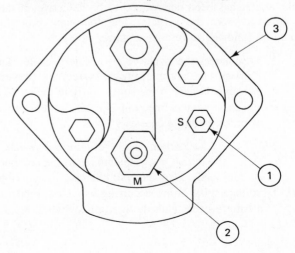

FIGURE 15-10 *GM solenoid ohmmeter check. Set the ohmmeter to the R × 1 scale. The reading between 1 and 3 should be 0.4 to 0.6 Ω (hold-in winding). The reading between 1 and 2 should be 0.2 to 0.4 Ω (pull-in winding).*

TO POSITVE(+) OF BATTERY

TO IGNITION SWITCH (START TERMINAL)

S

R

IGNITION RESISTOR BY-PASS TERMINAL (not used on all models)

'M'' (MOTOR) TERMINAL

FIGURE 15-11

signed to be "opened" when the plunger forces the solenoid disk into contact with the battery and starter terminals. A weaker winding, called the hold-in winding, keeps the circuit connected between the starter and the battery as long as the ignition switch is held in the start position. Using two jumper wires, a 12-V battery, and a plunger loosely installed in the solenoid, make the following connections:

1. Positive (+) post of the battery to the S terminal of the solenoid.

2. Negative (−) post of the battery to the M terminal of the solenoid. The M terminal connects the disk inside the solenoid with the starter motor field windings. See Figure 15-11.

3. When the last connection is made, a magnetic field is created that will rapidly pull the plunger into the solenoid.

4. If the connections are clean and tight and the plunger is not drawn into the solenoid, the solenoid should be replaced because there is an "open" in the pull-in windings.

TO CHECK HOLD-IN WINDINGS

The hold-in windings have the same number of turns of wire as the pull-in windings, but the wire gauge is smaller, resulting in less current draw from the battery to keep the starter drive engaged with the flywheel.

Using two jumper wires, a 12-V battery, and a plunger loosely installed in the solenoid, make the following connections:

1. Positive (+) post of the battery to the S terminal of the solenoid

2. Negative (−) post of the battery to the metal (ground) of the solenoid

The plunger should be forced into the solenoid; however, the force is less than the force created by the pull-in winding. In actual operation *both* windings are used to initially draw the plunger into the solenoid. At the instant the starter motor is engaged, the current for the pull-in winding stops.

OVERHAULING A STARTER SOLENOID

If the solenoid tests described above indicate that both the pull-in and the hold-in coils are okay, the solenoid can be disassembled and the contacts cleaned or reversed to reduce the voltage drop that occurs in used solenoids. See Figures 15-12 through 15-16.

SERVICING REMOTE-MOUNTED STARTER SOLENOIDS

A remote starter solenoid commonly used by Ford and other manufacturers uses only one winding. Using two jumper wires and a 12-V battery, make the following connections:

1. Positive (+) post of the battery to the S terminal of the solenoid

2. Negative (−) post of the battery to the metal mounting bracket (ground)

The solenoid should click whenever the last wire is connected.

STARTER MOTOR DISASSEMBLY

Before disassembly of any starter motor, mark with chalk or other suitable marker the relationship of the brush-end housing to the field housing. Also mark the location of the through bolts on the field housing. The drive-end housing usually has a locating notch or tab to locate it correctly on the field housing.

STARTER DRIVE REPLACEMENT PROCEDURE

The starter drive can easily be replaced without having to remove the armature or the brushes from a Ford starter. A defective starter drive unit will cause the starter to whine whenever engaged. The procedure for replacing the starter drive on a Ford starter includes:

1. Remove the negative (−) battery cable and remove the starter from the engine.

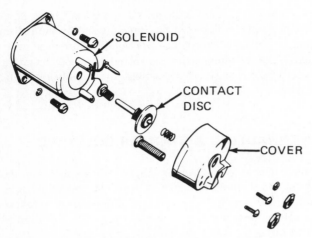

FIGURE 15-12 *To separate a typical GM-style solenoid, remove the inner nuts from the M and S terminals. Remove the two retaining screws and the plastic cover should pull straight off the terminals. (Courtesy of Chevrolet Motor Division, GMC.)*

FIGURE 15-13 *Cutaway view of a disk and one of the two terminal contacts.*

FIGURE 15-14 *The B terminal contact is worn and corroded.*

FIGURE 15-15 *The B terminal contact was reversed by loosening the retaining nut and turning the terminal 180°, then retightening the retaining nut.*

FIGURE 15-16 *The plunger disk can be sanded and cleaned or flipped upside down on some models.*

2. Use chalk or a scratch mark on the starter housing in line with the two long through bolts.

3. Remove the brush inspection band and cover plate.

4. Carefully remove both through bolts which hold the drive end housing to the field housing.

5. Carefully pull off the drive end housing *without* removing the armature.

NOTE: If the armature does come out accidently, the brushes must be repositioned before reinstalling the armature.

6. Remove the snap ring from the pivot pin. Remove the pin and starter drive yolk and the movable pole shoe assembly.

7. With the drive-end housing removed, the starter drive can be slipped off the end of the armature after removing a retainer and a C-clip fastener.

8. Inspect the replacement starter drive, being certain that the drive pinion gear is *exactly* the same length. There are several different Ford starter

FIGURE 15-17 *Excessively worn starter drive shift fork (yoke). Some forks can be flipped over (reversed) so that the round surface, now on the bottom in the photo, contacts the starter drive.*

drive units available and the major difference is in the length of the drive pinion gear.

9. Reinstall the replacement starter drive, yoke, C-clip, and drive-end housing.

10. Check the condition of the shift fork that operates the engagement of the starter drive. If the pins are worn, as in Figure 15-17, a replacement yoke (shift fork assembly) is required. Some shift forks can be reversed (flipped over) to provide a new surface against the starter drive.

NOTE: The drive-end housing includes a small notch that must be lined up with a locating tab on the starter field housing.

11. Align the chalk marks and install the through bolts.

A similar procedure is required to replace the starter drive units in other types of starters except that the armature usually has to be removed. The starter drive retaining clip also differs in design. See Figure 15-18 for the recommended method of reinstalling the retainer on a GM starter.

Bench test the starter before installing it in the vehicle to be assured of proper operation. Connect the red jumper cable leads to the positive (+) post of the battery and to the copper starter stud. Connect the black jumper cable lead to the negative (−) battery terminal and momentarily touch the other black jumper cable lead to the starter housing. Observe for correct starter engagement and operation.

CAUTION: When bench testing any starter, be certain that it is properly supported, because the starter will twist with a great force when engaged.

After bench testing the starter, check the tightness of the through bolts.

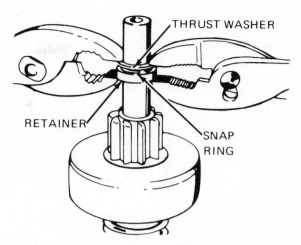

FIGURE 15-18 *Installing the starter drive retainer on a GM starter. (Courtesy of General Motors Corporation.)*

TESTING STARTER ARMATURES

Because the loops of copper wire are interconnected in the armature of a starter, an armature can be accurately tested only by use of a *growler*. A growler is a 110-V ac test unit that generates an alternating (60-H) magnetic field around an armature. A starter armature is placed into the V-shaped top portion of a laminated soft-iron core surrounded by a coil of copper wire. When the growler is plugged into a 110-V outlet and switched on, the moving magnetic field creates an alternating current in the windings of the armature.

GROWLER TEST FOR SHORTED ARMATURE WINDINGS

Place the armature on the growler and turn the growler ''on.''

CAUTION: Do not turn the growler on without an armature, or the growler will be damaged.

While rotating the armature by hand, gently place a hacksaw blade along the top of the armature. If any loop of the armature is *shorted*, the hacksaw blade will vibrate. See Figure 15-19. If an armature is shorted (copper-to-copper connection), it must be replaced or rewound by a specialist. The hacksaw vibrates because the alternating current creates an alternating electromagnet in the armature. If only one loop is shorted, it does not create the magnetic pull on the hacksaw blade in one direction and the blade will vibrate.

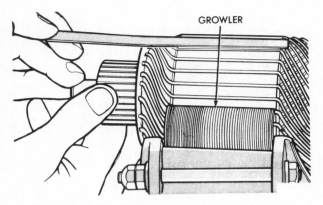

FIGURE 15-19 *Testing the armature for shorts using a hacksaw blade and a growler. If the hacksaw blade vibrates, the armature is shorted and must be replaced.*

TESTING THE ARMATURE FOR GROUNDS

Built into growlers is a 110-V test light with two test leads. Touch one lead to all segments (copper strips separated by mica insulation) of the commutator and touch the other test lead to the steel armature shaft or armature steel core. The test light should *not* light. If the test light is on, the armature is *grounded* (shorted to ground) and must be replaced. See Figure 15-20.

TESTING THE ARMATURES FOR OPENS

An "open" in an armature is usually observed visually as a break or unsoldered loop where it connects to the com-

FIGURE 15-20 *Checking the armature for grounds (copper to steel). If the test light on the growler lights when one probe is touched to the commutator (copper) and the other probe is touched to steel, the armature is grounded and must be replaced.*

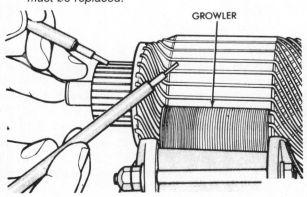

mutator segments. An open is usually caused by overheating of the starter due to excessive cranking time, or a shorted or grounded armature. A loose or broken solder connection can often be repaired by resoldering the broken connection using rosin-core solder. Many armatures can be tested on some growlers using a pickle-fork-shaped test probe. See Figure 15-21. With the growler "on," use the pickle-fork probe on the segments of the commutator. The loops of the armature are open if *no* current is indicated on the built-in ammeter on the growler. The current indicated on the ammeter is created by the moving magnetic field of the growler in loops of the armature. There are no exact specifications for the amount of current that should be indicted on the ammeter. If the ammeter reads *any* current between any two segments, the loops tested are *not* open. This test sequence should be repeated on each pair of segments. See Figure 15-22.

FIGURE 15-21 *Testing an armature for "opens."*

FIGURE 15-22 *An example of an "open" armature.*

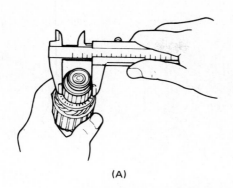

(A)

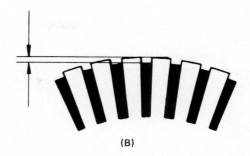

(B)

FIGURE 15-23   *(A) Measuring the diameter of a starter motor commutator to be certain that it is not below minimum specifications. (B) The end view of a commutator, showing the amount of undercut necessary. Standard undercut depth: 0.020 to 0.030 in. (0.5 to 0.8 mm). Undercutting can be accomplished on a special armature lathe attachment or by using a hacksaw blade to cut down between each copper segment of the commutator. (Courtesy of Toyota Motor Sales, USA, Inc.)*

NOTE: If the armature is open, shorted (copper to copper), or grounded (copper to steel), the armature must be repaired or replaced. The cost of a replacement armature often exceeds the cost of a replacement starter.

ARMATURE SERVICE

If the armature tests okay, the commutator should be measured as shown in Figure 15-23a and machined on a lathe, if necessary, to be certain that the surface is smooth and round. Some manufacturers recommend that the insulation between the segements of the armature (mica or hard plastic) should be undercut as shown in Figure 15-23b. Mica is harder than copper and will form raised "bumps" as the copper segments of the commutator wear. Undercutting the mica permits a longer service life for this type of starter armature.

TESTING STARTER MOTOR FIELD COILS

With the armature removed from the starter motor, the field coils should be tested for opens and grounds. A powered test light or an ohmmeter can be used. To test for a grounded field coil, touch one lead of the tester to a field brush (insulated or "hot") and the other end to the starter field housing. See Figure 15-24. The ohmmeter should indicate infinity (no continuity) and the test light should *not* light. If there is continuity, replace the field coil housing assembly.

NOTE: Many starters use removable field coils, and these coils must be rewound using the proper

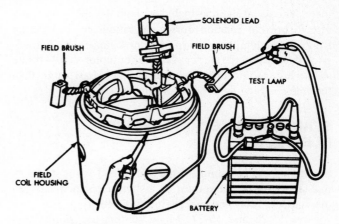

FIGURE 15-24   *Testing field coils with a battery and a test lamp (use a number 69 bulb). (Courtesy of Chrysler Corporation.)*

equipment and insulating materials. Usually, the cost involved in replacing defective field coils exceeds the cost of a replacement starter.

The ground brushes should show continuity to the starter housing.

STARTER BRUSH INSPECTION

Starter brushes should be replaced if the brush length is less than one-half of its original length (less than ¼ in. or 0.75 mm). See Figure 15-25. On some models of starter motors, the field brushes are serviced with the field coil assembly and the ground brushes with the brush holder. Many starters use brushes that are held in with screws and are easily replaced, whereas other starters may require soldering to remove and replace the brushes.

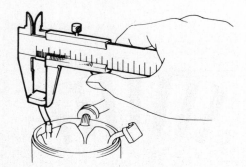

FIGURE 15-25 *Measuring the starter brush length. (Courtesy of Toyota Motor Sales, USA, Inc.)*

BENCH TESTING

Every starter should be tested before installation in a vehicle. The usual method includes clamping the starter in a vise to prevent rotation during operation and connecting heavy-guage jumper wires (minimum 4 guage) to a known-good battery and the starter. The starter motor should rotate as fast as specifications indicate and not draw more than the free-spinning amperage permitted.

STARTER DRIVE-TO-FLYWHEEL CLEARANCE

For the proper operation of the starter and without abnormal starter noise, there must be a slight clearance between the starter pinion and the engine flywheel ring gear. See Figures 15-26 and 15-27. Many starters use shims (thin metal strips) between the flywheel and the engine block mounting pad to provide the proper clearance.

FIGURE 15-26 *Use a screwdriver to pry the starter drive into engagement with the flywheel to measure the clearance. (Courtesy of General Motors Corporation.*

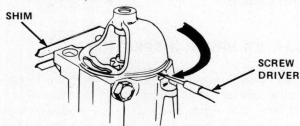

SHIM

SCREW DRIVER

A .015" SHIM WILL INCREASE THE CLEARANCE APPROXIMATELY .005". MORE THAN ONE SHIM MAY BE REQUIRED.

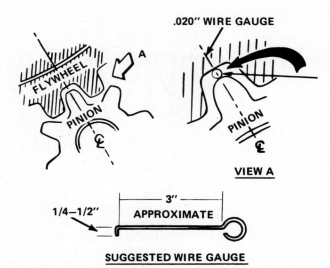

VIEW A

SUGGESTED WIRE GAUGE

FIGURE 15-27 *The clearance for a GM starter should be 0.020 in. A paper clip may be used, but paper clips are usually 0.025-to 0.030-in.-diameter wire. (Courtesy of General Motors Corporation.)*

NOTE: Some manufacturers use shims under the starter drive end housings during production. Other manufacturers *grind* the mounting pads at the factory for proper starter pinion gear clearance. If *any* GM starter is replaced, the starter pinion *must* be checked and corrected as necessary to prevent starter damage and excessive noise. If a different starter is to be installed, it is highly recommended that the *original* drive-end housing be transferred to the replacement starter to be assured of the proper clearance.

If the clearance is too great. The starter will produce a high-pitched whine during cranking.

If the clearance is too small. The starter will produce a high-pitched whine *after* the engine starts, just as the ignition key is released.

NOTE: The major cause of broken drive-end housings on starters is due to too small a clearance. If the clearance cannot be measured, it is best to put a shim between the engine block and the starter than to leave one out and chance breaking a drive-end housing.

If the clearance is excessive (0.060 in. or more) and there are not shims under the starter, the starter clearance can be reduced by shimming under only the outboard starter mounting bolt. A 0.015-in. shim used under the outboard bolt will decrease the clearance by approximately 0.010 in.

Troubleshooting charts for the starting system are shown in Figures 15-28 and 15-29.

RESULT	POSSIBLE CAUSE	PROBLEM SOURCE
1. Engine cranks slowly but does not start	• Battery discharged	Check battery.
	• Very low temperature	Battery must be fully charged; engine wiring and starting motor in good condition.
	• Undersized battery cables	Install proper cables.
	• Starting motor defective	Test starting motor.
	• Mechanical trouble in engine	Check engine.
2. Solenoid plunger chatters	• Low battery loose or corroded terminals	Charge battery, clean, and tighten terminals.
	• Hold-in winding of solenoid open	Replace solenoid.
3. Pinion disengages slowly after starting	• Sticky solenoid plunger	Clean and free plunger
	• Overrunning clutch sticks on armature shaft	Clean armature shaft and clutch sleeve.
	• Overrunning clutch defective	Replace clutch.
	• Shift-lever return return spring weak	Install new spring.
	• Tight alignment between flywheel and pinion	Realign cranking motor to flywheel.
4. Cranking motor turns but engine does not	• Pinion not engaged	Realign cranking motor to flywheel.
	• Pinion slips	Replace defective drive.

FIGURE 15-28 *This chart is to be used for complaints of slow cranking, solenoid chatter, and slow release of the pinion after starting. (Courtesy of General Motors Corporation.)*

RESULT	POSSIBLE CAUSE	PROBLEM SOURCE
1. No cranking, lights stay bright	· Open circuit in switch	Check switch contacts and connections.
	· Starting motor	Check commutator, brushes, and connections.
	· Open in control in circuit	Check solenoid, switch, and connections.
	· High resistance at battery connection	Clean and tighten terminal connections.
2. No cranking, lights dim heavily	· Battery discharged or malfunctioning	Recharge and test battery.
	· Very low temperature	Check wiring circuit and battery.
	· Pinion jammed	Poor alignment between cranking motor and flywheel, free pinion, check gear teeth.
	· Stuck armature	Frozen bearings, bent shaft, loose pole shoe.
	· Short in starting motor	Repair or replace as necessary.
	· Engine malfunction	Check engine for loss of oil, mechanical interference.
3. No cranking, lights dim slightly	· Loose or corroded battery terminal	Remove, clean, and reinstall.
	· Pinion not engaging	Clean drive and armature shaft, replace damaged parts.
	· Solenoid engages but no cranking	Clean commutator; replace brushes; repair poor connections.
	Excessive resistance or open circuit in starting motor	
4. No cranking, lights out	· Poor connection, probably at battery	Clean cable clamp and terminal; tighten clamp.
5. No cranking, no lights	· Open circuit	Clean and tighten connections; replace wiring.
	· Discharge or malfunctioning battery	Recharge and test battery.

FIGURE 15-29 *Troubleshooting chart for "no crank" situations. (Courtesy of General Motors Corporation.)*

REMANUFACTURED STARTERS

Remanufactured starters or "rebuilt" starters are disassembled and rebuilt. Even though many smaller rebuilders may not replace all worn parts, the major manufacturers totally remanufacture the starter, starter drive, and solenoid, if used. See Figure 15-30.

Most remanufacturers can replace just the commutator, if defective, and many rewind the armature windings. Field coils are tested and replaced or rewound if defective. All new bushings are installed and replacement brushes and brush holders are installed. *New* drive-end housings are installed, if needed, and each housing is designed to match the manufacturer's specifications exactly. For example, one particular starter may use 15 or more *different* drive-end housings, depending on its application. A four-wheel-drive vehicle may require the same starter as a passenger car, but the drive-end housing is constructed to enclose and seal the armature area of the starter from mud and water. Each starter is tested for proper operation, maximum current draw, and voltage drop before painting, boxing, and shipping. Most re-

FIGURE 15-30 *Starter drives being installed on armatures at a major electrical remanufacturer's plant.*

builders use the *Lester numbering system.* As each new electrical unit is introduced by a manufacturer, it is given a Lester number and listed in the Lester Rebuilders catalogs together with the original factory number. Therefore, the "part number" used by many rebuilders is often the Lester number.

SUMMARY

1. Cranking (starting) system testing must begin with testing the battery and battery cables.
2. A defective or discharged battery (less than 75% charged) may give false (higher-than-actual) starter amperage readings.
3. Battery voltage must be higher than 9.6 V during cranking.
4. Typical starting system problems and common causes include the following:

Starting Problem	*Common Causes*
Slow cranking	Low/discharged battery, corroded/high-resistance battery cables
Starter "whine"	Defective starter drive
"Nothing happens" on the start position	Open circuit in control circuit; loose or corroded S wire; starter solenoid; dead battery; loose or corroded battery cables; ignition switch or adjustment (if applicable)
Slow, jerky cranking when engine is warm	Overadvanced ignition timing, high resistance in battery cables

5. Before the starter is removed for service or replacement, the starter system should be tested using an ammeter.

6. After removal of the starter motor from the engine, the solenoid should be removed and tested separately (if equipped). The starter should be disassembled and component parts tested or inspected. The starter clutch should be replaced if starter whine is a problem. The starter brushes should be replaced if worn to less than one-half original new length. The starter bushings should be inspected, tested using the armature shaft for excessive play, and replaced as needed. The armature should be tested on a growler and replaced if needed. The field coils should be inspected and tested with an ohmmeter for opens and grounds.

7. After reassembly, the starter drive pinion-to-flywheel clearance must be checked and corrected when a GM pad type of starter is being serviced.

STUDY QUESTIONS

15-1. Describe the starter test using a voltmeter.

15-2. List four simple items to check when troubleshooting the starting circuit.

15-3. Explain how a weak battery can cause more current flow during cranking than that of a good battery.

15-4. Describe growler testing of the starter armatures.

15-5. Describe the reason that some starter motors must be shimmed for proper operation.

MULTIPLE-CHOICE QUESTIONS

15-1. Technician A says that a battery should be at least 75% charged for accurate starting system test results. Technician B says that the starter drive can be replaced without removing the starter from the engine. Which technician is correct?
(a) A only.
(b) B only.
(c) both A and B.
(d) neither A nor B

15-2. Slow cranking by the starter can be caused by all *except* the following:
(a) a low or discharged battery.
(b) corroded or dirty battery cables.
(c) engine mechanical problems.
(d) an ''open'' neutral safety switch.

15-3. A defective or misadjusted neutral safety switch will be indicated by what symptom?
(a) slow cranking, especially when warm.
(b) starter whine.
(c) ''no crank'' (no starter noise at all).
(d) solenoid chatter or clicking.

15-4. Technician A says that GM V-8 engines should not require more than 250 A to crank the engine. Technician B says that the overrunning clutch is part of a replaceable starter drive unit. Which technician is correct?
(a) A only.
(b) B only.
(c) both A and B.
(d) neither A nor B.

15-5. To check the pull-in windings of a GM solenoid, connect the battery as follows:
(a) positive (+) to S terminal, negative (−) to ground.
(b) positive (+) to BAT terminal, negative (−) to S terminal.
(c) positive (+) to S terminal, negative (−) to M terminal.
(d) negative (−) to S terminal, positive (+) to M terminal.

15-6. To check the hold-in windings of a GM solenoid, connect the battery as follows:
(a) positive (+) to S terminal, negative (−) to ground.
(b) positive (+) to BAT terminal, negative (−) to S terminal.
(c) positive (+) to S terminal, negative (−) to M terminal.
(d) negative (−) to S terminal, positive (+) to M terminal.

15-7. Armatures should be tested using:
(a) an ohmmeter on the high scale.
(b) an ohmmeter on the low scale.
(c) a growler.
(d) a test light.

15-8. Field coils should be tested using:
(a) a 110-V test light.
(b) an ohmmeter.
(c) a growler.
(d) either a or b.

15-9. If the starter pinion clearance with the engine flywheel is too great:

   (a) the starter will produce a high-pitched whine during cranking.

   (b) the starter will produce a high-pitched whine after the engine starts.

   (c) the starter drive will not rotate.

   (d) the solenoid will not operate.

15-10. To *decrease* the starter pinion clearance:

   (a) install a shim under the inboard bolt.

   (b) remove a long shim under both mounting bolts.

   (c) install a long shim under both mounting bolts.

   (d) replace the starter drive.

16

Alternators and the Charging Circuit

EVERY VEHICLE MUST BE EQUIPPED WITH A METHOD OF GENERATING ELEC-trical current to provide the electrical power needed for lighting, ignition, and other electrical units. The charging circuit must also be able to re-store the electrical power to the battery used for starting. The topics covered in this chapter include:

1. Basic dc generator operation
2. Dc generator controls
3. Alternator parts and operation
4. Charging system regulation
5. Electronic voltage regulator operation

PRINCIPLES OF OPERATION

All electrical generators use the principle of electromagnetic induction to generate electrical power from mechanical power. Electromagnetic induction (see Chapter 4 for details) involves the generation of an electrical current in a conductor when moved through a magnetic field. The amount of current generated can be increased by the following factors:

1. Increasing the *speed* of the conductor through the magnetic field
2. Increasing the *number* of conductors passing through the magnetic field
3. Increasing the *strength* of the magnetic field

DC GENERATORS

Dc generators are constructed similarly to a starter motor, but usually use only two brushes. Dc generators were used by car manufacturers until the 1960s.

The metal housing contains two field coils which create a strong magnetic field inside the generator. The generator is turned by the engine using a belt and drive pulley. The output current is generated in the copper wire loops of the *armature* as it rotates inside the magnetic field. This generated current is sent to the battery through the brushes at the A (armature) terminal of the generator. See Figure 16-1. The output of the generator is controlled by a three-part voltage regulator. Whenever the voltage produced by the generator exceeds a certain limit (approximately 14 V), a voltage-sensitive coil relay will "open" the circuit to the *field* windings (the terminal on the generator labeled "F"). The field coil will lose its magnetic field strength and the generator will stop generating because of the lack of a magnetic field required to produce current by electromagnetic

FIGURE 16-1 *Dc generator. Note the heavy steel case and the end cap.*

induction. The voltage regulator will then "close" the circuit to the field, restoring the magnetic field. Therefore, the electrical contact points of the voltage regulator will open and close many times per second to control the generator output voltage precisely. It is important to control the voltage of any generator to protect the battery from overcharging and to protect the lighting and accessories from possible burnout due to excessively high voltage.

The second part of the three-unit voltage regulator used on dc generators is a *current limiter*. The current limiter is a current-sensitive relay with heavy-gauge wire that "opens" the field circuit if the generator current output exceeds a maximum safe amperage. A dc generator is capable of producing extremely high outputs (over 100 A in some cases) if not controlled.

Excessively high amperage outputs create damaging heat in the generator, which can cause the solder (lead and tin alloy) to melt and be thrown out of the armature connections. This type of failure is called "throwing solder" and requires armature replacement. Therefore, the current to the field circuit can be opened and closed by either the voltage- or the current-sensitive relays of the voltage regulator. The third unit that is a part of a voltage regulator used for dc generators is called the *cutout relay*. The cutout relay "opens" the circuit between the battery (labeled "BAT" on the voltage regulator) whenever the battery voltage is higher than the generator voltage. This prevents the battery from being discharged through the armature of the generator whenever the engine is off or whenever the generator is not rotating fast enough to generate enough voltage to charge the battery.

Whenever a replacement generator or voltage regulator is installed on a vehicle or removed for service, the components have to be *polarized* after installation. "Polarity" means that the positive (+) connections of the generator and the positive (+) connections of the voltage regulator are the same. For an A-type field circuit (the field circuit is grounded externally), use a jumper wire connected for a fraction of a second between that BAT and the "armature" (ARM) terminals of the voltage regulator to make certain that the polarity is the same for both the generator and the voltage regulator. For a B-type circuit (internally grounded, usually Ford generators), connect F (field) to BAT.

AC GENERATORS (ALTERNATORS)

Even though dc generators worked well for many years, they could not produce current at low engine speeds. Since the early 1960s another type of generator has been used which can produce enough current at engine idle to allow the use of air conditioning, rear window defrosters, and other accessories that could not be used satisfactorily with a dc generator. An ac generator generates an alternating current when the current changes polarity during its rotation. However, a battery cannot "store" alternating current; therefore, this alternating current is changed to direct current (dc) by diodes inside the generator. Diodes are one-way electrical check valves that permit current to flow in only one direction. Most manufacturers call this type of ac generator an alternator, except for General Motors, which uses the brand name Delcotron. The name "alternator" is the most commonly accepted term for an automotive ac generator.

ADVANTAGES OF ALTERNATORS

1. Alternators can produce more current at engine idle.
2. Alternators do not require a cutout relay between the battery and the alternator to prevent the battery from discharging through the alternator when the engine stops. The diodes allow current to flow only *toward* the battery.
3. Alternators do not require a current-limited relay because the alternating current produced is self-limiting.
4. Alternators are lighter in weight and require less maintenance because the current generated does not flow through the brushes. The brushes of an alternator only conduct the current for the magnetic *field* (approximately 2 to 5 A maximum).

ALTERNATOR COMPONENTS

An alternator is constructed of a two-piece cast-aluminum housing. Aluminum is used because of its light weight, nonmagnetic properties, and heat-transfer properties needed to help keep the alternator cool. A front ball bearing is pressed into the front housing (drive-end housing) to provide the support and friction reduction necessary for the belt-driven rotor assembly. The rear housing (brush-end housing) usually contains a roller-bearing support for the rotor and mounting for the brushes, diodes, and internal voltage regulator (if equipped). See Figure 16-2.

ROTORS

The rotor creates the magnetic field of the alternator and produces a current by electromagnetic induction in the stationary stator windings. This differs from a dc generator, where the field current is created in the stationary field windings and the current is generated in the rotating armature. The alternator rotor is constructed of many turns of

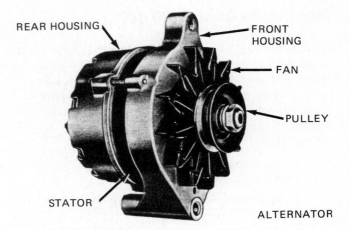

FIGURE 16-2 *Typical alternator, showing its major component parts. (Courtesy of Ford Motor Company.)*

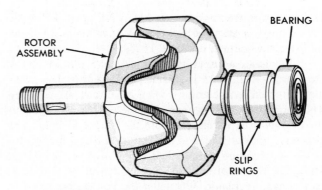

FIGURE 16-3 *Typical rotor assembly. (Courtesy of Chrysler Corporation.)*

copper wire coated with a varnish insulation wound over an iron core. The iron core is attached to the rotor shaft. See Figure 16-3.

At both ends of the rotor windings are heavy gauge-metal plates bent over the windings with trianglar fingers called poles. These pole "fingers" do not touch, but alternate or interlace. If current flows through the rotor windings, the metal pole pieces at each end of the rotor become electromagnets. The right-hand rule (see Chapter 4) indicates that whether a north or a south pole magnet is created depends on the *direction* the wire coil is wound. Since the pole pieces are attached to each end of the rotor, one pole piece will be a north pole magnet. The other pole piece is on the opposite end of the rotor, and therefore is viewed as wound in the opposite direction, creating a south pole. Therefore, the rotor fingers are alternating north and south magnetic poles. See Figure 16-4. The magnetic fields are created between the alternating pole piece fingers as shown in Figure 16-5. These individual magnetic fields produce a

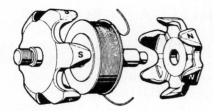

FIGURE 16-4 *Parts of a 12-pole rotor (six north and six south poles). (Courtesy of Robert Bosch Corporation.)*

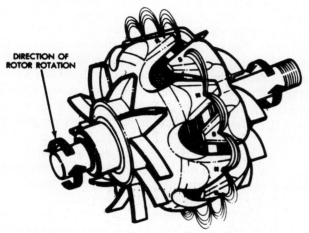

FIGURE 16-5 *Assembled rotor, showing alternating magnetic fields created by the alternating north and south poles. (Courtesy of Chrysler Corporation.)*

current by electromagnetic induction in the stationary stator windings.

The current necessary for the field (rotor) windings is conducted through slip rings. The current to the field is conducted through the slip rings with carbon brushes. The maximum rated alternator output in amperes is largely de-

pendent on the number and gauge of the windings of the rotor. Substituting rotors from one alternator to another can greatly affect its maximum output. Many commercially rebuilt alternators are tested and a sticker put on the alternator indicating its tested output. The original rating stamped on the housing is then ground off.

ALTERNATOR BRUSHES

The current for the field is controlled by the voltage regulator and is conducted to the slip rings through carbon brushes. The brushes conduct only the field current (approximately 2 to 5 A), and therefore tend to last longer than the brushes used on a dc generator, where all of the current generated in the generator must flow through the brushes. See Figure 16-6.

STATORS

Supported between the two halves of the alternator housing are three copper wire windings wound on a laminated metal core. See Figure 16-7. As the rotor revolves, its moving magnetic field induces a current in the windings of the stator.

DIODES

Diodes are constructed of a semiconductor material (usually silicon) and operate as a one-way electrical check valve that permits the current to flow in only one direction. See Figures 16-8 and 16-9. Alternators use six diodes (one positive and one negative pair for each of the three stator windings) to convert alternating current (ac) to direct current (dc). The symbol for a diode is shown in Figure 16-10.

FIGURE 16-6 *Typical brushes. (Courtesy of Chrysler Corporation.)*

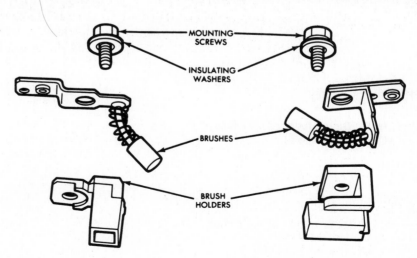

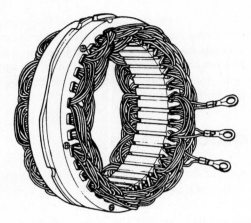

FIGURE 16-7 *Typical stator. (Courtesy of General Motors Corporation.)*

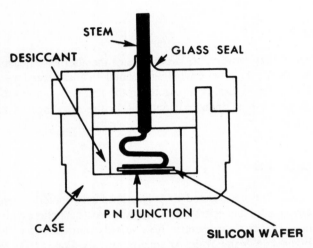

FIGURE 16-8 *Cutaway of a diode. The desiccant is installed inside the case of a diode to absorb moisture. See Chapter 5 for details on diodes. (Courtesy of General Motors Corporation.)*

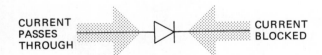

FIGURE 16-9 *Diode function. (Courtesy of General Motors Corporation.)*

CURRENT FLOW

FIGURE 16-10 *Diode symbol. (Courtesy of General Motors Corporation.)*

HOW AN ALTERNATOR WORKS

A rotor inside an alternator is turned by a belt and drive pulley by the engine. The magnetic field of the rotor generates a current in the windings of the stator by electromagnetic induction. See Figure 16-11.

Field current flowing through the slip rings to the rotor creates an alternating north and south pole on the rotor, with a magnetic field between each finger of the rotor. The induced current in the stator windings is an alternating current because of the alternating magnetic field of the rotor. The induced current starts to increase as the magnetic field starts to induce current in each winding of the stator. The current then peaks when the magnetic field is the strongest and starts to decrease as the magnetic field moves away from the stator winding. Therefore, the current generated is described as a sine-wave pattern (see Figure 16-12). As the rotor continues to rotate, this sine-wave current is induced in each of the three windings of the stator.

Since each of the three windings generates a sine-wave current, as shown in Figure 16-13, the resulting current combines to form a three-phase voltage output.

The current induced in the stator windings connects to diodes (one-way electrical check valves) which permit the alternator output current to flow in only one direction. All alternators contain six diodes, one pair of a positive and a negative diode for each of the three stator windings.

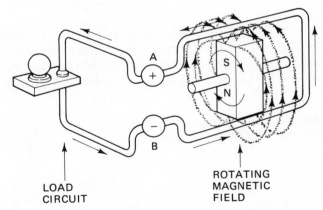

FIGURE 16-11 *Magnetic lines of force cutting across a conductor induce a voltage and current in the conductor.*

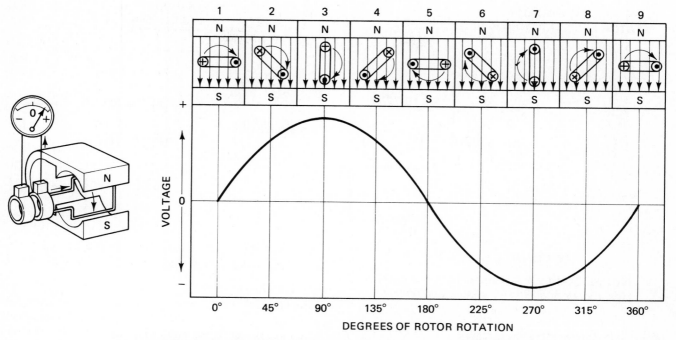

FIGURE 16-12 *Sine-wave voltage curve created by one revolution of a winding rotating in a magnetic field. (Courtesy of Robert Bosch Corporation.)*

FIGURE 16-13 *When three windings (u, v, w), are present in a stator, the resulting current generation is represented by the three sine waves. The voltages are 120° out of phase. The connection of the individual phases produces a three-phase alternating voltage. (Courtesy of Robert Bosch Corporation.)*

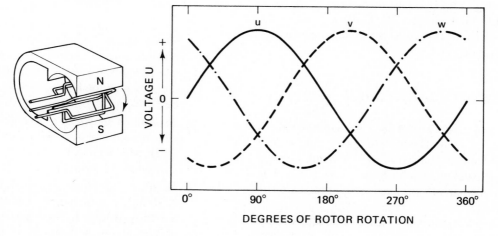

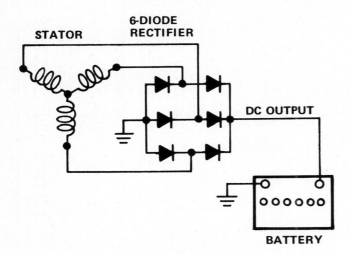

Y CONNECTION

FIGURE 16-14 Wye-connected stator windings. (Courtesy of General Motors Corporation.)

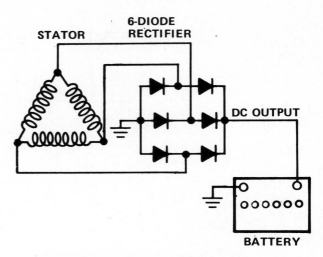

FIGURE 16-15 Delta-connected stator windings. (Courtesy of General Motors Corporation.)

WYE-CONNECTED STATORS

The Y-type (pronounced "wye" and generally so written) or star pattern (see Figure 16-14) is the most commonly used alternator stator winding connection. The output current with a wye-type stator connection is constant over a broad alternator speed range.

Current is induced in each winding by electromagnetic induction from the rotating magnetic fields of the rotor and produces a current in each winding. In a wye-type stator connection the currents must combine since two windings are always connected in series. The current produced in each winding is added to the other winding's current and then flows through the diodes to the alternator output terminal. One-half of the current produced is available at the neutral junction (usually labeled "STA." for stator). The voltage at this center point is used by some alternator manufacturers (especially Ford) to control the charge indicator light or is used by the voltage regulator to control the rotor field current.

DELTA-CONNECTED STATORS

The delta winding is connected in a trianglar shape, as shown in Figure 16-15. ("Delta" is a Greek letter shaped like a triangle.) Current induced in each winding flows to the diodes in a parallel circuit. More current can flow through two parallel circuits than can flow through a series circuit (as in a wye-type stator connection).

Delta-connected stators are used on alternators where high output at high alternator rpm (revolutions per minute) is necessary or required. The delta-connected alternator can produce 73% more current than the same alternator with wye-type stator connections. For example, if an alternator with a wye-connected stator can produce 32 A, the *same* alternator with delta-connected stator windings can produce 73% more current, or 55 A (32 × 1.73 = 55). The delta-connected alternator, however, produces lower current at low speed and must be operated at high speed to produce its maximum output. See Figure 16-16.

> **HINT:** General Motors delta-wound alternators can easily be identified by the location of the amperage output stamping on the *front* of the drive-end housing facing the drive pulley. Wye-wound GM alternators are stamped on top of the drive-end housing near the small pivot-end ear.

ALTERNATOR OUTPUT

The output voltage and current of an alternator depend on several factors:

1. *Speed of rotation.* Alternator output is increased with alternator rotational speed up to its maximum possible ampere output. Alternators normally rotate two to three times faster than engine speed, depending on the relative pulley sizes used for the belt drive.

2. *Number of conductors.* A high-output alternator contains more turns of wire in the stator windings. Stator winding connections (whether wye or delta) also affect the maximum alternator output.

3. *Strength of the magnetic field.* If the magnetic field is strong, a high output is possible because the current generated by electromagnetic induction is dependent on the number of magnetic lines of force that are cut.

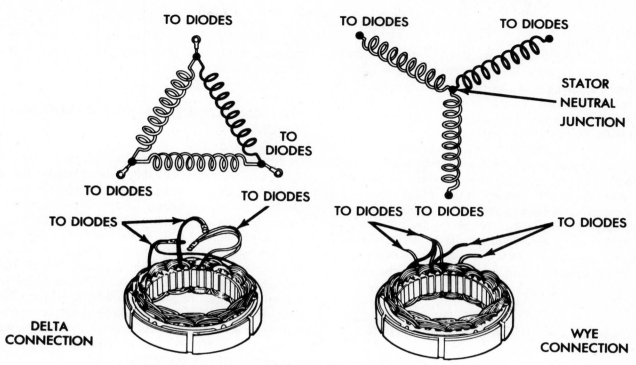

FIGURE 16-16 *Comparison between wye- and delta-connected stators. The wye-connected stator uses a junction connection, whereas the delta-connected stator has two wires connected at each connection. (Courtesy of Chrysler Corporation.)*

(a) The strength of the magnetic field can be increased by increasing the number of turns of conductor wire wound on the rotor. Higher output alternators have more turns of wire than does an alternator with a low-rated output.

(b) The strength of the magnetic field also depends on the current through the field coil (rotor). Since magnetic field strength is measured in ampere-turns, the greater the amperage or the number of turns, or both, the greater the alternator output.

ALTERNATOR CURRENT REGULATION

Alternators, unlike dc generators, do not require current limiters since they do not produce enough current to cause damage to the alternator itself. Whenever an alternating current is produced, the magnetic field is constantly changing and an opposing current is induced in the conductors (stator windings). This opposing current is called *inductive reactance*. The inductive reactance in an alternator limits the maximum current (amperes) that a particular alternator can produce.

An alternator, however, could produce over 250 V if the voltage is not controlled. All alternators require a regulator to control the voltage. See Figure 16-17. As the speed of the alternator increases, the voltage induced in the stator windings increases. Since it is not practical to maintain alternator speed at a constant rpm regardless of engine speed, the control of the field (rotor) current is used to control the output of the alternator.

At low alternator speed, the field current is maintained at its maximum to permit the alternator output to be as high as possible. If the alternator voltage is below the

FIGURE 16-17 *(Courtesy of General Motors Corporation.)*

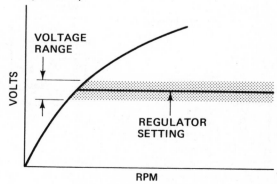

upper design limits, the voltage regulator does not limit the current to the field (rotor).

Alternator output is controlled (regulated) by increasing or decreasing the field current. No field current results in zero alternator output, and full field current allows the alternator to produce its maximum possible output. The field current of the rotor can be controlled by contact points or a transistor in a voltage regulator.

All voltage regulators must be able to "sense" or measure accurate battery voltage to control the alternator output. Therefore, all connecting wires and grounds must be secure for proper operation of the voltage regulator. A poor ground connection on the voltage regulator can cause *overcharging* because the voltage regulator "senses" a lower-than-actual battery voltage and allows the alternator output to exceed the normal charging voltage. See Chapter 17 for voltage-drop testing procedures for the charging circuit.

ALTERNATOR VOLTAGE REGULATION

An automotive alternator must be able to produce electrical pressure (voltage) higher than battery voltage to charge the battery. Excessively high voltage can damage the battery, electrical components, and the lights of a vehicle. If no (zero) amperes of current existed through the field coil of the alternator (rotor), alternator output would be zero because without field current a magnetic field does not exist. The field current required by most automotive alternators is under 3 A. It is the *control* of the *field* current that controls the output of the alternator. Current for the rotor flows from the battery through the brushes to the slip rings. After generator output begins, the voltage regulator controls the current flow through the rotor. See Figure 16-18.

The voltage regulator simply "opens" the field circuit if the voltage reaches a predetermined level, then "closes" the field circuit again as necessary to maintain the correct charging voltage.

BATTERY CONDITION AND CHARGING VOLTAGE

If the automotive battery is discharged, its voltage will be lower than the voltage of a fully charged battery. The alternator will supply charging current but may not reach the maximum charging voltage. For example, if a car is jump started and run at a fast idle (2000 rpm), the charging voltage may be only 12.0 V. As the battery becomes charged and the battery voltage increases, the charging voltage will also increase, until the voltage regulator limit is reached; then the voltage regulator will start to control the maximum charging voltage. A good, but discharged battery should be able to convert into chemical energy all the current the alternator could produce. As long as alternator voltage is higher than battery voltage, current will flow from high pressure (high voltage) to lower pressure (lower voltage). Therefore, if a voltmeter is connected to a discharged battery with the engine running, it may indicate lower-than-normal acceptable charging voltage. See Figure 16-19.

In other words, the condition and voltage of the battery *does* determine the charging rate of the alternator. It is often stated that the battery is the true "voltage regulator" and that the voltage regulator simply acts as the upper-limit voltage control. This is the reason that all charging system testing *must* be performed with a known-good battery, at least 75% charged, to be assured of accurate test results. If a discharged battery is used during charging system testing, tests could mistakenly indicate a defective alternator and/or voltage regulator.

FIGURE 16-18 *Mechanical-type voltage regulator. The regulator controls the current through the rotor (field) labeled "F." (Courtesy of General Motors Corporation.)*

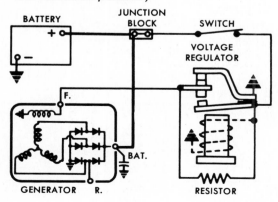

FIGURE 16-19 *(Courtesy of General Motors Corporation.)*

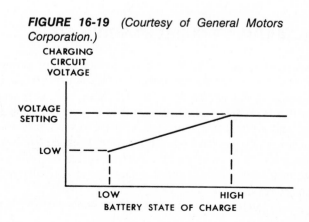

AN ALTERNATOR IS A POOR BATTERY CHARGER

Even though an alternator will eventually recharge a discharged battery, the charging rate will be low (less than rated output). The low charging rate causes the output current to flow continually through the stator windings. The voltage regulator is not switching the field current on and off because the output voltage is too low. Similar to starters being damaged with too low a voltage, the alternator *stator* is often damaged when operated using a defective or undercharged battery. A stator that has been operated in an alternator attempting to charge a bad battery is often black and smells of burned insulating varnish. It is best to charge a battery using a battery charger until at least 75% charged, then allow the charging system to recharge the battery fully. See Chapter 17 for complete and accurate charging system testing procedures.

MECHANICAL VOLTAGE REGULATORS

Older-style voltage regulators used tungsten steel contact points to control the alternator field (rotor) circuit. Most mechanical voltage regulators contain one or two units. See Figure 16-20. If the regulator has two units, one of the electromagnetic coils controls the field current and the other unit is a relay that controls the dash charging system failure light. See Figure 16-21. The voltage regulator contacts can switch the field current on and off as rapidly as 50 to 200 times per second.

All mechanical (point-type) voltage regulators must be mounted to a *vertical* surface. The operation of the points in the voltage regulator is, therefore, horizontal. The voltage regulation is not affected by the car's hitting bumps (or gravity), which would occur if the regulator is mounted with the points operating up and down.

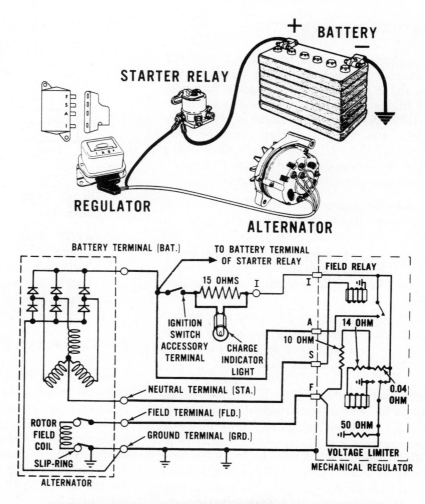

AUTOLITE ALTERNATOR SYSTEM— CHARGE INDICATOR LIGHT

FIGURE 16-20 (Courtesy of Ford Motor Company.)

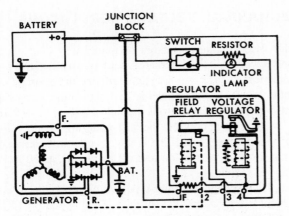

FIGURE 16-21 *Double-contact voltage regulator. By using double contacts, the voltage regulator can more accurately control the field current with increased point life. (Courtesy of General Motors Corporation.)*

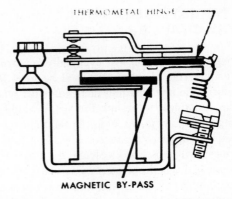

FIGURE 16-22 *Two methods of temperature compensation used on mechanical voltage regulators. (Courtesy of General Motors Corporation.)*

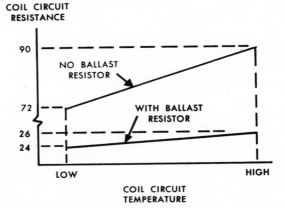

FIGURE 16-23 *(Courtesy of General Motors Corporation.*

TEMPERATURE COMPENSATION

All voltage regulators (mechanical or electronic) provide a method to increase the charging voltage slightly at low temperatures and to lower charging voltages at high temperatures. A battery requires a higher charging voltage at low temperatures because of the resistance to chemical reaction changes. However, the battery would be overcharged if the charging voltage were not reduced during warm weather. Three methods are used to provide temperature compensation in a *mechanical* voltage regulator:

1. A bimetallic hinge on the arm of the relay changes the tension on the movable arm of the voltage regulator as shown in Figure 16-22.

2. A magnetic shunt (or bypass) is mounted on top of a coil inside the voltage regulator, which changes its magnetic properties with temperature changes. See Figure 16-22.

3. A ballast resistor installed in series with the regulator coil provides constant resistance in the regulator circuit regardless of the temperature of the regulator coil. See Figure 16-23.

The three units listed above may be combined to provide consistent temperature-compensation control to the charging voltage. Electronic voltage regulators use a temperature-sensitive resistor in the regulator circuit. This resistor is called a *thermistor* and provides lower resistance as the temperature increases. A thermistor is used in the electronic circuits of the voltage regulator to control charging voltage over a wide range of under-the-hood temperatures.

NOTE: Voltmeter tests may vary according to temperature. Charging voltage tested at 32°F (0°C) will be higher than for the same vehicle tested at 80°F (27°C) because of the temperature-compensation factors built into voltage regulators.

A AND B FIELD CIRCUITS

When testing the charging circuit, most test equipment requires the technician to select either A or B field types. An A circuit is the most commonly used and means that the ground for the field circuit is *external* or grounded by the voltage regulator. See Figure 16-24. All electronic voltage regulators use A circuits because the controlling transistor(s) controls a lower-voltage current opening and closing the ground return path.

In a B-circuit field, the voltage regulator controls (open and closes) the power side of the field circuit and is

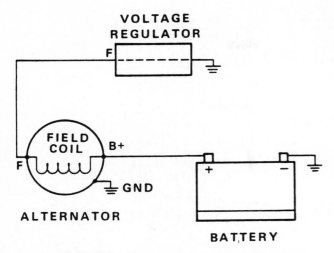

FIGURE 16-24 *Diagram of an A-type field circuit.*

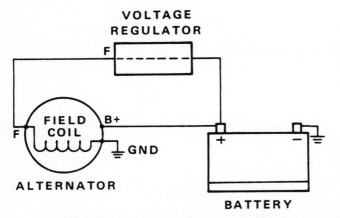

FIGURE 16-25 *Diagram of a B-type field circuit.*

grounded inside the generator or alternator. Therefore, a B circuit is internally grounded. See Figure 16-25.

ELECTRONIC VOLTAGE REGULATORS

Electronic voltage regulators have been used since the early 1970s. The electronic circuit of the voltage regulator cycles between 10 and 7000 times per *second* as needed to accurately control the field current through the rotor, and therefore the alternator output. The control of the field current is accomplished by "opening" and "closing" the *ground* side of the field current through the rotor of the alternator. Because the current flowing through the rotor drops the voltage, the controlling transistor switch functions at a lower operating voltage than if controlling the "hot" or the power side of the circuit. Electronic voltage regulators also use many resistors to help reduce the current and volt-

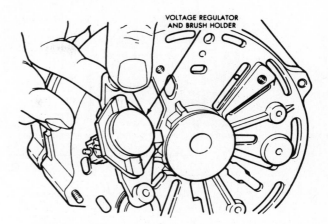

FIGURE 16-26 *Typical internal voltage regulator installation. (Courtesy of Chrysler Corporation.)*

age regulators, and this heat must be dissipated into the air to prevent damage to the diodes and transistors. Whether mounted inside the alternator or externally under the hood, electronic voltage regulators are mounted where normal airflow can keep the electronic components cool. See Figure 16-26.

HOW AN ELECTRONIC VOLTAGE REGULATOR WORKS

The zener diode is a major electronic component that makes voltage regulation possible. A zener diode blocks current flow until a specific voltage is reached, then it permits current to flow. Alternator voltage from the stator and diodes is first sent through a thermistor, which changes resistance with temperature, and then to a zener diode. Whenever the upper limit voltage is reached, the zener diode conducts current to a transistor, which then opens the field (rotor) circuit. All the current stops flowing through the alternator's brushes, slip rings, and rotor, and no magnetic field is formed. Without a magnetic field, an alternator does not produce current in the stator windings. When no voltage is applied to the zener diode, current flow stops and the base of the transistor is turned off, closing the field circuit. The magnetic field is thus restored in the rotor. The rotating magnetic fields of the rotor induce a current in the stator, which is again controlled if the output voltage exceeds the designed limit as determined by the zener diode breakdown voltage. Depending on the alternator rpm, vehicle electrical load, and state of charge of the battery, this controlled switching on and off can occur between 10 and 7000 times per second. See Figures 16-27 through 16-31.

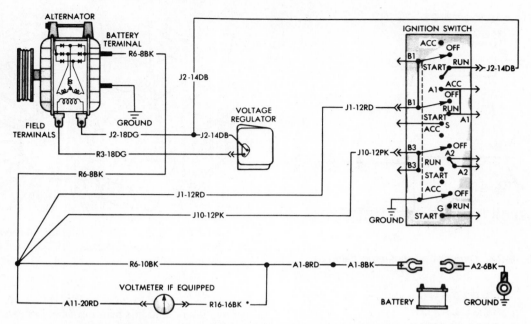

FIGURE 16-27   *Typical alternator and voltage regulator circuit. (Courtesy of Chrysler Corporation.)*

FIGURE 16-28   *Two diagrams of a typical General Motor's charging circuit with the voltage regulator inside the alternator housing. (A) and (B) show the same circuit. (Courtesy of Oldsmobile Division, GMC.)*

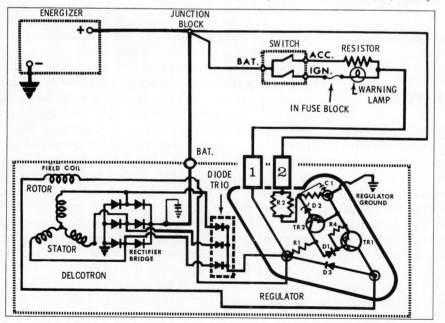

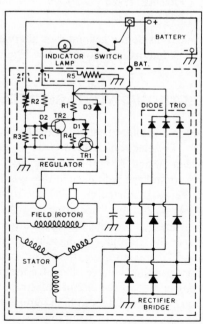

(a)

(b)

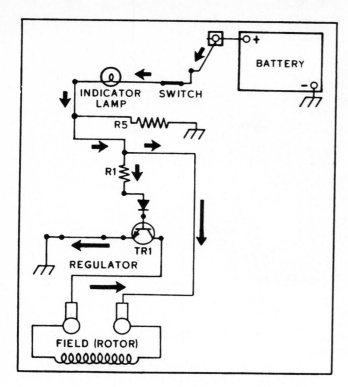

FIGURE 16-29 *Same diagram as in Figure 16-28, but showing the current flow to the rotor with the ignition switch "on" and the engine not running. (Courtesy of General Motors Corporation.)*

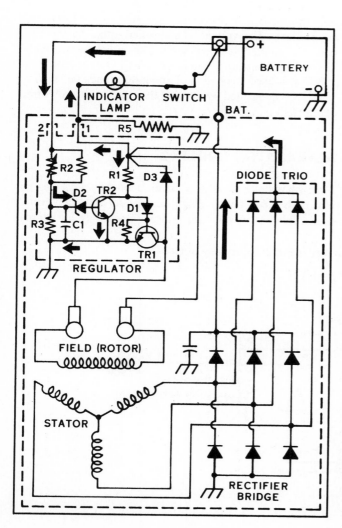

FIGURE 16-30 *Alternator output through the diode trio with the engine running. The "charge" light (indicator lamp) is put out because the voltage is being applied equally to both sides of the bulb. (Courtesy of General Motors Corporation.)*

FIGURE 16-31 *Alternator output being regulated. Notice that the voltage has reached a high-enough value to permit the zener diode (D_2) to turn on TR_2, permitting current to flow directly to ground instead of through the rotor (field). (Courtesy of General Motors Corporation.)*

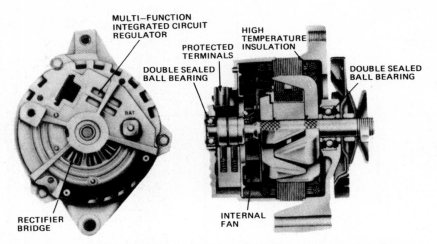

FIGURE 16-32 *General Motors CS (charging system) series of alternators. (Courtesy of General Motors Corporation.)*

COMPUTER-CONTROLLED ALTERNATORS

Beginning in the mid-1980s, General Motors introduced a smaller, yet high-output series of alternators. These alternators are called the CS (charging system) series. See Figure 16-32. After the letters CS are found numbers indicating the *outside diameter* in millimeters of the stator laminations. Typical sizes, designations, and outputs include:

CS-121, 5-SI	74 A
CS-130, 9-SI	105 A
CS-144, 17-SI	120 A

These alternators feature two cooling fans (one internal) and terminals designed to permit connections to an on-board body computer through terminals L and F (see Figure 16-33). See Chapter 27 for body computer details.

The reduced-size alternators also feature ball bearings front and rear and totally soldered internal electrical connections. The voltage is controlled either by the body computer (if equipped) or by the built-in voltage regulator. The voltage regulator switches the field voltage "on" and "off" at a fixed frequency of about 400 times per second. Voltage is controlled by varying the "on" and "off" time of the field current.

ALTERNATOR COOLING

Most automotive alternators are air cooled by a cooling fan next to the drive pulley. Some heavy-duty and many newer (mid-1980s) alternators are also equipped with an internal

FIGURE 16-33 *General Motors CS wiring diagram. (Courtesy of General Motors Corporation.)*

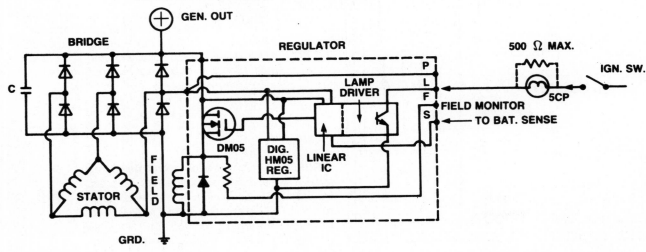

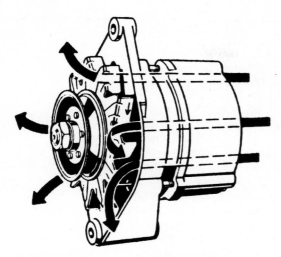

FIGURE 16-34 *Air flow through an alternator helps keep the diodes and internal voltage regulator (if equipped) cool. (Courtesy of Robert Bosch Corporation.)*

WILL USING A LARGER ALTERNATOR DAMAGE THE CAR?

Using a larger (higher-amperage output) alternator that will fit correctly may be possible. A higher-output alternator will be able to recharge the battery faster than will a lower-output alternator. Whenever the battery is fully charged, the alternator output is decreased almost to zero. Therefore, the higher-output alternator can help keep the battery charged, but as with any charging system, the alternator does not "work" unless required. A lower-output alternator would have to work longer than a higher-output alternator. The voltage output of all alternators, regardless of size, is controlled by the voltage regulator.

If a high-amperage accessory is added to a vehicle, such as safety lights for security, fire, or police use, a higher-output alternator may be required. Substituting a 100-A alternator for a stock 63-A alternator, for example, may be required to match the higher-amperage requirements. A higher-output alternator will not damage the battery or the electrical system. See Chapter 17 for the procedure for determining the *minimum* alternator amperage output required.

fan. These fans draw air in through the *rear* of the alternator and *out* the front (drive side) of the alternator. This assures that the air flows over the diodes (rectifiers) and voltage regulator (if equipped) before flowing over and around the stator and rotor. This airflow helps keep the most heat-sensitive components of the alternator cool. See Figure 16-34.

SUMMARY

1. All electrical generators produce electricity by using electromagnetic induction. Dc generators create electrical current in the armature rotating inside a strong magnetic field. The strength of the magnetic field is controlled by the voltage regulator. A dc generator voltage regulator unit also contains a current-limiter relay and a cutout relay.

2. A generator that produces alternating current is called an alternator. An alternator uses a rotating field (rotor), and current is induced in the stationary stator. A delta-wound alternator stator can produce greater current output than that produced by a wye-wound stator.

3. All voltage regulators control the field current of the alternator. If field current is at maximum, the alternator will produce its maximum. If the field current is zero, the alternator output is zero. All voltage regulators are temperature compensated to provide a higher charging voltage at low temperatures and a lower charging voltage during high-temperature operation.

STUDY QUESTIONS

16-1. List the major component parts of an alternator and describe how an alternator produces electrical current.

16-2. Explain the difference between wye- and delta-wound alternator stators.

16-3. Describe how the voltage regulator controls the alternator output.

16-4. Explain why all voltage regulators have temperature compensation.

MULTIPLE-CHOICE QUESTIONS

16-1. Technician A says that the diodes regulate the alternator output voltage. Technician B says that dc generators require a current limiter to protect the generator. Which technician is correct?
   (a) A only.
   (b) B only.
   (c) both A and B.
   (d) neither A nor B.

16-2. The magnetic field is created in the _____ in an alternator.
   (a) stator.
   (b) diodes.
   (c) rotor.
   (d) bearings.

16-3. The voltage regulator controls the current:
   (a) through the alternator brushes.
   (b) through the rotor.
   (c) through the alternator field.
   (d) all of the above.

16-4. Technician A says that a wye-wound stator in an alternator can produce greater amperage output at low speed than can a delta-wound stator. Technician B says that excessive amperage output can damage the alternator. Which technician is correct?
   (a) A only.
   (b) B only.
   (c) both A and B.
   (d) neither A nor B.

16-5. Technician A says that two diodes are required for each stator winding lead. Technician B says that diodes change alternating current into direct current. Which technician is correct?

   (a) A only.
   (b) B only.
   (c) both A and B.
   (d) neither A nor B.

16-6. Alternator output voltage:
   (a) is controlled by the voltage regulator.
   (b) is controlled by the diode.
   (c) can be too high due to a poor ground on the voltage regulator.
   (d) both (a) and (c) are correct.

16-7. Mechanical voltage regulators:
   (a) use A field circuits only.
   (b) use B field circuits only.
   (c) use either A or B field circuits.
   (d) vibrate at 7000 times per second.
   (e) none of the above.

16-8. Field circuit types are classified as follows:
   (a) ''A'' means internally grounded.
   (b) ''B'' means internally grounded.
   (c) all electronic regulators are A circuits.
   (d) both (b) and (c) are correct.

16-9. Alternator brushes must carry *approximately* _____ amperes.
   (a) 1 to 3.
   (b) 2 to 5.
   (c) 10 to 36.
   (d) 36 to 108.

16-10. Using an alternator for an extended period in an attempt to charge a defective battery can cause:
   (a) burned diodes.
   (b) an overheated stator.
   (c) burned rotor slip rings.
   (d) an overheated voltage regulator.

17

Alternator and Charging Circuit Testing and Service

A PROPERLY OPERATING CHARGING CIRCUIT IS NECESSARY FOR THE CORrect operation of an automobile's entire electrical system. After the battery has been tested and known to be capable of supplying its rated capacity, and has been tested to be at least 75% charged, the charging system can be tested in a sequence designed to pinpoint exact problems or faults in the charging system. The topics covered in this chapter include:

1. On-the-car alternator and voltage regulator testing with and without testing equipment
2. Off-the-car alternator and voltage regulator testing
3. Alternator disassembly and service procedures for the most commonly used alternators
4. Alternator component testing
5. Reassembly procedures

GENERAL CHARGING CIRCUIT TROUBLESHOOTING

If the "charge" light is on while the engine is running, the alternator output voltage is below the battery voltage. The charge light (could also be labeled "GEN" or "ALT") should be "on" when the ignition is "on" without the engine running, and should be "off" whenever the engine is running. If the "charge" light is "on" when the engine is running, the problem could be:

1. Defective alternator component(s)
2. A defective voltage regulator
3. A defect in wiring between the alternator and the voltage regulator or indicator light
4. A defective electric choke relay (if applicable)

If the alternator light is dim with the engine running, there is a possible problem with the diode trio (if equipped) or other problems requiring further charging circuit testing.

If the vehicle is equipped with an ammeter, the *normal* charging system operation should indicate "charging" after an engine start and slowly return to zero charge as the battery reaches full charge. If the ammeter indicates discharge or no charge all the time, the charging circuit should be tested.

If the vehicle is equipped with a voltmeter, the normal charging voltage should indicate between 13 and 15 V. Battery voltage below 12 V with the engine running indicates a "no charging" or limited charging. Excessive current drain caused by high-amperage-drawing accessories could be causing the low charging voltage. To avoid discharging the battery during extended idling periods, turn off electrical accessories to relieve the electrical load. A voltmeter reading higher than 16 V usually indicates a failure of the voltage regulator.

GENERAL ALTERNATOR TESTING AND PRECAUTIONS

The charging system should be checked using a series of tests that will pinpoint the cause(s) of improper charging system operation. The importance of the basic items that need to be inspected cannot be overemphasized. The following list includes tests and safety procedures that can prevent injury to the technician and/or components of the charging system:

1. Always check that the alternator belt tension is proper and in good condition.
2. Check all fusible links for signs of overheating damage. If in doubt, check for continuity with an ohmmeter.
3. To avoid a possible spark or arc that could cause damage, always disconnect the negative (−) bat-

tery cable before performing any service on any electrical component.

4. Never disconnect a battery cable when the engine is running. This was a popular test of a *dc generator*. If the engine continued to run, the generator was producing current. However, an alternator could produce 250 V when disconnected from the battery and can cause damage to any electrical component that is "on" at the time of the procedure. Also, this high voltage could cause battery damage or a battery explosion if the battery cable was reconnected to the battery with the engine running. Some electronic voltage regulators are designed to prevent the alternator from charging, by opening the field circuit if either battery cable is disconnected.

5. Never ground the alternator output terminal (usually labeled "BAT" or battery terminal). Alternators usually have plastic or rubber protective covers surrounding this terminal, to prevent accidental grounding.

6. Never attempt to polarize an alternator unless instructed by the alternator manufacturer (usually only on heavy-duty truck applications).

7. If a booster battery is used for starting, be certain to connect the jumper cables positive (+) lead to the positive (+) post and the negative (−) lead to the negative (−) post. See Chapter 13 for the proper jump-starting procedure.

VOLTMETER TESTING ON THE CAR

The simplest basic test that can be performed on the charging circuit is to connect a voltmeter to the battery and then measure the charging voltage at 2000 engine rpm. See Figure 17-1. The charging voltage should be 13.5 to 15.0 V or a minimum of 0.5 V higher and a maximum of 2.0 V higher than basic battery voltage. If the charging voltage is too *high* (over 15.0 V or over 2 V higher than basic battery voltage), there could be a problem with one or more of the following:

1. Defective voltage regulator
2. Poor ground on the voltage regulator
3. Defective (high-resistance) wiring between the alternator and the voltage regulator

If the charging voltage is too *low* (less than 13.5 V), there could be a possible problem with one of the following:

1. Loose alternator belt (see Figure 17-2)
2. Defective voltage regulator
3. Discharged battery
4. Defective alternator

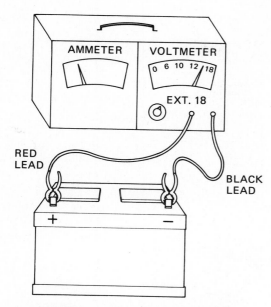

FIGURE 17-1 *Typical voltmeter hookup.*

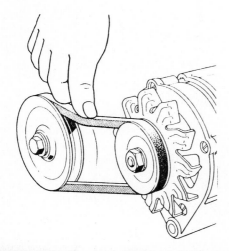

FIGURE 17-2 *The maximum deflection of an alternator drive belt should be ¹/₂ in. when the belt is depressed in the middle between the pulleys. (Courtesy of Robert Bosch Corporation.)*

VOLTAGE-DROP TESTING THE CHARGING CIRCUIT

Voltage-drop testing is a method used to test wires and connections for excessive resistance. The voltage drop of all wires and connections combined should not exceed 3% of the system voltage or 0.4 V with a 20-A charging rate for a 12-V system.

> **NOTE:** Accurate voltage-drop testing must be performed with a known current flow in amperes to accurately determine excessive voltage drop.

A poor ground connection on the voltage regulator can create overcharging. The voltage regulator must be able to measure battery voltage accurately. If the voltage drop in the connecting wires or ground connections is excessive (over 0.4 V total for all wires and connections), the voltage regulator will not "sense" the correct battery voltage.

To test for charging circuit voltage drops, connect a carbon-pile load unit to the battery. With the engine running, adjust the carbon pile to obtain a 20-A charging rate. While maintaining a constant 20-A rate, check every wire and connection by touching the positive (+) voltmeter test lead temporarily to the *most* positive connection (closest to the alternator) and the negative (−) lead to the most negative end of the same wire or cable. See Figure 17-3.

Any particular wire or connection should not exceed 0.2 V and total system voltage drops should be less than 0.4 V. If the voltage drop exceeds 0.2 V, remove and clean or replace the affected wire or connection. A voltage drop between the voltage regulator base and the battery negative (−) post indicates a high-resistance ground connection.

ALTERNATOR OUTPUT TESTING ON THE CAR

If reduced charging or no charging is suspected, the amperage output of the alternator should be tested. An alternator will only produce enough current (amperes) to keep the battery charged. Only if the battery is discharged or the electrical load is high will an alternator produce its maximum rated output. Therefore, the alternator must be loaded using a variable carbon-pile electrical load tester with the engine running at a high enough speed (approximately 2000 rpm) to permit the alternator to produce its maximum rated output. The output of an alternator should be within 10% of its rated output as stamped or indicated on the alternator housing. Most General Motors alternators are stamped near the pivot bolt side, while Chrysler and Ford use tag colors for identification.

Ford ID Colors

Orange	40 A	Rear terminal
Black	65 A	Rear terminal
Green	60 A	Rear terminal
Black	70 A	Side terminal
Red	100 A	Side terminal

Chrysler ID Colors

Violet	41 A	
Natural	50 A	
Yellow	60 A	
Brown	65 A	
Yellow	100 A	(larger housing than 60 A)

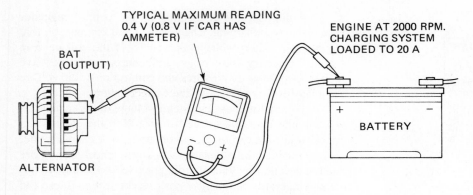

VOLTAGE DROP — INSULATED CHARGING CIRCUIT

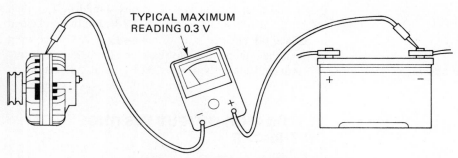

VOLTAGE DROP — GROUND CHARGING CIRCUIT

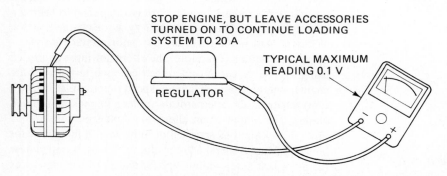

VOLTAGE DROP — ENTIRE GROUND CIRCUIT

FIGURE 17-3 *Voltmeter hookup to test the voltage drop of the charging circuit.*

HOW TO DETERMINE MINIMUM REQUIRED ALTERNATOR OUTPUT

All charging systems must be able to supply the electrical demands of the electrical system. If lights and accessories are used constantly and the alternator cannot supply the necessary ampere output, the battery will be drained. To determine the minimum electrical load requirements, connect an ammeter in series with either battery cable.

> **NOTE:** If using an inductive-pickup ammeter, be certain that the pickup is over *all* the wires leaving the battery terminal. Failure to include the small body ground wire from the negative (−) battery

terminal to the body or the small positive (+) wire (if testing from the positive side) will *greatly* decrease the true current flow readings.

After connecting an ammeter correctly in the battery circuit, turn on all lights and accessories (with the engine off) that are likely to be used continuously, as follows:

1. Turn ignition to "on" ("run") (do not start the engine).
2. Turn the heat selector to air conditioning (if equipped).
3. Turn the blower motor to high speed.
4. Turn the headlights on bright.

5. Turn on the radio.

6. Turn on the windshield wipers.

7. Turn on any other accessories that may be used continuously (do not operate the horn, power door locks, or other units that are not used for more than a few seconds).

Observe the ammeter. The current indicated (normally 30 A or more) is the electrical load that the alternator must be able to exceed to keep the battery fully charged. The minimum acceptable alternator output should be 5 A greater than the accessory load. For example, if the measured continuous electrical load was 36 A, the *minimum* acceptable alternator output should be 41 A (36 + 5 = 41 A).

NO-TOOL TESTING ON THE CAR

If the voltage regulator and the rotor are functioning correctly, the entire rotor shaft and the front and rear alternator bearings will be magnetized. The rotor shaft should be magnetized because the voltage regulator completes the circuit for the field (rotor) through the alternator brushes. This current flow creates a strong magnetic field in the rotor and magnetizes the rotor shaft and bearings. To *safely* check for a properly operating alternator field circuit, use any steel object to check if the rear bearing is magnetized with the engine running. See Figure 17-4. If the rear bearing is *not* magnetized, there is a possible problem with one or more of the following items:

1. The voltage regulator (not completing the field rotor) circuit

2. Stuck, excessively worn, or defective alternator brushes

FIGURE 17-4 *If the rear bearing is magnetized, the voltage regulator, alternator brushes, and the rotor are functioning.*

3. Defective (open, shorted, or grounded) alternator rotor

To confirm or eliminate the voltage regulator as the cause of no alternator output (no field current), *full fielding* can be used to allow full battery voltage to be connected to the field windings (rotor). "Full fielding" means that full battery voltage is sent to the rotor. With full battery voltage, the rotor should produce its full magnetic field strength. If the rear bearing *is* magnetized, the "no charging" problem is due to defective components inside the alternator. Because the rotor shaft (and bearings) are magnetized, the voltage regulator is controlling the field current.

ALTERNATOR OUTPUT TESTING PROCEDURE

To test the amperage output of any alternator, a carbon-pile tester must be connected to the positive (+) and negative (−) terminals of the battery. An ammeter must be connected in series between the battery and the alternator to measure the alternator's amperage output. A voltmeter also must be connected to the battery terminals to measure the battery voltage. Most automotive starting and charging testers are designed with all three components necessary in one unit. See Figure 17-5.

1. Determine the output of the alternator being tested.

2. Connect the carbon pile, ammeter, and voltmeter according to the test equipment manufacturer's instructions.

FIGURE 17-5 *Typical hookup of a starting and charging tester (Sun Electric VAT-40 shown). (Courtesy of Sun Electric Corporation.)*

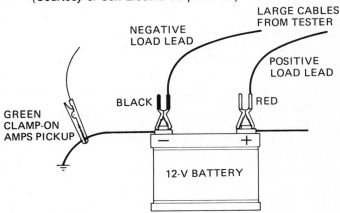

Test lead connections for testing the starting system, charging system, voltage regulator, and diode stator.

3. Turn the ignition switch "on" ("run") (do not start the engine). Observe and record the ammeter reading. This is the current required to operate the ignition system and any other electrical component that is "on" with the ignition "on" ("run"): for example, the clock, radio, and computer.

4. Start the engine and hold it at a fast idle (2000 rpm).

5. Turn the load control knob until the highest reading possible is indicated on the ammeter.

NOTE: Do not increase the load too much where the battery voltage drops below 12.0 V. The maximum alternator output is always developed at a battery voltage higher than 12.0 V.

6. Watch the ammeter scale while rotating the load knob until the highest reading is observed. The ammeter reading indicates the maximum alternator output.

7. Add the maximum tested output to the current reading observed with the ignition "on" ("run").

8. This total should be within 10% of the rated output. For example:

Current reading with ignition "on" =   3.0 A

Maximum observed alternator output = 56.0 A

Total = 59   A

Rated output = 63   A

10% of the rated output =   6.3 A

The alternator is okay because 59 A is within 6.3 A of the rated capacity.

If the output is 2 to 8 A below specifications, the usual cause is an *open* diode or a slipping drive belt. If the output is 10 to 15 A below specifications, the usual cause is a *shorted* diode or a slipping alternator belt. A shorted diode will often cause a whining noise at engine idle and could cause the charge indicator light to remain on with the ignition switch "off."

If the alternator output is below 10% of the rated output, perform the same alternator output test again, with a full field.

CAUTION: Do not operate any alternator with a full field (by passing the voltage regulator) without a load applied to the battery. Since the voltage regulator is bypassed, there is no control of the alternator's output voltage. The output voltage can reach high levels that could cause damage to any electrical unit that is operating when the test is performed.

FULL FIELDING GM ALTERNATORS

To full field a GM SI alternator (Delcotron) with an *internal* voltage regulator, insert a screwdriver momentarily into the test hole (usually D-shaped) in the rear of the alternator housing with the ignition switch "on" (engine not running). See Figures 17-6 and 17-7. Attached to the internal voltage regulator is a metal tang which is electrically connected to the ground rotor brush. See Figure 17-8. Battery voltage is present at the "hot" rotor brush whenever the ignition switch is "on" and the voltage regulator completes the circuit by providing a path to ground. With the screwdriver touching the side of the alternator housing and the grounding tab, the field (rotor) circuit is completed to ground and current will flow through the rotor and create a magnetic field if the brushes are contacting the slip rings

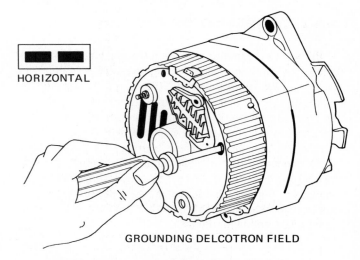

GROUNDING DELCOTRON FIELD

FIGURE 17-6   *A GM alternator with an internal voltage regulator can be identified by the horizontal plug-in connector.*

FIGURE 17-7   *To full-field an* internal *regulator GM alternator, insert a screwdriver through the test hole (usually D shaped).*

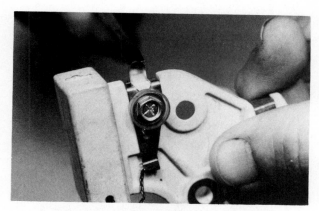

FIGURE 17-8 *The screwdriver grounds the tab on the ground brush, thereby providing full current flow through the brushes and rotor.*

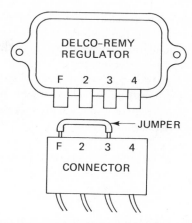

FIGURE 17-10 *Connections required to full-field a GM alternator with an external voltage regulator.*

of the rotor. With the screwdriver inserted, the rear bearing should be magnetized. If the bearings are not magnetized, the problem is due to defective brushes or a defective rotor inside the alternator. If the bearings *are* magnetized with the voltage regulator bypassed, the voltage regulator is defective and must be replaced.

To full field a GM alternator (Delcotron) with an *external* voltage regulator, a jumper wire can be connected directly to the F (field) terminal of the alternator. See Figure 17-9. This procedure may be difficult because of the close clearance between the F terminal of the alternator and the alternator case. A better method involves removing the connector at the external voltage regulator. Then connect a jumper lead between the F connector lead and connector lead 3 with the ignition switch "on" (engine not running), as shown in Figure 17-10. With the engine running and the regulator bypassed, the alternator output should be within 10% of specifications.

If the alternator bearings are *not* magnetized, the problem is due to defective brushes or rotor, requiring that the alternator be removed from the vehicle and disassembled for additional testing. If the alternator bearings *are* magnetized with the jumper wire connected, the voltage regulator is defective and should be replaced.

FULL-FIELDING FORD ALTERNATORS (REAR OR SIDE TERMINAL)

If the charging system is equipped with an external voltage regulator (mechanical or electronic), unplug the connector from the voltage regulator and use a jumper wire between the A and F terminals of the connector, as shown in Figure 17-11. The terminals are lettered on the connector or on

FIGURE 17-11 *Connections required to full-field a Ford alternator with an external voltage regulator.*

SEPARATE RELAY-TYPE REGULATORS

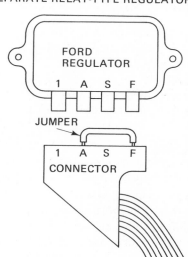

FIGURE 17-9 *A GM alternator with an external voltage regulator can be identified by the vertical plug-in connector.*

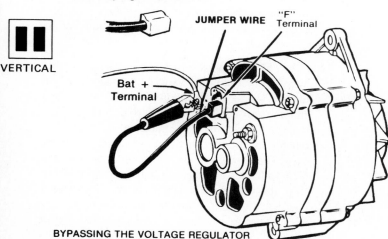

the voltage regulator. See Figure 17-12. Start the engine and immediately apply the carbon-pile load to determine the maximum alternator amperage output and to prevent output voltage from exceeding 16.0 V.

If the alternator is now within 10% of specifications, the problem is a defective voltage regulator or defective wiring between the regulator and the alternator. If the rear bearing is not magnetized after full fielding, the brushes or rotor are defective.

Newer Ford products use an integral electronic voltage regulator. To full-field these alternators, ground the F terminal.

FULL-FIELDING CHRYSLER ALTERNATORS

To full field a Chrysler with an externally mounted voltage regulator (mechanical or electronic), disconnect the *green* field wire from the alternator with the ignition ''off.'' Then connect a jumper wire from the alternator field terminal to ground. With carbon pile, ammeter, and voltmeter connected, start the engine and run at 2000 rpm with a load being applied. If the output is now normal (amperage within 10% of specifications), the problem is a defective voltage regulator. See Figures 17-13 and 17-14.

FULL-FIELDING OTHER ALTERNATORS

All other types of alternators operate similarly; however, exact testing procedures and test specifications may vary. If the exact procedures or specifications are unknown, check with a service manual for the exact charging system being tested.

ALTERNATOR NOISE

A noisy alternator can be caused by defective alternator bearings, a defective diode, a defective or loose belt, or loose mounting brackets and bolts. Check all mounting bolts and drive belts to be certain of the location of the noise. An ''open'' diode can cause a howling or squealing noise similar to a defective bearing. When disassembling an alternator, the front pulley may *not* have to be removed to check and replace parts, *except* for the front bearing.

> NOTE: Some delta-wound alternators are noisy during normal operation. The noise may be objectionable to many owners, and the noise often does sound as if the bearings are defective. Before replacing or servicing a delta-wound alternator for a noise complaint, check the operating noise of a similar alternator in another vehicle.

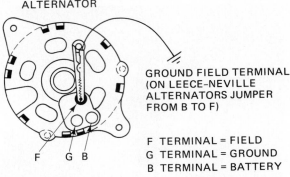

BACK OF FORD ALTERNATOR

GROUND FIELD TERMINAL (ON LEECE-NEVILLE ALTERNATORS JUMPER FROM B TO F)

F TERMINAL = FIELD
G TERMINAL = GROUND
B TERMINAL = BATTERY

FIGURE 17-12 *Jumper wire connections required to full-field a Ford alternator (or Leece-Neville) with an internal electronic voltage regulator.*

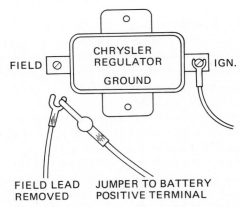

FIELD | CHRYSLER REGULATOR GROUND | IGN.

FIELD LEAD REMOVED

JUMPER TO BATTERY POSITIVE TERMINAL

FIGURE 17-13 *Connections required to full-field a Chrysler alternator with a mechanical voltage regulator.*

ALTERNATOR DISASSEMBLY

If testing has confirmed that there are alternator problems, remove the alternator from the vehicle *after* disconnecting the *negative* (−) battery cable. This will prevent the occurrence of damaging short circuits. Mark the case with a scratch or with chalk to assure proper reassembly of the alternator case. See Figure 17-15.

> NOTE: Most alternators of a particular manufacturer can be used on a variety of cars, which may require wiring connections placed in various locations. For example, a Chevrolet and an Oldsmobile alternator may be identical except for the position of the rear section containing the electrical connections. The four through bolts that hold the two halves together are equally spaced; therefore, the rear alternator housing *can* be reassembled in any one of four positions to match the wiring needs of various models. See Figure 17-16.

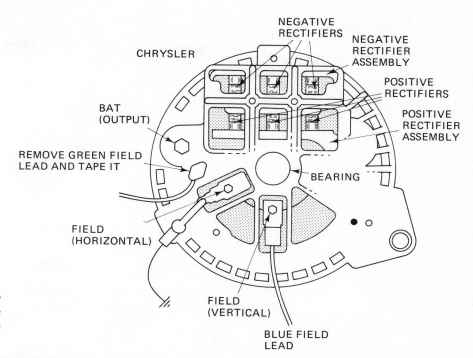

FIGURE 17-14 *Connections required to full-field a Chrysler alternator with an electronic voltage regulator.*

FIGURE 17-15 *Always mark the case of the alternator before disassembly to be assured of correct reassembly. Note that the wire connectors indicate that this GM alternator has an internal voltage regulator.*

After the through bolts have been removed, carefully separate the two halves; the stator windings stay with the rear case. The rotor can be inspected and tested while attached to the front housing.

TESTING THE ROTOR

The slip rings on the rotor should be smooth and round (within 0.002 in. of being perfectly round). If grooved, the slip rings can be machined to provide a suitable surface for the brushes. Do not machine beyond the minimum slip ring dimension as specified by the manufacturer.

If the slip rings are discolored or dirty, they can be cleaned with 400-grit or fine emery (polishing) cloth. The rotor must be turned while being cleaned to prevent flat spots on the slip rings.

The field coil continuity in the rotor can be checked by touching one test lead of a 110-V (15-W bulb) tester on each slip ring. The test light should light. A more accurate method is to measure the resistance between the slip rings using an ohmmeter. Typical resistance values include:

GM	2.4 to 3.5 Ω
Ford	3.0 to 5.5 Ω
Chrysler	3.0 to 6.0 Ω

The resistance values listed above are typical only; exact specifications for the alternator being tested should be consulted before condemning a rotor. Ohmmeters can also vary in accuracy.

1. If the resistance is below specification, the rotor is shorted.
2. If the resistance is above specification, the rotor connections are corroded or open.

Rotor inspection specifications often include an acceptable amperage current. With an ammeter connected in series with the test leads, connect the jumper leads directly to the positive (+) post of the battery and one slip ring. Connect another test lead from the negative (−) post of the battery and the other slip ring. Current will flow through

EXPLANATION OF DELCO "CLOCK" POSITIONS

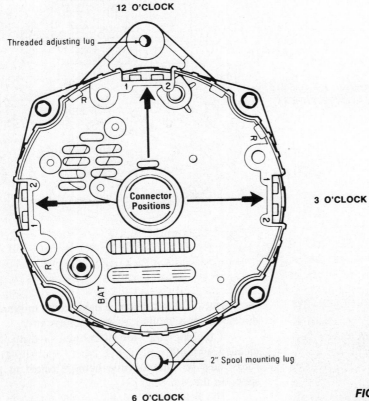

The connector position is determined by viewing the alternator from the diode end with the threaded adjusting lug in the up or 12 o'clock position. Select the 3 o'clock, 9 o'clock or 12 o'clock position to match the unit being replaced.

FIGURE 17-16   *The 9 o'clock position usually fits Pontiacs, the 12 o'clock position usually fits Buicks and Cadillacs, and the 3 o'clock position usually fits Chevrolets and Oldsmobiles. (Courtesy of Lester Catalog Company.)*

FIGURE 17-17   *If the ohmmeter reads infinity between the slip rings, the rotor is open. If the ohmmeter reads lower than specifications or zero, the rotor is shorted. If the ohmmeter is connected between a slip ring and steel and reads less than infinity, the rotor is grounded. (Courtesy of Chevrolet Motor Division, GMC.)*

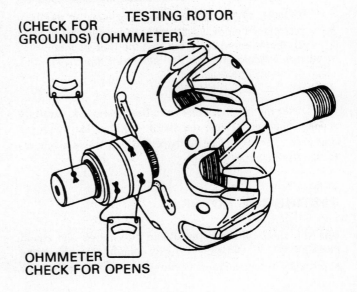

TESTING ROTOR
(CHECK FOR GROUNDS) (OHMMETER)

OHMMETER
CHECK FOR OPENS

the field windings of the rotor and create a magnetic field. Compare the ammeter reading with the manufacturer's specifications (usually between 1.8 and 4.5 A). If the current draw is above specification, the rotor is shorted. If the current draw is below specification, the rotor has high resistance or corroded connections or an open rotor winding. See Figure 17-17.

If the rotor is found to be open, shorted (copper to copper), or grounded (copper to steel), the rotor must be replaced or repaired by a specialized shop. Loose connections at the rotor slip rings may be repaired by resoldering.

> **NOTE:** The cost of a replacement rotor may exceed the cost of an entire rebuilt alternator. Be certain, however, that the rebuilt alternator is rated at the same or higher output than the original.

TESTING THE STATOR

The stator must be disconnected from the diodes (rectifiers) before testing. Since all three windings of the stator are electrically connected (either wye or delta), a powered test light (110- or 12-V) or an ohmmeter can be used to check a stator. There should be low resistance among all three stator leads (continuity) and the test light *should* light. There should *not* be continuity (infinity ohms, or the test light should not light) when tested between any stator lead and the metal stator core. If there is continuity, the stator is *grounded* (short to ground) and must be repaired or replaced. See Figure 17-18. Since the resistance is very low for a normal stator, it is generally *not* possible to test for a *shorted* (copper-to-copper) stator.

A shorted stator will, however, greatly reduce alternator output. If all alternator components test okay and the output is still low, substitute a known-good stator and retest. If the stator is black or smells burned, check the vehicle for a discharged or defective battery. If battery voltage never reaches the voltage regulator cutoff point, the alternator will be continuously producing current in the stator windings. This continuous charging often overheats the stator.

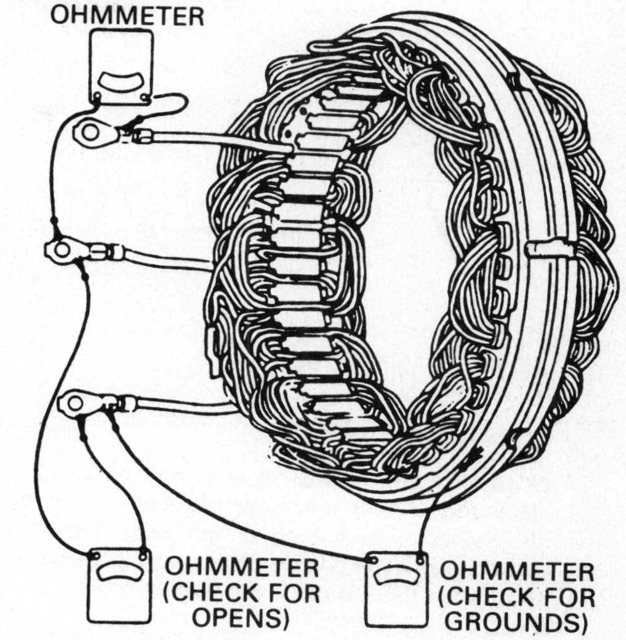

TESTING THE DIODE TRIO

Many alternators are equipped with a diode trio. A diode is an electrical one-way check valve which permits current to flow in only one direction. "Trio" means "three." A diode trio is three diodes connected together. See Figure 17-19. The diode trio is connected to all three stator windings. The current generated in the stator flows through the diode trio to the internal voltage regulator. The diode trio is designed to supply current for the field (rotor) current

TESTING STATOR
(CHECK FOR OPENS)
OHMMETER

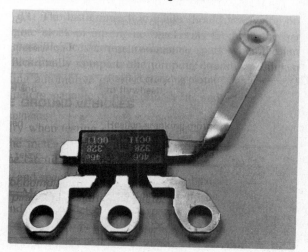

OHMMETER (CHECK FOR OPENS)   **OHMMETER (CHECK FOR GROUNDS)**

FIGURE 17-18 *If the ohmmeter reads infinity between any two of the three stator windings, the a stator is open and defective. The ohmmeter should read infinity between any stator lead and the steel laminations. If the reading is less than infinity, the stator is grounded.*

FIGURE 17-19 *Diode trio.*

and turns off the charge indicator light whenever the alternator voltage equals or exceeds the battery voltage. If one of the three diodes in the diode trio is defective (usually "open"), the alternator may produce close to normal output; however, the charge indicator light will be on dimly.

A diode trio should be tested with an ohmmeter or 12-V self-powered test light (continuity light). The ohm-

TESTING TRIO

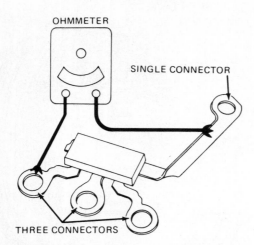

FIGURE 17-20 *As with most diodes, an ohmmeter should read high ohms between the single connector and each of the three stator connections when connected one way and low ohms when the test leads are reversed.*

meter should indicate low ohms (less than 300 Ω) one way and high ohms (more than 300 Ω) after reversing the test leads and touching all three connectors of the diode trio. See Figure 17-20. The test light should light one way and not light when the leads are reversed.

CAUTION: Do not test with a 110-V test light because the high voltage can damage the diodes.

TESTING THE RECTIFIER BRIDGE (DIODES)

Alternators are equipped with six diodes to convert the alternating current (ac) generated in the stator windings into direct current (dc) for use by the vehicle's battery and electrical components. The six diodes include three positive (+) diodes and three negative (−) diodes [one positive (+) and one negative (−) for each winding of the stator]. These diodes can be individual diodes or grouped into a positive and negative "rectifiers" each containing three diodes. All six diodes can be combined into one replaceable unit called a *rectifier bridge*. The rectifier(s) (diodes) should be tested using an ohmmeter or a 12-V test light.

Since a diode (rectifier) should allow current to flow in only one direction, each diode should be tested to check if the diode allows current flow in one direction and blocks current flow in the opposite direction. To test many alternator diodes, it may be necessary to unsolder the stator connections. Accurate testing is not possible unless the diodes are separated electrically from other alternator components. See Figure 17-21. Using an ohmmeter, one test lead should

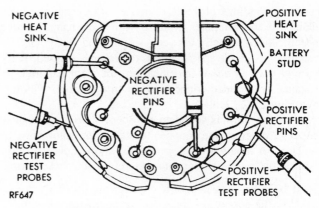

FIGURE 17-21 *Some diodes, such as in this Chrysler alternator, must be unsoldered before testing. (Courtesy of Chrysler Corporation.)*

be connected to the diode lead wire and the other connected to the diode case (base of the diode). Read the ohmmeter. Reverse the test leads. A good diode should have high resistance one way (reverse bias) and low ohms the other way (forward bias).

If the ohmmeter reads low ohms in both directions, the diode is shorted. If the ohmmeter reads high ohms in both directions, the diode is open. Open or shorted diodes must be replaced. Most alternators group or combine all positive and all negative diodes in one replaceable component called a *rectifier*. General Motor's Delcotron alternators use a replaceable *rectifier bridge* containing all six diodes in one unit combined with a finned heat sink. See Figure 17-22. Some Ford and other alternators use six diodes in a single replaceable bridge.

FIGURE 17-22 *Testing a GM alternator rectifier bridge containing six diodes in one unit. Notice that the test leads are connected to the top finned heat sink and the copper diode connector (not the stud). The test leads should be reversed and all six diodes tested.*

END FRAME VIEW

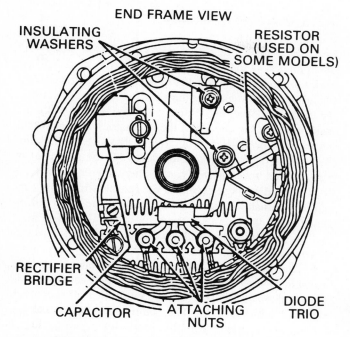

FIGURE 17-23 *Correct location of all components and insulated screws in a typical GM alternator. (Courtesy of Cadillac Motor Car Division, GMC.)*

TESTING THE GM INTERNAL VOLTAGE REGULATOR

Even though the voltage regulator can be tested on the car with the engine running, the internal voltage regulator can also be tested using a special tester.

> **NOTE:** If the current insulated screw is not installed in the correct location, either unregulated (maximum) output or zero output is the result. See Figure 17-23.

BRUSH HOLDER REPLACEMENT

Alternator carbon brushes often last many years and require no scheduled maintenance. The life of the alternator brushes is extended because they conduct only the field (rotor) current, which is normally only 2 to 5 A. The alternator brushes should be inspected whenever the alternator is disassembled and should be replaced whenever worn to less than ½ in. in length. Alternator brushes are spring loaded and if the springs are corroded or damaged, the brushes will not be able to keep constant contact with the slip rings of the rotor. If the brushes do not contact the slip rings, field current cannot create the magnetic field in the rotor which is necessary for current generation. Brushes are commonly purchased assembled together in a brush holder.

BRUSHES RETAINED IN HOLDER

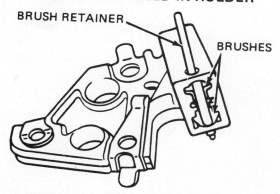

FIGURE 17-24 *GM alternator brush holder assembly. Replacement brushes come assembled with the brush holder. (Courtesy of Chevrolet Motor Division, GMC.)*

FIGURE 17-25 *A brush holder assembly shown assembled in the alternator. The brush retainer is actually a straightened-out paper clip.*

See Figure 17-24. After the brushes are installed (usually retained by two or three screws) and the rotor installed in the alternator housing, a brush retainer pin can be pulled out through an access hole in the rear of the alternator, allowing the brushes to be pressed against the slip rings by the brush springs. See Figure 17-25.

BEARING SERVICE AND REPLACEMENT

The bearing of an alternator must be able to support the rotor and reduce friction. An alternator must be able to rotate up to 15,000 rpm and withstand the forces created by the drive belt. The front bearing is usually a ball-bearing type and the rear is a smaller roller bearing. The front bearing is located under a retainer and pressed into the front alternator case. The pulley must be removed before the

FIGURE 17-26 *Remove the drive pulley from a GM alternator by holding a ⁵/₁₆ in. hex (allen) wrench in a vise and remove the retaining nut with a ¹⁵/₁₆ in. box end wrench.*

FIGURE 17-27 *Alternative method of removing the front pulley from a GM alternator. Gently place the rotor in a vise after removing the brush end frame housing. The vise will hold the rotor while removing the front nut.*

rotor can be separated from the case. Chrysler alternator press-fit pulleys must be removed using a puller, while Ford and General Motors alternators use a nut to hold the drive pulley to the rotor. The front pulley can be removed from most GM alternators by using a ¹⁵/₁₆-in. wrench while holding the rotor shaft from rotating using a ⁵/₁₆-in hex (Allen) wrench. See Figures 17-26 and 17-27.

The old or defective bearing can be pushed out and the replacement pushed in the front housing by applying pressure with a socket or pipe against the outer edge of the bearing (outer race). Replacement bearings are usually pre-lubricated and sealed. Adding additional grease could cause overheating of the bearing by reducing heat transfer from the bearing surfaces to the alternator housing. Many alternator front bearings must be removed from the rotor using a special puller.

ALTERNATOR ASSEMBLY

Reassemble the alternator rectifier(s), regulator, stator, and brush holder. If the brushes are internally mounted, insert a wire through the holes in the brush holder and the alternator rear frame to retain the brushes for reassembly. Install the rotor and front-end frame properly aligned with the mark made on the outside of the alternator housing. Install the through bolts. Before removing the wire pin holding the brushes, spin the alternator pulley. If the alternator is noisy or not rotating freely, the alternator can easily be disassembled again to check for the cause. After being certain the alternator is free to rotate, remove the brush holder pin and spin the alternator again by hand. The noise level may be slightly higher with the brushes released onto the slip rings.

Alternators should be tested on a bench tester, if available, before reinstalling on a vehicle. When installing the alternator on the vehicle, be certain that all mounting bolts and nuts are tight. The battery terminal should be covered with a plastic or rubber protective cap to help prevent accidental shorting to ground, which could seriously damage the alternator.

REMANUFACTURED ALTERNATORS

Remanufactured or rebuilt alternators are totally disassembled and rebuilt. Even though there are many smaller rebuilders who may not replace all worn parts, the major national remanufacturers *totally* remanufacture the alternator. Old alternators (called cores) are totally disassembled and cleaned. Both bearings are replaced and all components tested. Rotors are rewound to original specifications if required. The rotor windings are not counted but are rewound on the rotor "spool" using the correct-gauge copper wire to the *weight* specified by the original manufacturer. New slip rings are replaced as required and soldered to the rotor spool windings and machined. The rotors are also balanced and measured to be certain that the outside diameter of the rotor meets specifications. An undersized rotor would produce less alternator output because the field must be close to the stator windings for maximum output. *Individual* diodes (within the rectifiers) are replaced if required. See Figure 17-28. Every alternator is then assembled and tested for proper output, boxed, and shipped to warehouses. Individual parts stores (called jobbers) purchase parts from various regional or local warehouses.

FIGURE 17-28 *Diodes being soldered to the stator on a Ford alternator on the assembly line at a national electrical remanufacturer.*

SUMMARY

1. Charging system testing requires that the battery be at least 75% charged to be assured of accurate test results. The charge indicator light should be "on" with the ignition switch "on," but should go out whenever the engine is running. Normal charging voltage (at 2000 engine rpm) is 13.5 to 15.0 V.

2. If the charging system is not charging properly, the rear bearing of the alternator should be checked for magnetism. If the rear bearing *is* magnetized, the voltage regulator, brushes, and alternator rotor are functioning correctly. If the rear bearing is not magnetized, the voltage regulator, alternator brushes, or rotor are not functioning. Bypass the voltage regulator by supplying battery voltage to the field. If the rear bearing is now magnetized and the charging system output is normal, the voltage regulator is at fault.

3. To check for excessive resistance in the wiring between the alternator and the battery, a voltage-drop test should be performed.

4. Electricity is lazy because it always travels the path of least resistance. Alternators are also lazy because they do not produce their maximum rated output unless required by circuit demands. Therefore, to test for maximum alternator output, the battery must be loaded to force the alternator to produce its maximum output.

5. Each alternator should be marked across its case before disassembly to be assured of proper "clock position" during reassembly. After disassembly, all alternator internal components should be tested using a continuity light or an ohmmeter. The components that should be tested include:
   (a) Stator
   (b) Rotor
   (c) Diodes
   (d) Diode trio (if equipped)
   (e) Bearings
   (f) Brushes (more than ½ in. in length)

6. Electronic voltage regulators can be tested either off the car using a special tester or on the car using the full-field bypass procedure.

STUDY QUESTIONS

17-1. Describe the normal operation of the dash charging system units, including the charge light, the ammeter, and the voltmeter.

17-2. Explain the proper procedure for testing the amperage output of an alternator.

17-3. What is the procedure for determining the minimum required alternator output?

17-4. Describe the procedure used to test alternator rotors, stators, and diodes.

MULTIPLE-CHOICE QUESTIONS

17-1. An average charging circuit voltage on a 12-V system is:
(a) 7.5 to 8.5 V.
(b) 11.5 to 13.3 V.
(c) 14.9 to 16.1 V.
(d) 13.5 to 15.0 V.

17-2. Technician A says that by full fielding the alternator, you are bypassing the voltage regulator. Technician B says that voltage regulators control the alternator output by controlling the field current. Which technician is correct?
(a) A only.
(b) B only.
(c) both A and B.
(d) neither A nor B.

17-3. Technician A says that a voltage-drop test of the charging circuit can be performed only if current is flowing through the circuit being tested. Technician B says that if the ground for the voltage regulator is corroded, the charging voltage could be too high. Which technician is correct?
(a) A only.
(b) B only.
(c) both A and B.
(d) neither A nor B.

17-4. An alternator output test:
(a) reads the ammeter and the voltmeter at 2000 rpm.
(b) must be within 10% of specifications.
(c) must include the ignition amperage in the reading.
(d) all of the above.

17-5. To obtain maximum output from a GM alternator with an external voltage regulator:
(a) ground the field.
(b) supply the battery voltage to the field terminal.
(c) connect a resistor to the output terminal.
(d) ground the diodes.

17-6. When checking an alternator rotor, if an ohmmeter shows 0 Ω between the slip rings and the rotor shaft, the rotor is:
(a) okay—normal.
(b) defective—shorted rotor windings.
(c) defective—grounded rotor windings.
(d) okay—open rotor windings.

17-7. When checking an alternator stator, if an ohmmeter shows low ohms between all three stator connections, the stator is:
(a) okay—normal.
(b) okay—open.
(c) defective—grounded.
(d) defective—shorted.

17-8. When checking an alternator diode trio, if an ohmmeter reads low ohms in both directions on one leg, the diode trio is:
(a) okay—normal.
(b) defective—shorted.
(c) okay—open.
(d) defective—grounded.

17-9. When checking an alternator rectifier bridge, if an ohmmeter reads 5 Ω on all connections, the part is:
(a) okay—normal.
(b) defective—open.
(c) defective—grounded.
(d) defective—shorted.

17-10. Noisy alternators can be caused by:
(a) defective bearing(s).
(b) a shorted diode.
(c) loose mounting brackets.
(d) all of the above.

18

Basic Ignition Operation

THE IGNITION SYSTEM INCLUDES COMPONENTS AND WIRES NECESSARY TO create a high-voltage spark (over 20,000 V) and distribute this high voltage to the correct spark plug at an exact time. This high-voltage spark can occur at over 200 times per *second* when the car is traveling at highway speeds. This high voltage is produced in the ignition coil and is required to ignite the gas and air inside the engine. The topics covered in this chapter include:

1. High-voltage uses
2. Ignition coils
3. Electromagnetic induction
4. Basic ignition operation
5. Ballast resistors/bypass
6. Dwell
7. Coil energy
8. Capacitance and induction spark

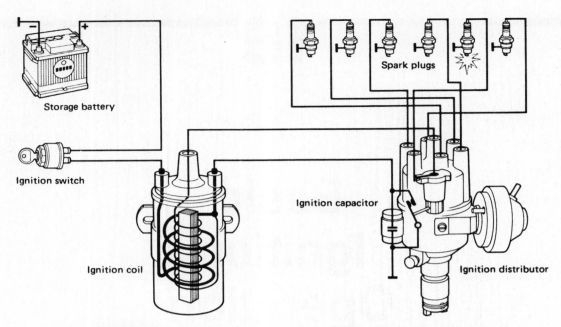

FIGURE 18-1 *Basic point-type ignition system. (Courtesy of Robert Bosch Corporation.)*

HIGH-VOLTAGE USES

All ignition systems require a high-voltage spark to fire the spark plugs. See Figure 18-1. The automotive battery can supply approximately 12 V, but the voltage required to jump the gap of a spark plug can range from 5000 V to more than 20,000 V. The voltage required to fire the spark plug depends on the following engine operating conditions:

1. The wider the gap of the spark plug, the greater the voltage required.
2. The leaner the fuel mixture in the cylinder, the greater the voltage required. Droplets of fuel in the engine provide ''stepping stones'' for the current. Therefore, the richer the mixture is, the easier the mixture can conduct current. A lean mixture (less fuel) contains fewer droplets of fuel and the mixture is less conductive to current flow.
3. The greater the compression in the engine, the greater the voltage required.

The greatest voltage required by the ignition system occurs at part-throttle acceleration. This condition requires the highest voltage because the fuel system is supplying a lean fuel mixture with increased combustion pressures.

IGNITION COILS

The heart of any ignition system is the ignition coil. The coil creates a high-voltage spark by electromagnetic induction. Most ignition coils contain two separate but electrically connected windings of copper wire. Other coils are true transformers in which the primary and secondary windings are not electrically connected. See Figure 18-2.

The center of an ignition coil contains a core of laminated soft iron (thin strips of soft iron). This core increases the magnetic strength of the coil. Surrounding the laminated core are approximately 20,000 turns of fine wire (approximately 42 gauge). These windings are called the *secondary* coil windings. Surrounding the secondary winding is approximately 150 turns of heavy wire (approximately 21 gauge). These windings are called the *primary* coil windings. In many coils, these windings are surrounded with a thin metal shield and insulating paper and placed into a metal container. The metal container and shield help retain the magnetic field produced in the coil windings. The primary and secondary windings produce heat because of the electrical resistance in the turns of wire. Many coils contain oil to help cool the ignition coil. Other coil designs, such as those used on GM's High Energy Ignition (HEI) systems and others, use an air-cooled,

FIGURE 18-2 *Typical ignition coil. This style of ignition coil is usually filled with transformer oil to help control internal coil temperatures. (Courtesy of Chrysler Corporation.)*

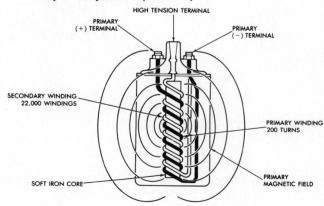

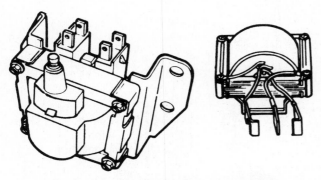

FIGURE 18-3 *Two examples of E-style ignition coils. E coils do not contain oil but are usually constructed of an epoxy material over the primary and secondary windings. (Courtesy of General Motors Corporation.)*

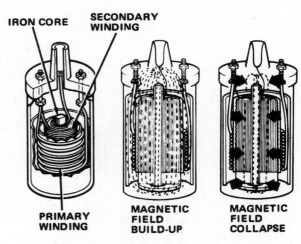

FIGURE 18-4 *Whenever a current starts to flow into any coil, it requires a certain amount of time before the magnetic field reaches full strength due to the self-induction of the coil windings. (Courtesy of Ford Motor Company.)*

epoxy-sealed *E coil*. It is called an E coil because the laminated soft iron core is E-shaped, with the coil wire turns wrapped around the center "finger" of the E and the primary winding wrapped inside the secondary winding. See Figure 18-3.

The primary windings of the coil extend through the case of the coil and are labeled positive (+) and negative (−). The positive (+) terminal of the coil attaches to the ignition switch, which supplies current from the positive (+) battery terminal. The negative (−) terminal is attached to contact points or an electronic ignition module which "opens" and "closes" the primary ignition circuit by opening or closing the ground return path of the circuit. With the ignition switch "on," current should be available at *both* the positive (+) and the negative (−) terminals of the coil if the primary windings of the coil have continuity. The labeling of positive (+) and negative (−) of the coil indicates that the positive (+) terminal is *more* positive (closer to the positive terminal of the battery) than the negative (−) terminal of the coil. This is called the coil *polarity*. The polarity of the coil must be correct to be certain that electrons flow from the hot center electrode of the spark plug. The polarity of an ignition coil is determined by the direction of rotation of the coil windings. The correct polarity is then indicated on the primary terminals of the coil. If the coil primary leads are reversed, the voltage required to fire the spark plugs is increased by 40%. The coil output voltage is directly proportional to the ratio of primary versus secondary turns of wire used in the the coil.

SELF-INDUCTION

Whenever current starts to flow into a coil, an opposing current is created in the windings of the coil. This opposing current generation is caused by self-induction and is called *inductive reactance*. Inductive reactance is similar to re-

sistance because it opposes any increase in current flow in a coil. Therefore, whenever an ignition coil is first energized, there is a slight delay of approximately 0.01 second until the ignition coil has reached its maximum magnetic field strength. See Figure 18-4. The point where a coil's maximum magnetic field strength is reached is called *saturation*. See Chapter 4 for details on self-induction.

MUTUAL INDUCTION

In an ignition coil, there are two windings, a primary and a secondary winding. Whenever there is a *change* in the magnetic field of one coil winding, there is a change in the other coil winding. Therefore, if the current is stopped from flowing (circuit is "opened"), the collapsing magnetic field cuts across the turns of the secondary winding and creates a high voltage in the secondary winding. The collapsing magnetic field also creates a voltage of up to 250 V in the *primary* winding. See Chapter 4 for details on mutual induction.

HOW IGNITION COILS CREATE 40,000 VOLTS

All ignition systems use electromagnetic induction to produce a high-voltage spark from the ignition coil. Electromagnetic induction means that a current can be created in a conductor (coil winding) by a moving magnetic field. The magnetic field in an ignition coil is produced by current flowing through the primary windings of the coil. The current for the primary windings is supplied through the igni-

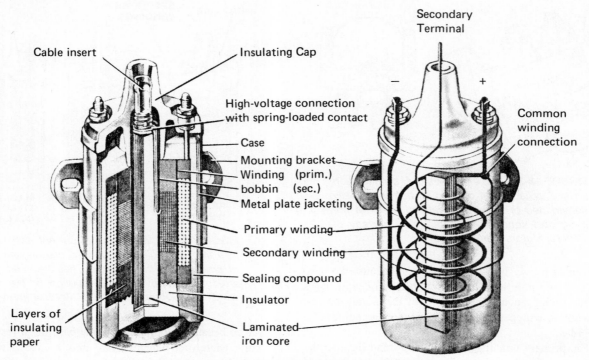

Cable insert

Insulating Cap

Secondary
Terminal

High-voltage connection
with spring-loaded contact

Case

Mounting bracket

Winding (prim.)

bobbin (sec.)

Metal plate jacketing

Primary winding

Secondary winding

Sealing compound

Insulator

Laminated
iron core

Layers of
insulating
paper

Common
winding
connection

− +

FIGURE 18-5 *A typical ignition coil with proper triggering can change low voltage current (10 to 12 V at 2 to 8 A) into very high voltage current (up to 40,000 V at 20 to 80 mA). (Courtesy of Robert Bosch Corporation.)*

tion switch to the positive (+) terminal of the ignition coil. The negative (−) terminal is connected to the ground return through the use of movable mechanical ignition points or through an electronic ignition module. See Figure 18-5.

If the primary circuit is completed, current (approximately 1 to 5 A) can flow through the primary coil windings. This creates a strong magnetic field inside the coil. When the primary coil winding ground return path connection is "opened," the magnetic field *collapses* and induces a high-voltage (20,000 to 40,000 V) low-amperage (20 to 80 mA) current in the secondary coil windings. This high-voltage pulse flows through the coil wire (if equipped) distributor cap, rotor, and spark plug wires to the spark plugs. For each spark that occurs, the coil must be charged with a magnetic field and then discharged. The ignition components, which regulate the current in the coil primary winding by turning it "on" and "off," are known collectively as the *primary ignition circuit*. All of the components necessary to create and distribute the high voltage produced in the secondary windings of the coil are called the *secondary ignition circuit*. The components included in each circuit include (see Figure 18-6):

Primary Ignition Circuit	Secondary Ignition Circuit
1. Battery	1. Secondary windings of coil
2. Ignition switch	2. Distributor cap and rotor (if equipped)
3. Primary windings of the coil	3. Spark plug wires
4. Points and condenser (if point-type)	4. Spark plugs
5. Pulse generator and module (if electronic ignition)	

POINT-TYPE IGNITION OPERATION

The point-type ignition system worked well for many years but was replaced in the mid-1970s by electronically triggered ignition systems. The point-type ignition is commonly referred to as the Kettering ignition, named for its inventor, Charles F. Kettering (1876–1958), a Dayton, Ohio, inventor and industrialist. See Figure 18-7.

Located under the distributor cap and rotor is a set

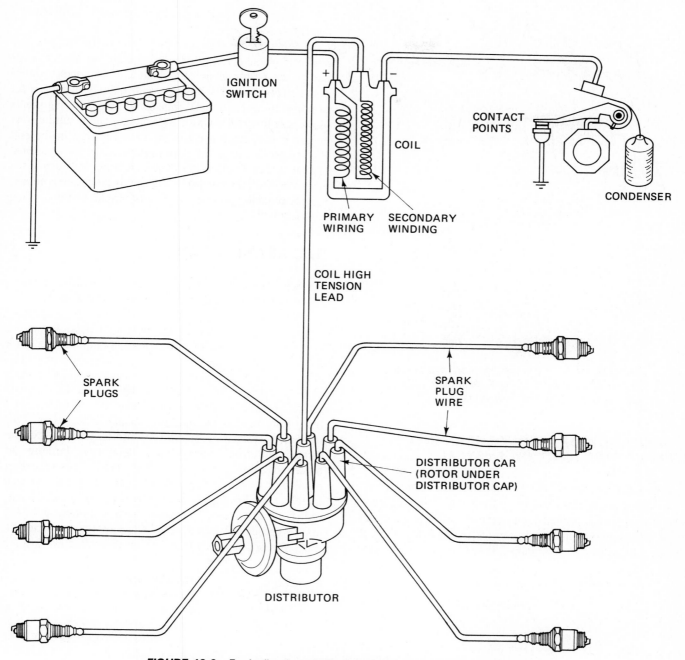

FIGURE 18-6 *Basically, the primary ignition circuit is located below the distributor rotor and the secondary above the rotor.*

of contact points. These tungsten steel points act as an electrical switch that controls the current flowing through the ignition coil. The points open and close the ground return path of the primary coil winding. The points are opened by a small cam lobe located on the shaft of the distributor (four lobes for four cylinder, six lobes for six cylinder, etc.) and closed by a built-in spring.

When the points are closed. Current flows through the coil and through the closed points and to ground, which completes the primary ignition circuit. Current (10 V and 5 A) flows through the coil's primary wiring, creating a magnetic field. See Figure 18-8.

When the points are open. Whenever the engine operation rotates the distributor cam lobe enough for the points to just open, the primary circuit is *opened* and the stored energy that has built up inside the coil discharges out of the tower of coil (20,000 V or more) and is sent to the distributor cap and rotor through the coil wire.

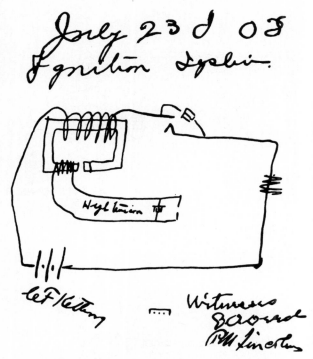

July 23d 08
Ignition System.

High tension

C.F. Kettering

Witnesses
ground
R M Lincoln

FIGURE 18-7 *Kettering ignition system as drawn in 1908 by Charles F. Kettering, a Dayton, Ohio, inventor and industrialist.*

OPERATION OF A CONDENSER IN A POINT-TYPE IGNITION SYSTEM

Attached to the points there is another wire in addition to the one coming from the side of the coil. This lead goes to a small silver-colored round canister called a condenser. An automotive condenser is constructed of many alternate layers of paper and aluminum foil and is capable of "stor-

ing" an amount of electrons. See Chapter 4 for details on capacitors (condensers).

When the ignition points are just starting to open, the tendency of the current flow is to "jump" or arc the small gap of the points in order to maintain a complete closed circuit. However, this arcing would quickly destroy a set of points by causing metal transfer (pitting). Arcing points also cause low coil output (low voltage) and resulting poor engine performance because the stored energy in the coil is released slowly rather than rapidly as occurs in normal operation. See Figure 18-9. The condenser "absorbs" the electrons that would normally arc across the opening points, thereby reducing point arcing to a minimum and improving coil output.

BALLAST RESISTORS

Between the ignition switch and ignition coil, a heat-sensitive variable resistor or calibrated resistor wire is in the primary circuit, as shown in Figure 18-10. The purpose of this resistor is to provide the maximum possible current flow through the coil for maximum coil output to the spark plugs at all driving speeds, and yet provide acceptable ignition point life. At lower engine speeds, the points remain closed a relatively long time and the coil can easily reach saturation. The variable resistor heats up due to this relatively long "on" time, and its resistance is increased. This increased resistance effectively limits primary current through the points and prevents them from burning. At higher engine speeds, the points are closed a relatively short time, so that the ballast resistor does not get as hot and therefore has lower resistance. The resulting increased current flow is necessary to assure coil saturation at higher engine speeds. Therefore, the ballast resistor tends to stabilize the primary magnetic field in the coil, thus keeping

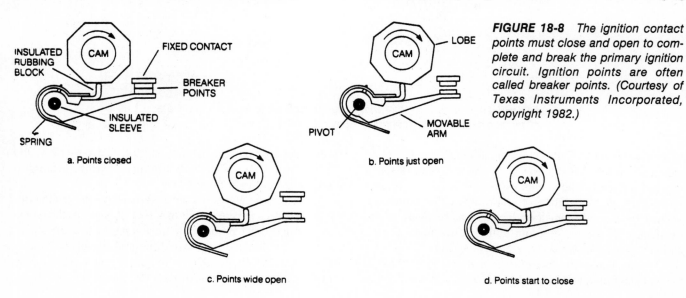

a. Points closed

b. Points just open

c. Points wide open

d. Points start to close

FIGURE 18-8 *The ignition contact points must close and open to complete and break the primary ignition circuit. Ignition points are often called breaker points. (Courtesy of Texas Instruments Incorporated, copyright 1982.)*

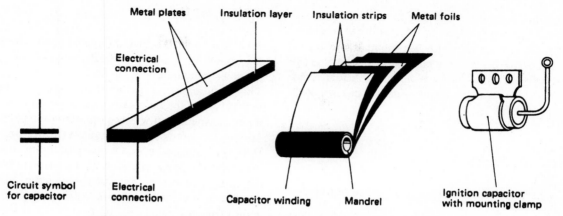

Circuit symbol
for capacitor

Metal plates   Insulation layer   Insulation strips   Metal foils

Electrical
connection

Electrical
connection   Capacitor winding   Mandrel   Ignition capacitor
with mounting clamp

FIGURE 18-9 *Typical automotive ignition condenser (also called a capacitor). The condenser (capacitor) helps prevent the ignition points from burning and provides for rapid shutoff of the primary ignition circuit necessary for high voltage from the coil. (Courtesy of Robert Bosch Corporation.)*

FIGURE 18-10 *Basic point-type ignition system. Note the resistance wire between the ignition switch and the positive of the ignition coil. (Courtesy of Ford Motor Company.)*

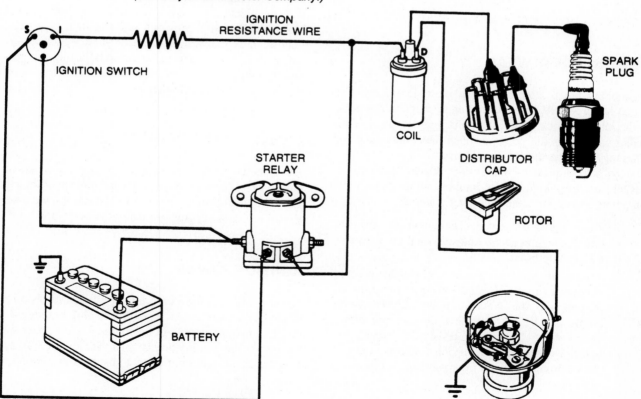

IGNITION
RESISTANCE WIRE

IGNITION SWITCH

SPARK
PLUG

COIL

STARTER
RELAY

DISTRIBUTOR
CAP

ROTOR

BATTERY

the points from arcing by controlling the induced primary voltage. Some ignition coils use an internal ballast resistor. Some electronic ignition cars also use a ballast resistor to stabilize and control the primary current. A ballast resistor (or wire) usually measures between 1.2 and 2.5 Ω at room temperature. Primary resistance wire usually measures 0.40 Ω per foot of length.

RESISTOR BYPASS

The current-limiting ballast resistor is bypassed during engine cranking because:

1. The battery source voltage is lower during cranking due to the electrical load of the cranking circuit (approximately 10 V instead of the typical 14 V during running).
2. This bypass allows maximum current to reach the ignition coil during starting.
3. The cranking time is usually of such short duration that point life is unaffected.

   NOTE: If a car will start on the "start" position but fails to keep running when switched to the "on" position, the problem is usually traceable to an open ballast resistor, resistor wire, or on some cars, an open ignition circuit (ignition fuse).

IGNITION POINT WEAR

The rubbing block often wears the most during the first several hundred miles after installing new ignition points. As the rubbing block wears, the point gap decreases. This smaller gap causes the points to open later, thus delaying (retarding) the time the sparks will occur.

To help prevent as many point-related problems as possible, use only high-quality ignition points. Such ignition points should have the following features:

1. They should be vented. These hole or cross-cut features help keep the tungsten steel points below the melting point.
2. A brass or copper strip is used to provide a better conductor (lower resistance) between the coil connection and the movable point. Lower-quality points use the current flow through the return spring itself. The lower resistance through the brass or copper conductor helps increase the primary current flow and improves coil output.

POINT GAP

Point gap is the distance the points open, measured when the points are fully opened. Point gap specifications are given in thousandths of an inch, for example: 0.019 in. This is an important specification since point gap will determine when the points *open*, and this in turn determines when the spark plug will fire (ignition timing). If the points are set too far apart (for example, 0.022 in. instead of 0.016 in.), the coil may not reach saturation at high engine speed. If the points are set too close (for example, 0.010 in. instead of 0.016 in.), premature point burning could occur because the points are conducting current for too long. The coil could also be overheated if the points are set too close.

If used points are *blue,* this indicates excessive primary circuit voltage. Check the voltage regulator setting, primary circuit resistance, or condenser. If the points are *black,* dirt, oil, or grease have accumulated on the points. Check for excessive distributor cam lube. A defective PCV system can also cause crankcase pressure to force engine oil into the distributor.

DWELL

Dwell is the electrical method of measuring point gap and is the the "number of degrees of distributor cam rotation that the points are *closed.*" A wide gap will result in a lower dwell, and a narrower gap will increase dwell. As long as there is dwell, a magnetic field is "dwelling" in the ignition coil. When replacing or adjusting the ignition points, a dwell meter is more accurate than setting the point gap with a thickness (feeler) gauge. See Figure 18-11.

DWELL METER HOOKUP

Dwell meters are connected to the negative (−) side of the coil so as to "sense" the opening and closing of the points. The other lead (usually black) of most dwell meters is connected to a good engine ground.

IGNITION TIMING

Ignition timing is the *exact time* in relation to piston position when a spark is sent to the spark plug to ignite the air and gas mixture already in the cylinder. This spark occurs the instant the ignition points *open.* See Figure 18-12. See Chapter 20 for ignition timing adjustment procedures.

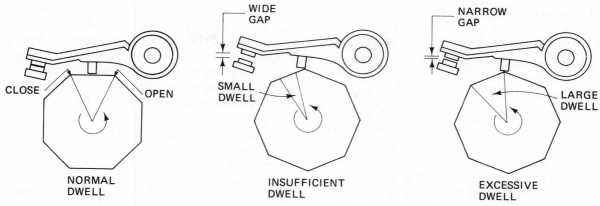

FIGURE 18-11 *Note that as the rubbing block wears, the point gap decreases and the dwell angle increases. As the dwell increases, the ignition timing is retarded 1° for every 1° that dwell is increased.*

FIGURE 18-12 *For the engine to develop its maximum power and economy, the spark must occur several degrees of crankshaft rotation before the piston reaches the top of the compression stroke.*

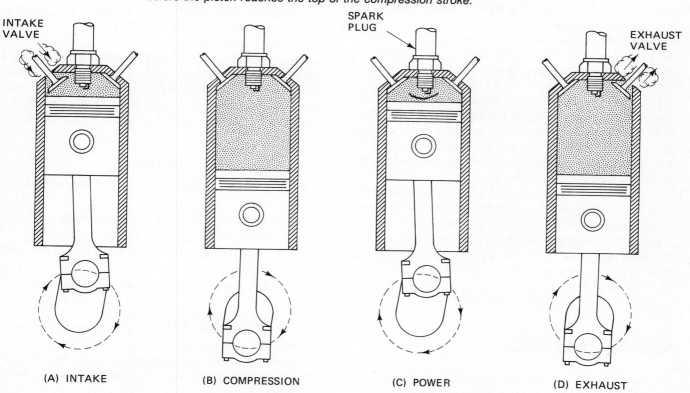

DWELL VERSUS TIMING

The spark occurs when the points just open, and when the points open is determined by point gap (the narrower the gap, the later the point opening). If dwell (point gap) is not correct, the timing may not be correct.

> **NOTE:** For every degree that dwell is increased, timing is retarded 1°. For every degree that dwell is decreased, timing is advanced 1°. Therefore, dwell changes timing and must be set before adjusting ignition timing. Adjusting ignition timing does *not* change dwell (point gap). Ignition timing should be changed only after making certain that the point gap (dwell) is set correctly.

COIL ENERGY

The energy of an ignition system is expressed as the product of $\frac{1}{2} Li^2$, where L is the coil inductance (and is constant for each coil) and i is the primary current existing at the time the primary circuit is opened. The total energy of an ignition system is determined largely by the amount of primary circuit current (in amperes). With a point-type ignition, the maximum current possible is determined by the service life of the breaker points. Electronic ignition systems are capable of controlling higher primary ignition current than is breaker-point ignition. (Electronic ignition systems can generally handle about 8 amperes, compared to 3–5 amperes for a point-type ignition.) The higher the primary circuit current, the higher the energy in the ignition coil. General Motors named their electronic ignition HEI (High Energy Ignition) because without breaker points, the ignition system was designed for high energy. A high-energy ignition system is necessary to ignite lean fuel mixtures properly by providing the necessary spark *duration*. Electronic ignition systems currently in use produce the necessary high energy required to ignite lean mixtures.

Coil energy is expressed in an electrical unit for energy called a "joule," named after an English physicist, James Prescott Joule (1818-1889). One joule (J) is also equal to 1 watt-second. Coil energy can be as low as 0.001 J to more than 0.100 J (1 to 100 milliwatt-sec.). This means that a typical ignition coil will provide 20 to 80 mA (0.020 to 0.080 A) of current across the spark plug electrodes with a typical voltage of 10,000 to 40,000 V.

HIGH-VOLTAGE COILS

Electronic ignition systems are capable of producing up to 40,000 V (40 kV), yet ignition coils are seldom manufactured to produce over 40 kV. According to automotive de-sign engineers, if coils were able to produce over 40 kV, the insulation materials used in the construction of the ignition coil would have to be greatly improved. Also, if higher voltage is produced, all insulating materials, including spark plug wires, spark plugs, and distributor caps, would require a greater dielectric strength.

Replacing the original coil with a high-voltage coil may not improve the voltage to the spark plugs or engine performance. The voltage output of any coil is limited to the actual voltage *required* by the spark plug to arc the gap inside the engine (remember, electricity is lazy). Therefore, a higher-voltage output coil will be able to provide this higher voltage, but only if the higher voltage is required to fire the spark plug. Original-equipment ignition coils are usually designed to provide a maximum voltage of 38 to 40 kV.

SECONDARY IGNITION CAPACITANCE

Until the voltage is high enough to overcome the millions of ohms of resistance that the spark plug gap represents, no spark occurs and no current flows through the secondary circuit. The air gaps between the rotor and distributor cap and the spark plug gaps represent a capacitor, with the air/fuel mixture acting as the dielectric. Moisture (especially condensation) on spark plug wires greatly increases the capacitance and the required voltage necessary for a spark to occur at the spark plug. If the coil voltage is high enough, the air/fuel mixture between the spark plug electrodes will be *ionized* (become electrically conductive), allowing a spark to occur at the spark plug. The voltage required to overcome the capacitive resistance of the secondary circuit can vary from 5000 to 15,000 V or more. Factors that increase the capacitive load and require higher voltages to overcome include:

1. Wide-gapped spark plugs
2. Rounded spark plug center electrodes (normal with usage)
3. Moisture on the distributor caps and/or spark plug wires
4. Lean fuel mixtures (rich fuel mixtures are more conductive)
5. Increase in cylinder pressure (compression)

If the condition of the secondary ignition system offers too high a capacitive resistance, no spark will occur at the spark plugs. The type of spark plug wire does not control the voltage at the plug. The spark plug *gap* actually determines the firing voltage as long as the plug wires are not open. See Figure 18-13.

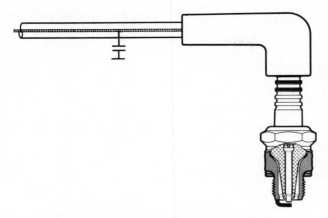

FIGURE 18-13 *Electrical components of the ignition system such as spark plug wires have the effect of a capacitor and reduce the ignition energy if dirty or moist. (Courtesy of Robert Bosch Corporation.)*

The spark energy, created by induction, remaining in the ignition coil after the capacitive resistance has been overcome is available to provide a spark long enough in duration (at least 1 millisecond) to ignite the fuel mixture in the cylinder. This means that for an ignition system to perform properly, all secondary ignition components should be kept clean and dry for best performance. Many automobile manufacturers provide rubber or plastic covers over the ignition coil and/or distributor to help prevent accumulation of dirt or moisture on these components.

WHAT IS RISE TIME?

Various types of ignition systems have their own operating characteristics. One of the factors that varies among ignition system types is rise time. See Figure 18-14. According to SAE, rise time is the time, measured in microseconds (millionths of a second), for the output voltage of a coil to rise from 10% to 90% of its maximum output. The faster the rise time of an ignition system, the easier it is to fire a fouled spark plug. An ignition spark that reaches its maximum voltage quickly (short rise time) has no time to flow through fouled spark plug deposits toward ground. Typical rise times for various types of ignitions are:

1. Point type: 200 microseconds (μs) (relatively slow)
2. Electronic ignitions: 20 to 50 μs
3. Capacitor discharge ignition: 1 to 3 μs (extremely fast)

Capacitor discharge (CD) ignitions charge a capacitor, then discharge the capacitor through the primary windings of the ignition coil. This creates a high-voltage (short-rise-time) spark which is short in duration. This short-duration spark makes a CD ignition unacceptable for use with lean fuel mixtures. CD ignition systems were popular in the 1960s and 1970s as an added ignition accessory for point-type ignition systems. CD ignition systems are currently used for small gasoline engines and many outboard motors.

Point-type ignition systems with pitted or arcing points have a longer-than-normal rise time, and therefore spark plugs are more likely to foul. Each ignition system design requires its own spark plug gap, rotor gap, and other factors for proper operation.

FIGURE 18-14 *From the time the points open until the time of ignition is called the rise time. In this example, the rise time is 30 μs. (Courtesy of Robert Bosch Corporation.)*

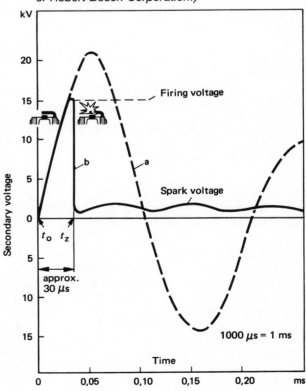

SUMMARY

1. All ignition systems must be able to produce over 20,000 V and distribute this high-voltage (up to 40,000 V), low-amperage (0.020 to 0.080 A) current to the correct spark plug at the instant needed.

2. All ignition systems use a low-voltage primary circuit that controls (pulses) the primary circuit of the ignition coil to produce high-voltage pulses out of the coil.

3. Point-type ignition systems use a set of points to open and close the primary ignition ground return path circuit. The polarity of any coil is determined by the direction in which the coil windings are wound. When the points close, current starts to flow through the primary coil windings (approximately 21-gauge wire) and starts to build a strong magnetic field. After about 0.01 second, the magnetic field has reached its greatest strength, called saturation. When the points open, the magnetic field collapses and induces a high-voltage spike in the secondary winding of the coil. The condenser prevents the points from arcing and burning and permits a more rapid shutoff of the primary ignition current. The more rapid this cutoff, the higher the induced voltage. A ballast resistor is used on point-type ignition systems to provide the greatest possible current flow through the coil at high engine speeds, yet provide for lower current flow at low engine speed where the maximum current is not required. Reduction of primary current flow at low engine speed also helps preserve coil and breaker point life.

4. Dwell is an electrical method of measuring breaker point gap. Dwell is the number of degrees of distributor cam rotation that the points are closed. The dwell *must* be adjusted before checking or adjusting ignition timing.

5. Original ignition coils can produce approximately 40,000 V and enough energy to provide for a long-duration (1 millisecond) spark required to ignite a lean mixture. Dirt and/or moisture on the distributor cap or spark plug wire can greatly increase the voltage required to fire the spark plugs.

STUDY QUESTIONS

18-1. Describe how high voltage is produced in the ignition coil.

18-2. Explain the operation of the point-type ignition system and why a condenser must be used in the system to create high secondary voltage.

18-3. What is the purpose of the ballast resistor?

18-4. Describe the relationship between point gap and dwell.

18-5. Why is it important that the secondary circuit components be clean and dry for proper ignition system operation?

MULTIPLE-CHOICE QUESTIONS

18-1. Cam dwell or cam angle means:
   (a) the length of *time* the points are closed.
   (b) the angle at which the rubbing block contacts the cam.
   (c) the difference between the lobes on the cam.
   (d) the number of degrees of distributor cam rotation that the points are closed.
   (e) none of the above.

18-2. Excessive point gap results in:
   (a) retarded timing.
   (b) reduced dwell.
   (c) both (a) and (b).
   (d) none of the above.

18-3. Ignition coils reach saturation in approximately:
   (a) 10 seconds.
   (b) 1 second.
   (c) 0.1 second.
   (d) 0.01 second.

18-4. With the ignition switch in the "start" or "crank" position, the engine starts, but as soon as the switch is released to run position, the engine stalls. What is the probably cause?
(a) a defective bypass circuit.
(b) a defective coil.
(c) a defective primary ignition resistor.
(d) a defective condenser.

18-5. The maximum coil output is approximately:
(a) 20,000 V and 5 A.
(b) 40,000 V and 0.080 A.
(c) 25,000 V and 20 mA.
(d) 80,000 V and 60 mA.

18-6. Technician A says that dwell must only be adjusted at the specified idle RPM. Technician B says that if the dwell is adjusted 2°, the timing will be changed by 2°. Which technician is correct?
(a) A only.
(b) B only.
(c) both A and B.
(d) neither A nor B.

18-7. Coil energy is measured in units of:
(a) volts.
(b) amperes.
(c) watt-seconds.
(d) coulombs.

18-8. As ignition points wear:
(a) the point gap increases.
(b) the points remain the same.
(c) the dwell decreases.
(d) the point gap decreases.

18-9. The spark occurs when:
(a) the points open.
(b) the points close.
(c) the ballast resistor is "on."
(d) the ignition switch is "on."

18-10. The items that increase the voltage required to fire a spark plug include:
(a) a rich mixture; a wide plug gap.
(b) dirty spark plug wires; a lean fuel mixture.
(c) a narrow spark plug gap; a rich mixture.
(d) a rich mixture; a wide spark plug gap.

19

Electronic Ignition Operation

ELECTRONIC IGNITION SYSTEMS HAVE BEEN STANDARD EQUIPMENT SINCE the mid-1970s. The topics covered in this chapter include:

1. Electronic ignition operation
2. Magnetic pulse generators
3. GM HEI electronic ignition
4. Ford electronic ignition systems
5. Chrysler electronic ignition systems
6. Hall-effect electronic ignition systems
7. Direct-fire (no distributor) ignition

IGNITION BACKGROUND

The point-type ignition system functioned well for many years, but had the following disadvantages:

1. High maintenance requirements, due to rubbing block and point wear
2. Timing changes due to point gap changes
3. Decreased gas mileage and increased exhaust emissions, due to resulting retarded timing changes
4. Limited ability to ignite lean fuel mixtures, due to lower spark energy

The key to solving the problems noted above was to eliminate the mechanical contact points and replace them with non-wearing electronic components that could perform the same function as points.

TRANSISTOR BASICS

The type of electronic switch now in use is called a transistor. A transistor is made of three sections or layers referred to as P (positive)- or N (negative)-type material. The center layer determines its polarity (for example; PNP or NPN). The three elements (layers) are:

E = emitter
C = collector
B = base

If current can flow from E to B, the transistor "turns on" C to E.

A transistor is like two diodes back to back. A diode allows current flow in one direction only. However, a transistor will allow current flow if the electrical conditions allow it to switch "on." See Figure 19-1.

FIGURE 19-1 *(Courtesy of General Motors Corporation.)*

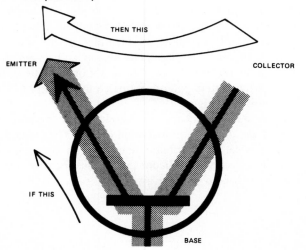

The electrical conditions are determined or "switched" by means of the *base* or *B*. The base will carry current only when the proper voltage range and polarity is applied.

The base current can be turned "off" or "on" (and therefore the transistor can be turned "off" or "on") by means of mechanical switches *or* by reversing the current flow in the base circuit. (See Chapter 5 for detailed information on transistors.)

Electronic ignition does this by sending a voltage signal or "timing pulse" generated in the distributor (called a *pulse generator*) which "opens" the transistor. This "opens" the ignition primary circuit, which causes the coil to discharge a high voltage.

PULSE GENERATORS

A pulse generator consists of:

1. A permanent magnet (stationary)
2. A pole piece (stationary)
3. A reluctor (rotating)

All three units occupy the area in the distributor below the rotor that would normally be for the points and condenser.

The various component parts of a pulse generator differ among various manufacturers. See Figure 19-2. The terms commonly used include the following:

Manufacturer	Stationary Pickup Coil	Rotating Trigger Wheel
AMC	Sensor	Trigger wheel
Chrysler	Pickup coil	Reluctor
Ford	Stator	Armature
GM	Pole piece and magnetic pickup	Timer core

FIGURE 19-2 *Typical electronic primary ignition system. (Courtesy of Chrysler Corporation.)*

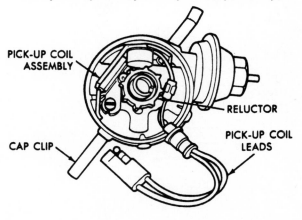

HOW ELECTRONIC IGNITION WORKS

Surrounding the pickup coil is a magnetic field formed by the permanent magnet. When a conductor comes close to the magnetic lines of force, the *reluctance* (resistance) to the magnetic lines of force decreases. Therefore, the movable armature is often called a *reluctor* because it decreases the reluctance of the magnetic field surrounding the pickup coil. See Figure 19-3.

As the armature tooth approaches the center of the stator, a strong magnetic field begins to build around the magnetic pickup, permitting the primary coil current to flow across the emitter–base circuit through the magnetic pickup to ground. This allows the ignition coil to build up a strong magnetic field. When the reluctor tooth passes the center of the magnetic pickup and starts away from it, it reverses the induced voltage and turns "off" the "base" of the switching transistor. This, in turn, switches "off" the

FIGURE 19-3 *Operation of a typical pulse generator. (Courtesy of Texas Instruments Incorporated, copyright 1982.)*

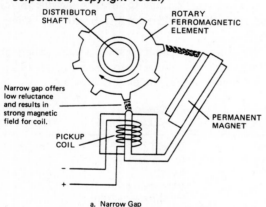

a. Narrow Gap

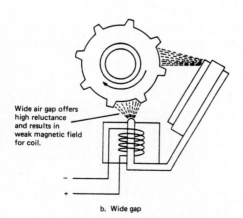

b. Wide gap

emitter–collector of the transistor and turns "off" the primary ignition coil current. The very rapid cutoff of primary current "collapses" the magnetic field in the ignition coil to produce the high secondary voltage required to fire the spark plug.

GM HEI ELECTRONIC IGNITION OPERATION

High Energy Ignition (HEI) has been the standard equipment ignition system on General Motors vehicles since the 1975 model year. Most V-6 and V-8 models use an ignition coil inside the distributor cap. Some V-6, in-line six-cylinder, and four-cylinder models use an externally mounted ignition coil. The operation of both styles is similar. The large-diameter distributor cap provides additional space between the spark plug connections to help prevent cross-fire. HEI distributors also use 8-mm-diameter spark plug wires which use female connections to the distributor cap towers. See Figure 19-4. (See Chapter 21 for details on HEI distributor caps, rotors, and spark plug wires.) HEI ignition coils must be replaced (if defective) with the exact replacement style. HEI coils differ and can be identified by the colors of the primary leads. The primary coil leads can be either white and red, or yellow and red. The correct color lead coil must be used for replacement. The colors of the leads indicate the direction the coil is wound, and therefore its polarity.

HEI PULSE UNIT

Beneath the rotor is a pole piece and magnetic pickup which are stationary. Attached to the distributor shaft is a timer core. Whenever the "teeth" of the timer core and magnetic pickup rotate pass each other, a small current is generated. This small electrical pulse is sent to the module. Replacement pickup coil assemblies are identified by the color of the plug connector. Replacement pickup coil assemblies must have the same-color wire connector or cable tie to be assured of the correct polarity to match the polarity of the ignition coil. See Figure 19-5.

The module contains various electronic components that open and close the primary coil circuit. The module is installed inside the distributor housing with two screws. Special heat-dissipating silicone grease must be used under the module to help conduct away the heat created in the module.

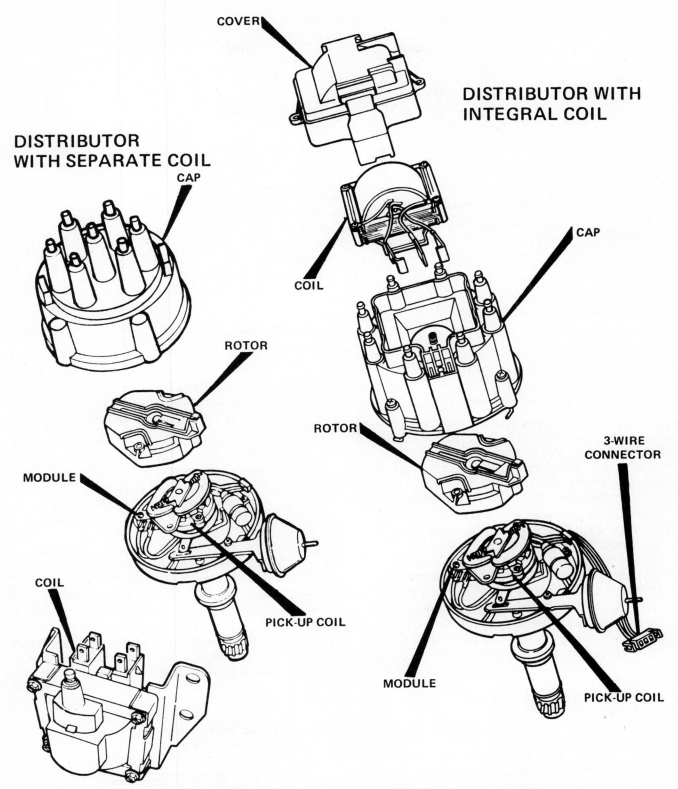

COVER

DISTRIBUTOR WITH
INTEGRAL COIL

DISTRIBUTOR
WITH SEPARATE COIL

CAP

COIL

CAP

ROTOR

ROTOR

3-WIRE
CONNECTOR

MODULE

PICK-UP COIL

COIL

MODULE

PICK-UP COIL

FIGURE 19-4   *General Motors uses ignition systems with remote mounted coils and coils mounted inside the distributor cap. (Courtesy of General Motors Corporation.)*

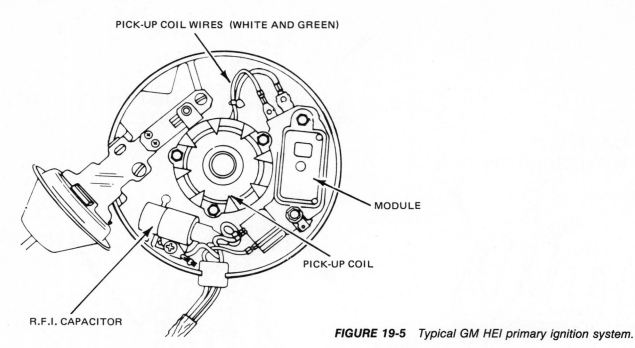

PICK-UP COIL WIRES (WHITE AND GREEN)

MODULE

PICK-UP COIL

R.F.I. CAPACITOR

FIGURE 19-5 *Typical GM HEI primary ignition system.*

MYSTERY SPARKS!

GM HEI electronic ignitions use two different ignition coils wound in opposite directions. An ignition coil wound clockwise, for example, must use a correspondingly wrapped pickup coil. The identifying part of the pickup coil is the color of the cable connector on the white and green pickup coil leads. The ignition coils are identified by a white/red or yellow/red primary wire leads. See Figure 19-6. If the wrong pickup coil (or ignition coil) is used and the polarities do not match, the moving magnetic field from the ignition coil could cause the pickup coil to create a "pulse" at the wrong times. Whenever the pickup creates a pulse, the module receives this pulse as a voltage signal and turns off the primary coil current and the coil discharges. Loose or partially broken pickup coil wires could also result in a "no start," "poor running," or other mystery engine operation.

NOTE: Using the wrong pickup coil could also cause severe and damaging detonation (ping) or failure to start. The current flow through the large battery cable to the starter could cause a strong enough magnetic field to "trigger" the pickup coil and fire the spark plug at any time. Be certain that replacement parts are the correct parts, to prevent mystery sparks.

Many different electronic ignition control modules are used in General Motors vehicles. In addition to a rare three-pin unit, these modules include:

FIGURE 19-6 *The ignition coil and pickup coil must have the same polarity. A yellow wire coil must be used with a yellow connector pickup coil. (Courtesy of Chevrolet Motor Division, GMC.)*

IGN. COIL PICK-UP COIL

RED WIRE WHITE WIRE

CLEAR, BLACK,

IGN. COIL PICK-UP COIL

YELLOW WIRE RED WIRE

YELLOW

1. *Four-pin, either white or black color, modules.* These modules function in the same way and are interchangeable. However, the scope pattern created by these modules differs slightly depending on module color. See Figure 19-7. (See Chapter 24 for scope pattern differences.)

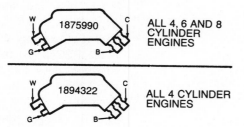

FIGURE 19-7 *(Courtesy of General Motors Corporation.)*

2. *Five-pin module.* This module includes an extra circuit which retards the ignition timing electronically whenever the engine is cold (below 120°F engine temperature). This module is called the *electronic module retard* (EMR) system. The five-pin module is also used on early turbocharged Buick V-6 engines. See Figure 19-8.

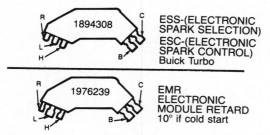

FIGURE 19-8 *(Courtesy of General Motors Corporation.)*

3. *Seven-pin module.* This module is used on computer command control (CCC) vehicles and includes three additional pins for electronic spark timing (EST). See Chapter 25 for computer operation. There are numerous seven-pin (EST) modules. Double check that a replacement module is correct for the vehicle. See Figure 19-9.

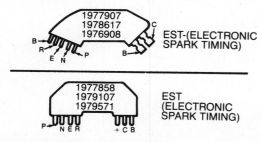

FIGURE 19-9 *(Courtesy of General Motors Corporation.)*

The pickup coil assembly creates an electronic pulse (pulse generator) which is received by the electronic control module. The module then opens the ignition coil primary circuit, which creates the high voltage from the secondary windings of the coil. See Figure 19-10.

After the spark has occurred, the module circuit reconnects the coil primary circuit. The number of degrees of distributor rotation that the primary current flows through the coil (dwell) is variable and is controlled by the module. As engine speed is increased, the magnetic pulse created in the pickup coil assembly occurs earlier. This increase is caused by the increase in induced current in the pickup coil with increasing engine speed. The module advances the ignition timing with increasing engine speed approximately 1 to 11 distributor degrees (2 to 22° on the crankshaft). This spark advance is called *module timing*. The module also increases dwell "on" time as engine speed is increased. The increased dwell "on" time promotes coil saturation at high engine speeds, yet decreases dwell time at lower engine speeds to help protect the ignition coil.

FIGURE 19-10 *(Courtesy of Oldsmobile Division, GMC.)*

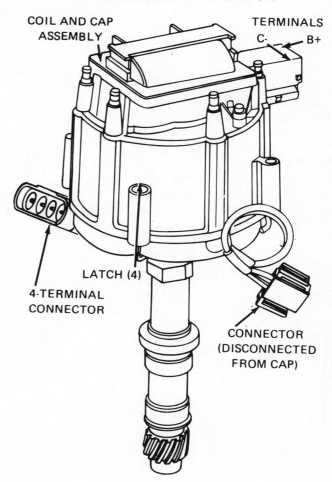

"COIL IN CAP" DISTRIBUTOR

FORD ELECTRONIC IGNITION

Ford electronic ignition systems all function similarly, even though the name for Ford electronic ignition system has changed many times since 1974.

1974–1976	Ford Solid State Ignition System
1977–1978	Duraspark II—larger distributor cap and 8-mm-diameter spark plug wires were major changes from earlier systems. (Duraspark I was used only in 1977–1979 and only for California's V-8s.) See Figure 19-11.

NOTE: The Duraspark II system can be modified to allow either altitude compensation or economy calibration. Modules designed for this ''dual-mode'' modification are equipped with an added three-wire connector.

1978	EEC I (Electronic Engine Control) uses the Duraspark III ignition, which is basically the same as the Duraspark II except some circuits have been eliminated that the EEC system performs instead of the ignition module.
1979	EEC II is basically the same as the EEC I system except for some added components, and the circuits were changed.
1980	EEC III changes were again mostly internal electronic circuits. EEC III used the basic Duraspark III ignition system.
1982–present	EEC IV—The newest generation of computer engine control was first used in 1982.

FIGURE 19-11 *Typical Ford electronic ignition system. Notice the use of resistance wire. Most newer Ford electronic ignition systems do not use a ballast resistor (resistance wire). (Courtesy of Ford Motor Company.)*

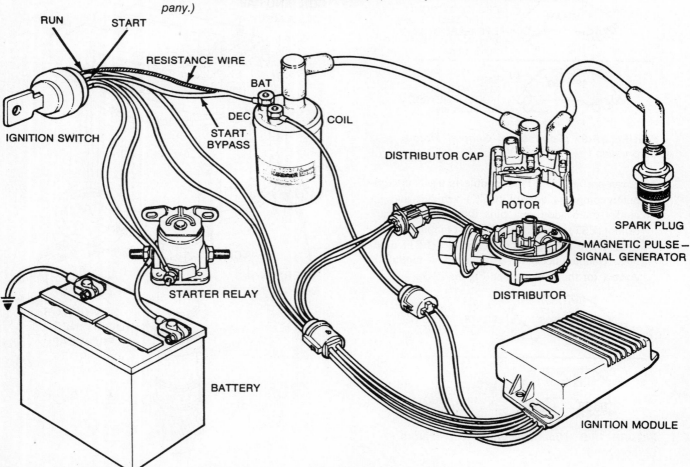

TFI IGNITION

The EEC IV system uses the TFI ignition system. *Thick-film-integrated* (TFI) *ignition systems* were first used on the Escort/Lynx and similar models. This system uses a smaller control module attached to the distributor and uses an air-cooled epoxy E-style coil. "Thick-film ignition" means that all the electronics are manufactured on small built-up layers to form a thick film. See Figure 19-12. Construction includes using pastes of different electrical resistances which are deposited on a thin, flat ceramic material by a process similar to silk-screen printing. These resistors are connected by tracks of palladium silver paste. Then the chips that form the capacitors, diodes, and integrated circuits are soldered directly to the palladium silver tracks. The thick-film manufacturing process is highly automated.

FORD MODULE IDENTIFICATION

Ford electronic ignition control units (modules) are identified by the color of the sealing block (grommet). Replacement modules should be the correct part number *and* have the same color sealing block. The colors available include black, blue, brown, green, red, and white, in addition to one yellow seal or two yellow seals. The module is simply an electronic switching circuit which turns the primary coil circuit "on" and "off." The module can get hot during normal operation and is therefore mounted where normal airflow should keep the temperature from getting too high. Rustproofing materials or other added under-the-hood accessory should be kept away from the ignition module. See Figures 19-13, 19-14, and 19-15.

OPERATION OF FORD ELECTRONIC IGNITION

Ford electronic ignition systems function basically the same way regardless of year and name (Duraspark, EEC, etc.). Under the distributor cap and rotor is a magnetic pickup assembly. This assembly produces a small alternating electrical pulse (approximately 1.5V) whenever the distributor armature rotates past the pickup assembly (stator). See Figures 19-16 and 19-17. This low-voltage pulse is sent to the ignition module. The ignition module then switches

FIGURE 19-12 *TFI system used with a vacuum advance distributor. (Courtesy of Ford Motor Company.)*

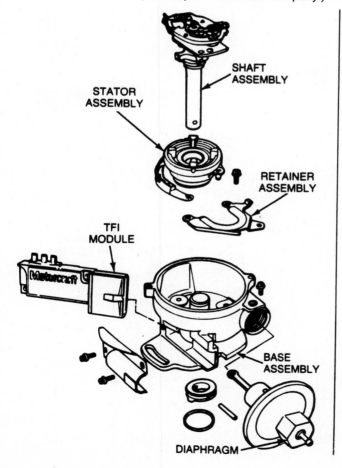

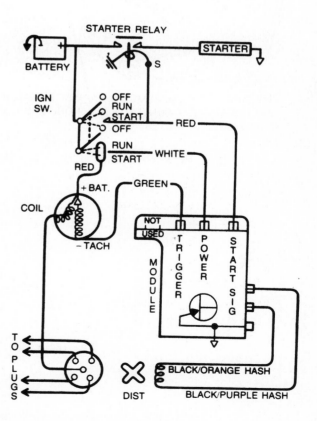

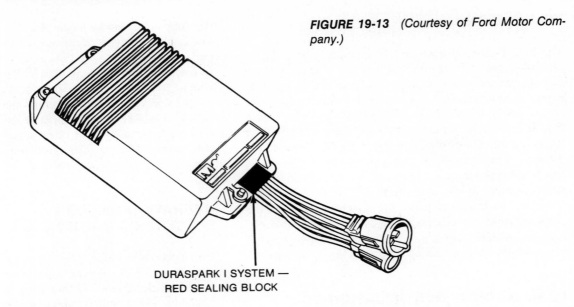

FIGURE 19-13 (Courtesy of Ford Motor Company.)

DURASPARK I SYSTEM —
RED SEALING BLOCK

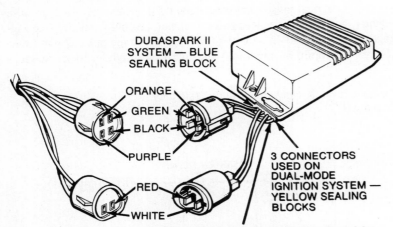

FIGURE 19-14 (Courtesy of Ford Motor Company.)

DURASPARK II
SYSTEM — BLUE
SEALING BLOCK

ORANGE
GREEN
BLACK
PURPLE

RED
WHITE

3 CONNECTORS
USED ON
DUAL-MODE
IGNITION SYSTEM —
YELLOW SEALING
BLOCKS

1979 CRANKING RETARD MODULE — WHITE SEALING BLOCK
AND WHITE 4-PIN CONNECTOR (2.3L NON-TURBOCHARGED
ENGINE ONLY 49 STATES)

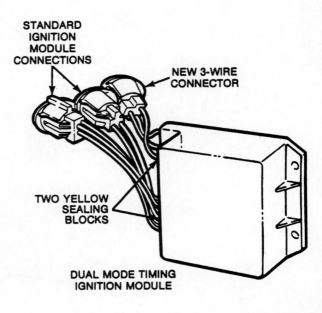

STANDARD
IGNITION
MODULE
CONNECTIONS

NEW 3-WIRE
CONNECTOR

TWO YELLOW
SEALING
BLOCKS

DUAL MODE TIMING
IGNITION MODULE

FIGURE 19-15 (Courtesy of Ford Motor Company.)

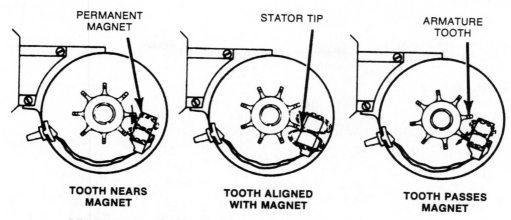

PERMANENT
MAGNET

STATOR TIP

ARMATURE
TOOTH

**TOOTH NEARS
MAGNET**

**TOOTH ALIGNED
WITH MAGNET**

**TOOTH PASSES
MAGNET**

FIGURE 19-16   The rotating armature (reluctor) and the magnet of the stator (pickup) coil combine to create a signal for the module. (Courtesy of Ford Motor Company.)

FIGURE 19-17   The armature and magnetic pickup assembly can be replaced with the distributor installed in the engine. (Courtesy of Ford Motor Company.)

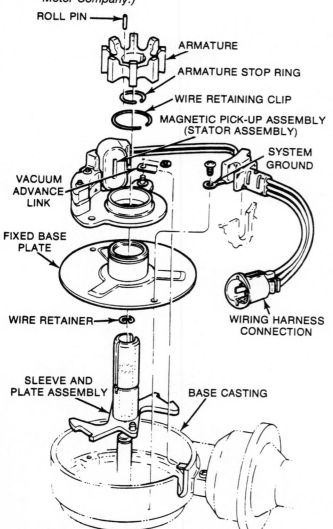

ROLL PIN

ARMATURE

ARMATURE STOP RING

WIRE RETAINING CLIP

MAGNETIC PICK-UP ASSEMBLY
(STATOR ASSEMBLY)

SYSTEM
GROUND

VACUUM
ADVANCE
LINK

FIXED BASE
PLATE

WIRE RETAINER

WIRING HARNESS
CONNECTION

SLEEVE AND
PLATE ASSEMBLY

BASE CASTING

(through transistors) "off" the primary ignition coil current. When the ignition coil primary current is stopped quickly, a high-voltage "spike" discharges from the coil secondary winding. Some Ford electronic ignition systems used a ballast resistor to help control the primary current through the ignition coil in the run mode (position); other Ford systems do not use a ballast resistor. The coil current is controlled in the module circuits by decreasing dwell (coil charging time) depending on various factors determined by operating conditions. See Chapter 23 for diagnostic procedures and troubleshooting hints.

CHRYSLER ELECTRONIC IGNITION

Chrysler was the first domestic manufacturer to produce electronic ignition as standard equipment. The Chrysler system consists of a pulse generator unit in the distributor (pickup coil and reluctor). Chrysler's name for their electronic ignition is EIS (Electronic Ignition System) and the control unit (module) is called the ECU (Electronic Control Unit). See Figure 19-18. The ECU used on 1971–1979 units used a five-pin connector, and 1980 and later models use a four-pin connector. See Figure 19-19.

The pickup coil in the distributor generates (pulse generator) the signal to open and close the primary coil circuit. Some engines used two (dual) pickup coils. One pickup coil is called the *starting pickup* and provides slightly retarded ignition timing to aid in starting. The other pickup is called the *run pickup*. See Figure 19-20.

CHRYSLER BALLAST RESISTOR

The five-pin ECU used a dual-ballast resistor. The dual-ballast resistor uses a 0.5-Ω resistance to limit the voltage in the ignition primary circuit to help protect the ignition

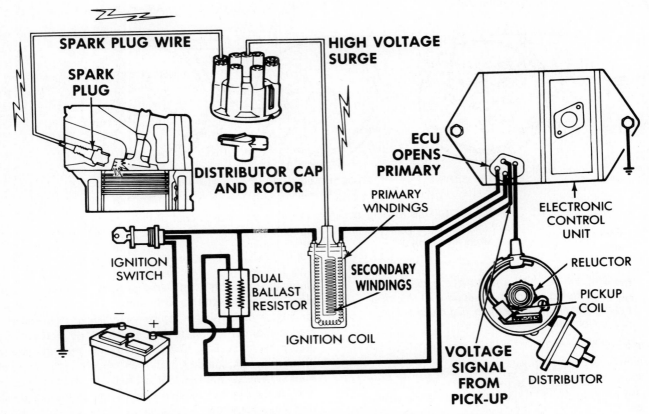

FIGURE 19-18 *Typical Chrysler electronic ignition system (EIS). (Courtesy of Chrysler Corporation.)*

FIGURE 19-19 *A five-pin module was used until 1980, when it was replaced with a four-pin unit. (Courtesy of Chrysler Corporation.)*

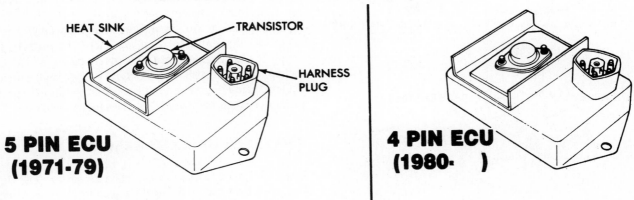

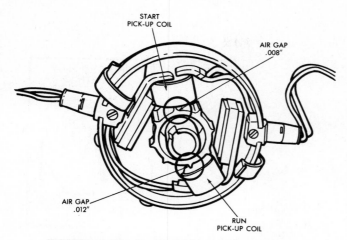

FIGURE 19-20 *Dual pickup coils were used in some Chrysler models. (Courtesy of Chrysler Corporation.)*

coil from overheating at low engine speeds. The *second* resistance of 5 Ω is used to limit the voltage to the ECU only. A single 0.5-Ω resistor is used on 1980 and later four-pin ECU ignitions to control coil primary voltage. Whenever a "no start" or an intermittent spark occurs during cranking, the ballast resistor should be the first item checked on Chrysler cars. See Figure 19-21. See Chapter 23 for complete ignition system troubleshooting procedures.

FIGURE 19-21 *The single resistors used on 1980 and newer four-pin module ignitions are 0.5 Ω.*

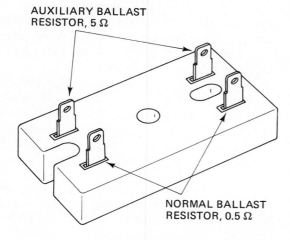

DUAL-BALLAST RESISTOR

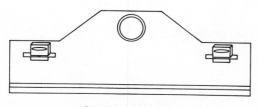

IGNITION RESISTOR

HALL-EFFECT IGNITION SYSTEMS

The *Hall effect* was discovered by Edward H. Hall in 1879. He found that when a thin rectangular gold conductor, carrying a current, is crossed at right angles by a magnetic field, a difference of potential is produced at the edges of the gold conductor. Modern Hall-effect units use semiconductor material (usually silicon) instead of gold. See Figure 19-22.

The Hall effect can be used as a very accurate electronic switch when a moving metallic shutter blocks the magnetic field from striking the semiconductor. Whenever the opening of the shutter allows the magnetic field to strike the sensor, a small voltage is produced and is sent to the electronic control unit. As the distributor rotates, a blocking shutter blocks the magnetic field and the current stops flowing from the sensor. The electronic control unit can be designed to either turn "on" or turn "off" the ignition coil primary current when the shutter blades are blocking. See

FIGURE 19-22 *(Courtesy of Ford Motor Company.)*

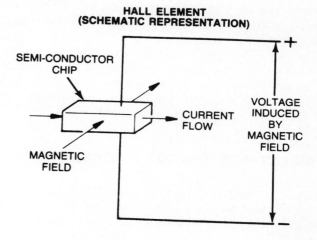

HALL ELEMENT (SCHEMATIC REPRESENTATION)

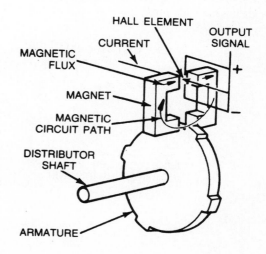

SIGNAL GENERATION

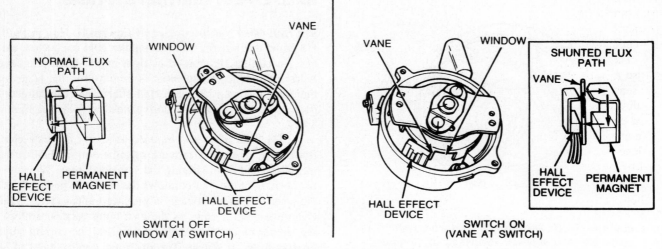

FIGURE 19-23 *Typical Hall-effect ignition sensor. (Courtesy of Ford Motor Company.)*

Figure 19-23. Hall-effect ignition is used by many manufacturers, including Ford, GM, Chrysler, and many imports. The advantage of the Hall effect over the magnetic pulse generator is its accuracy. The Hall effect can easily trigger the ignition to within $\pm\frac{1}{4}°$ of distributor rotation. The Hall-effect ignition unit is commonly used on electronic fuel injected engines to trigger the "on" time accurately for the fuel injector nozzles. This accuracy is important for maximum performance, gas mileage, and lowest emissions on computer-equipped engines.

DIRECT-FIRE IGNITION SYSTEMS

Direct-fire ignition refers to ignition without using a distributor (distributorless). Direct-fire ignition was introduced in the mid-1980s and uses the on-board computer to fire the ignition coils. Direct-fire ignition systems were first used on some Saabs and some General Motors engines. Some four-cylinder engines use four coils, but usually a four-cylinder engine uses two ignition coils and a six-cylinder engine uses three ignition coils. Each coil is a true transformer, where the primary winding and secondary winding are not electrically connected. Each end of the secondary winding is connected to a cylinder exactly opposite the other in the firing order. See Figure 19-24. This means that *both* spark plugs fire at the same time! When one cylinder (for example, number 6) is on the compression stroke, the other cylinder (number 3) is on the exhaust stroke. The voltage required to jump the spark plug gap on cylinder 3 (the exhaust stroke) is only 2 to 3 kV and provides the *ground circuit* for the secondary coil circuit. The remaining coil energy is used by the cylinder on the compression stroke. One spark plug of each pair fires straight polarity and the other cylinder fires reverse polarity. Spark plug life is not greatly affected by the reverse polarity. If there is only one defective spark plug wire or spark plug, two cylinders may be affected.

Direct-fire ignitions require sensors to trigger the coils at the correct time. The direct-fire system on 3.8 liter V-6 Buick-built engines uses a camshaft and a crankshaft Hall-effect sensor for the computer and ignition module to accurately determine the correct spark plug firing time. See Figure 19-25. Neither the camshaft sensor nor the crank-

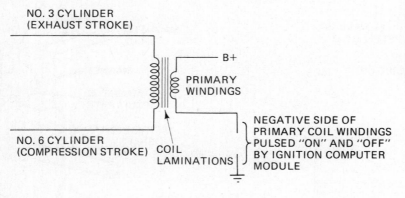

FIGURE 19-24 *Ignition coils used in most direct-fire ignition systems are true transformers. There is no electrical connection between the primary and secondary windings.*

shaft sensor may be moved to adjust ignition timing. Ignition timing is not adjustable. The slight adjustment of the crankshaft sensor is designed to position the sensor exactly in the middle of the rotating metal disk for maximum clearance. The Buick 3.0 liter V-6 and some other engines do not use a cam sensor, but uses double Hall-effect crankshaft sensors. Again, ignition timing is not adjustable. The GM direct-fire system is commonly called *computer-controlled coil ignition* (C3I) or *direct ignition system* (DIS). There are also two types of direct ignition systems found on General Motors vehicles. Type I, which is manufactured by Magnavox, can be identified as having the spark plug wire connections on both sides of the coils. Type II, manufactured by Delco Remy Division of General Motors, features coils that have all of their spark plug wire connections on the same side of the coil assembly. See Figure 19-26.

Some direct-fire ignition systems such as those on some Swedish Saab cars and other newer engines, place the ignition coils directly on the spark plug. This type of system eliminates the problems of high-voltage losses caused by spark plug wires.

TROUBLESHOOTING DIRECT-FIRE IGNITION SYSTEMS

If ignition-related problems are suspected on a direct-fire ignition system, check for spark at several spark plugs while cranking the engine.

If No Spark at Any Plug

1. Check the ignition fuse.
2. Check the ignition feed wire(s) to the ignition module and the ignition coils.
3. Check the resistance between the secondary terminals of each coil (generally less than 15,000 Ω).

If Spark at Some Plugs

1. Check the spark plug for proper resistance (should be less than 30,000 Ω for any one wire).
2. Check the spark plugs for excessive cracks, fouling, or other faults.
3. Check the resistance between the secondary terminals of each coil (generally less than 15,000 Ω).

FIGURE 19-25 *Typical direct-fire ignition system found on the 3.8-liter Buick V-6. The unit shown is manufactured by Magnavox. (Courtesy of General Motors Corporation.)*

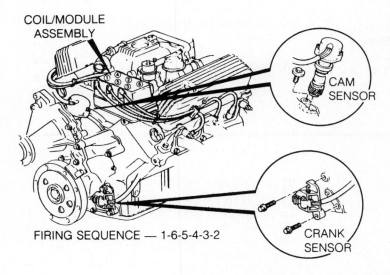

COIL/MODULE ASSEMBLY

CAM SENSOR

CRANK SENSOR

FIRING SEQUENCE — 1-6-5-4-3-2

FIGURE 19-26 *(Courtesy of General Motors Corporation.)*

IGNITION SYSTEM IDENTIFICATION

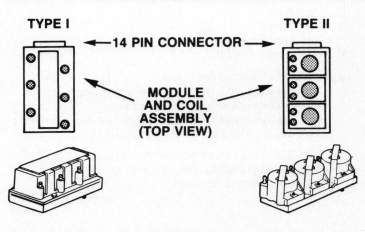

TYPE I

TYPE II

← 14 PIN CONNECTOR →

MODULE AND COIL ASSEMBLY (TOP VIEW)

SUMMARY

1. Electronic ignition systems use electronic components that do not wear, to trigger the negative (ground return path) side of the ignition coil to produce a high-voltage spark. Electronic ignitions use either a pulse generator (pickup coil) or Hall-effect switch to trigger the electronic ignition control unit (module). The control unit (module) electronically turns the primary current of the coil ''on'' and ''off.'' Some electronic control units use a ballast resistor, while others use a varying dwell time to protect the ignition coil from overheating.

2. Direct-fire ignition systems use the on-board computer and sensors on the camshaft and/or crankshaft to trigger coils. A four-cylinder engine usually uses two coils and a six-cylinder engine uses three coils. Whenever each coil fires, a high-voltage spark is sent to two cylinders opposite each other in the firing order. The spark occurring on the cylinder on the exhaust stroke provides the ground for the coil secondary winding, and the other spark plug ignites the air/fuel mixture on the cylinder that is on the power stroke. The ignition timing is controlled by the computer and is not adjustable.

STUDY QUESTIONS

19-1. What is a pulse generator?

19-2. What is the purpose of the electronic control module in the operation of an electronic ignition system?

19-3. Why do only some electronic ignition systems require the use of a ballast resistor?

19-4. Describe the operation of a Hall-effect-triggered electronic ignition system.

19-5. Explain how an ignition system can function without a distributor.

MULTIPLE-CHOICE QUESTIONS

19-1. The pulse generator:
(a) fires the spark plugs directly.
(b) signals the electronic control unit (module).
(c) replaces the electronic control unit (module).
(d) fires the distributor rotor.

19-2. Technician A says that dwell cannot be adjusted on cars equipped with pointless electronic ignition. Technician B says that dwell cannot be determined at all on pointless electronic ignition. Which technician is correct?
(a) A only.
(b) B only.
(c) both A and B.
(d) neither A nor B.

19-3. Most direct-fire ignition systems:
(a) use one coil to fire all cylinders at the same time.
(b) fire two cylinders at the same time.
(c) fire four cylinders at the same time.
(d) use two coils for a six-cylinder engine.

19-4. The Ford thick-film-integration (TFI) module:
(a) is mounted on the inner fender.
(b) is mounted on the distributor.

(c) does not use a pickup coil.
(d) both (b) and (c).

19-5. Technician A says that a defective pickup coil can cause a ''no start'' situation. Technician B says that a defective pickup coil will cause the module to fail. Which technician is correct?
(a) A only.
(b) B only.
(c) both A and B.
(d) neither A nor B.

19-6. Ballast resistors are:
(a) used on some GM HEI units and all Chrysler electronic ignitions.
(b) used on most older-style Ford and Chrysler electronic ignition systems.
(c) not used on electronic ignition systems.
(d) only used on direct-fire ignition systems.

19-7. Chrysler electronic ignition systems (EIS) use:
(a) either a three- or four-pin module.
(b) either a four- or five-pin module.
(c) either a five- or six-pin module.
(d) either a six- or seven-pin module.

19-8. GM HEI systems:

(a) all use integral coil (ignition coil in distributor).

(b) do not use an ignition coil.

(c) all use a four-pin module.

(d) none of the above.

19-9. Electronic ignition systems:

(a) use special 80,000-V ignition coils.

(b) "fire" the spark plug by charging, then discharging the ignition coil.

(c) have lower spark energy than that of point-type systems.

(d) require unleaded fuel to burn the spark plugs correctly.

19-10. Hall-effect ignition systems:

(a) use a Hall-effect switch instead of a pickup coil to trigger the module (electronic control unit).

(b) use a blade shutter to open and close a special set of contact points.

(c) must have the shutter blades grounded for proper engine operation.

(d) both (a) and (c).

20

Ignition Timing
and
Timing Advance

IGNITION TIMING MUST BE CORRECT FOR AN ENGINE TO DELIVER ITS MAX-
imum power and economy with the lowest possible exhaust emissions.
Normal engine wear (timing chain) and distributor point wear (point-type
ignition) require that the ignition timing be checked and adjusted (if necessary) at least every year. The topics covered in this chapter include:

1. Ignition timing background
2. Characteristics of improper timing
3. Ignition timing procedure
4. Need for timing advance
5. Mechanical advance
6. Vacuum advance
7. Distributor testing:
    (a) On the car
    (b) Using a distributor machine
8. Magnetic timing
9. Average timing methods

IGNITION TIMING

Ignition timing is the *exact time* a spark is sent to the spark plug relative to piston position on the compression stroke to ignite the gas and air mixture already in the cylinder. See Figures 20-1 and 20-2. This spark occurs the instant the ignition primary circuit is opened (either by ignition points or electronic ignition transistor). Most cars use the number 1 cylinder as the reference cylinder. If the timing of the number 1 cylinder is correct, the timing of all other cylinders will be correct because of engine design.

When the spark occurs exactly at *top dead center* (TDC), it is said to be zero degrees advanced. Most spark ignition occurs slightly *before top dead center* (BTDC or BTC) because it takes some time (even if it is measured in

FIGURE 20-1 *Piston coming up on the compression stroke. (Courtesy of General Motors Corporation.)*

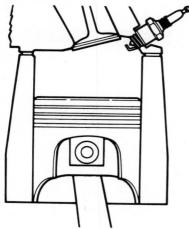

FIGURE 20-2 *When the piston approaches top dead center (TDC), the spark occurs and the expanding burning gases push the piston downward. (Courtesy of General Motors Corporation.)*

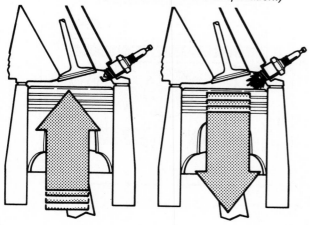

thousandths of a second) for the fuel to burn. If the fuel is ignited slightly before the piston reaches the top, the fuel will be burning as the piston is starting down. This slight ''spark advance'' results in proper performance and maximum gas mileage because the fuel mixture is burning and exerting maximum pressure on the piston as it starts downward. If the spark occurs after the piston has reached the top, it is referred to as firing *after top dead center* (ATDC or ATC).

RETARDED TIMING

Retarded timing means that a spark occurs *after* it should. For example, retarded timing means that the timing is set at 4° BTDC instead of 10° BTDC.

Characteristics of excessively retarded timing
1. Long cranking before starting
2. Poor or reduced performance and gas mileage
3. Possible slow, rough idle

ADVANCED TIMING

Advanced timing means that a spark occurs *before* it should. For example, advanced timing is timing that is at 10° BTDC instead of 4° BTDC.

Characteristics of excessively advanced timing
1. Pings on acceleration (see below)
2. Slow, ''jerky'' cranking when engine is warm

PING

Ping (also known as *spark knock* or *detonation*) is caused by a second explosion occurring inside the combustion chamber during engine operation. This second explosion is ignited by the heat of compression in the cylinder. See Figure 20-3.

Ping is caused by one or more of several reasons, including:

1. Low-octane-rated fuel
2. Advanced ignition timing
3. Excessive engine temperature
4. Excessive compression for the type of fuel

If engine ping (spark knock) is permitted to continue regularly or if the engine is driven hard for a short time, serious engine damage can result. See Figures 20-4 and 20-5.

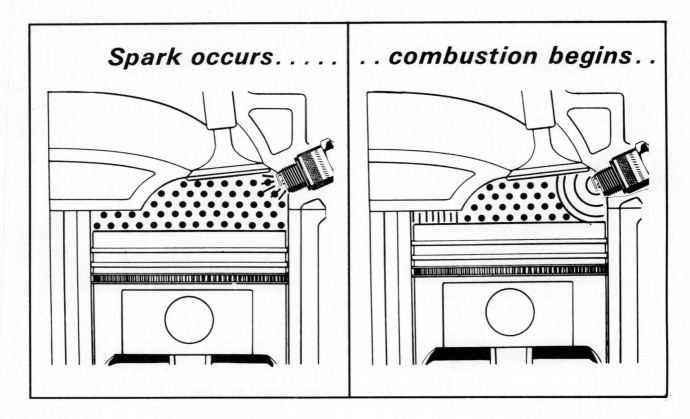

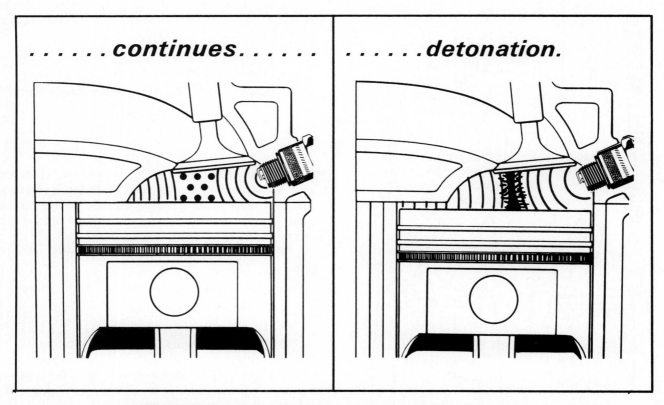

FIGURE 20-3 *How detonation occurs inside an engine. (Courtesy of Champion Spark Plug Company.)*

FIGURE 20-4 *Piston damage caused by detonation. (Courtesy of Champion Spark Plug Company.)*

FIGURE 20-5 *Close-up view of piston damage caused by detonation. (Courtesy of Champion Spark Plug Company.)*

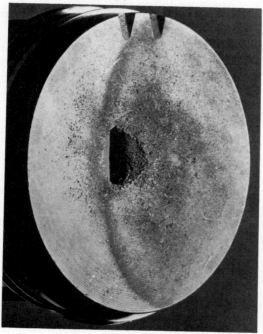

Serious engine damage can result because when the two flame fronts stroke each other, the cylinder pressure and temperature are greatly increased. The sound of spark knock can be a mild sound like marbles in a coffee can or

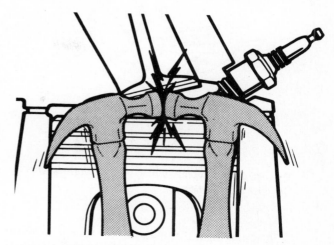

FIGURE 20-6 *Spark knock (detonation) sounds as if two hammers were striking together. (Courtesy of General Motors Corporation.)*

a severe sound like two hammers striking together. See Figure 20-6.

Technicians often report that their customers tell them to adjust their engine's valves because they are noisy during acceleration. This valve noise is usually spark knock (ping). Ignition timing is the primary item that should be checked while attempting to correct a ping problem. Other nonelectrical causes of ping include:

1. EGR valve or control system not functioning correctly
2. Lean fuel mixture
3. Excessive engine temperature caused by an engine or cooling system problem

NOTE: Preignition is caused by a hot object in the combustion chamber igniting the air/fuel mixture before the ignition occurs. The factors noted above have limited influence in preventing preignition.

TIMING LIGHTS

Timing lights are often called *stroboscopic lights* because they permit the viewing of the moving timing marks by lighting the marks every time the number 1 cylinder fires. The bright light of the stroboscope is usually provided by an electrical current through a xenon-gas-filled tube. Three basic types of stroboscopes include:

1. A timing light that is powered just by the current flowing through the spark plug wire. The flashing light is very dim light, and this type of timing light is not recommended.

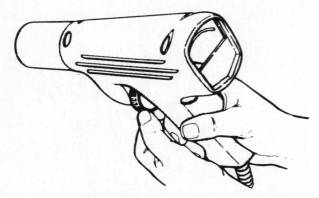

FIGURE 20-7 *Typical timing light. (Courtesy of Chrysler Corporation.)*

2. A timing light that uses the car battery to power the light. This portable timing light is the most commonly used type.

3. A timing light that is powered by a 110-V outlet. Such 110-V-powered timing lights are usually part of electronic test analyzers and are often the brightest. See Figure 20-7.

IGNITION TIMING PREPARATION

Since accurate ignition timing depends on many factors, it is important that the manufacturer's specifications be checked before attempting to set the timing on any engine. If specifications are not adhered to, poor engine performance, poor fuel economy, overheating, or damage to the engine could result. Since 1968 the correct ignition timing has been listed on the under-the-hood emission decal of all cars and light trucks.

PRETIMING CHECKS

Before checking or adjusting the ignition timing, the following items should be done to ensure accurate timing results:

1. The dwell must be correct (point-type ignition systems only).

2. The engine should be at normal operating temperature (upper radiator hose hot and pressurized).

3. The engine should be at the correct timing rpm (check specifications).

4. The vacuum hoses should be removed and the hose plugged from the vacuum advance unit (if equipped) on the distributor unless otherwise specified.

5. If the engine is computer equipped, check the *exact* timing procedure specified by the manufacturer. This may include disconnecting a "set

FIGURE 20-8 *Typical timing light hookup. (Courtesy of ALLTEST INC.)*

timing'' connector wire, grounding a diagnostic terminal, disconnecting a four-wire connector, or similar ignition timing procedures.

TIMING LIGHT CONNECTIONS

1. Connect the timing light battery leads to the vehicle battery: the red to the positive (+) terminal and the black to the negative (−) terminal.

2. Connect the timing light high-tension lead to the number 1 spark plug cable. See Figure 20-8.

CYLINDER NUMBER 1

Four- or six-cylinder engines. On most in-line four- and six-cylinder engines, the number 1 cylinder is the *most forward* cylinder.

V-6 or V-8 engines. Most V-type engines use the *left-front* cylinder (driver's side) as the number 1 cylinder, *except* Ford engines and some Cadillacs, which use the right-front (passenger's side) cylinder.

Sideways (transverse) engines. Most front-wheel-drive cars with engines installed sideways use the cylinder to the far right (passenger's side) as the number 1 cylinder (plug wire closest to the fan belts).

Rule of thumb. If the number 1 cylinder or the type of engine is unknown, the number 1 cylinder is the cylinder *most forward* as viewed from above (except Pontiac V-8 engines). See Figure 20-9.

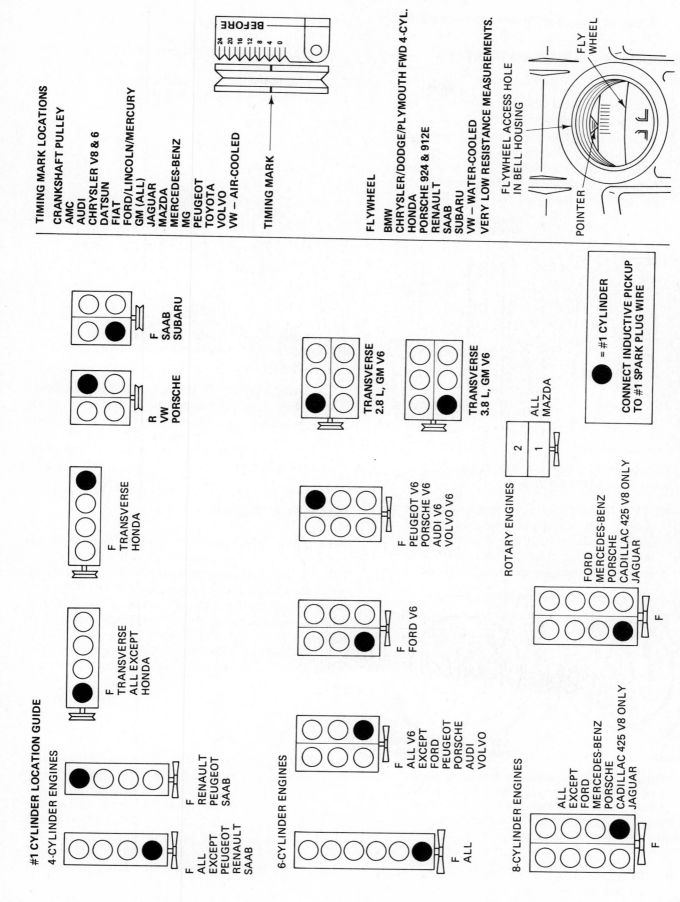

FIGURE 20-9 *(Courtesy of ALLTEST INC.)*

#1 CYLINDER LOCATION GUIDE

4-CYLINDER ENGINES

F
ALL
EXCEPT
PEUGEOT
RENAULT
SAAB

F
RENAULT
PEUGEOT
SAAB

F
TRANSVERSE
ALL EXCEPT
HONDA

F
TRANSVERSE
HONDA

R
VW
PORSCHE

F
SAAB
SUBARU

6-CYLINDER ENGINES

F
ALL

F
ALL V6
EXCEPT
FORD
PEUGEOT
PORSCHE
AUDI
VOLVO

F
FORD V6

F
PEUGEOT V6
PORSCHE V6
AUDI V6
VOLVO V6

TRANSVERSE
2.8 L, GM V6

TRANSVERSE
3.8 L, GM V6

8-CYLINDER ENGINES

F
ALL
EXCEPT
FORD
MERCEDES-BENZ
PORSCHE
CADILLAC 425 V8 ONLY
JAGUAR

F
FORD
MERCEDES-BENZ
PORSCHE
CADILLAC 425 V8 ONLY
JAGUAR

ROTARY ENGINES

ALL
MAZDA

● = #1 CYLINDER

CONNECT INDUCTIVE PICKUP
TO #1 SPARK PLUG WIRE

TIMING MARK LOCATIONS

CRANKSHAFT PULLEY
AMC
AUDI
CHRYSLER V8 & 6
DATSUN
FIAT
FORD/LINCOLN/MERCURY
GM (ALL)
JAGUAR
MAZDA
MERCEDES-BENZ
MG
PEUGEOT
TOYOTA
VOLVO
VW – AIR-COOLED

TIMING MARK

BEFORE

24 20 16 12 8 4 0

FLYWHEEL
BMW
CHRYSLER/DODGE/PLYMOUTH FWD 4-CYL.
HONDA
PORSCHE 924 & 912E
RENAULT
SAAB
SUBARU
VW – WATER-COOLED
VERY LOW RESISTANCE MEASUREMENTS.

FLYWHEEL ACCESS HOLE
IN BELL HOUSING

FLY
WHEEL

POINTER

CHECKING OR ADJUSTING IGNITION TIMING

1. Start the engine and adjust the speed to that specified for ignition timing.

2. With the timing light aimed at the stationary timing pointer, observe the position of the timing mark with the light flashing. Refer to the manufacturer's specifications for the correct setting on the under-the-hood decal. See Figure 20-10.

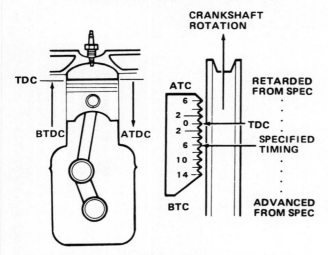

FIGURE 20-10 *Typical timing marks. (Courtesy of Ford Motor Company.)*

NOTE: If the timing mark appears ahead of the pointer, in the direction of crankshaft rotation, the timing is advanced. If the timing mark appears behind the pointer, in the direction of crankshaft rotation, the timing is retarded. See Figure 20-11.

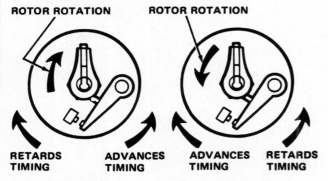

FIGURE 20-11 *To advance timing on point-type (shown) or electronic ignition, turn the distributor housing opposite the rotor rotation. To retard timing, turn it the same direction as the rotor rotation. (Courtesy of Ford Motor Company.)*

3. To adjust, loosen the distributor locking bolt or nut and turn the distributor housing until the tim-

ing mark is in correct alignment. Turn the distributor housing in the direction of rotor rotation to retard the timing and against rotor rotation to advance the timing. See Figure 20-12.

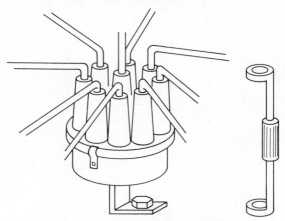

FIGURE 20-12 *Typical distributor wrench designed to reach hold-down bolts or nuts on most distributors.*

4. After adjusting the timing to specifications, carefully tighten the distributor locking screw. It is sometimes necessary to readjust the timing after the initial setting because the distributor may have rotated slightly when the hold-down bolt was tightened.

MAGNETIC TIMING

Since the late 1970s, many manufacturers have provided a receptacle for use with a magnetic ignition timing meter. Magnetic timing involves a magnetic pickup sensor that is inserted in a holder that is located close to TDC. The number of degrees that the sensor is offset from true TDC (0°) is called the *magnetic offset angle*. See Figures 20-13 and 20-14.

The magnetic offset angle varies according to the manufacturer. Listed below are typical magnetic offset angles.

GM	−9.5°
AMC	−10°
Chrysler	−10°
Ford	Magnetic timing is not recommended by Ford (offset angles vary according to engine plant, engine, and year) (some typical Ford engine offset angles include 26°, 52°, 68°, 135°, 314°, and 334°). Magnetic timing on Ford engines is designed to be used only on the assembly line during engine production.

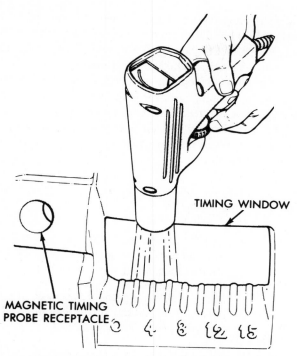

FIGURE 20-13 *(Courtesy of Chrysler Corporation.)*

FIGURE 20-14 *Magnetic timing receptacle offset from zero degrees. A magnetic timing pickup probe "senses" the notch in the crankshaft pulley.*

Mercedes	−15°
BMW	−20°
Renault	−20°
VW/Audi	−20°
Porsche	−20°
Volvo	−20°

Magnetic timing units (meters) are available through most auto-supply stores. When using the magnetic sensor, the harmonic balancer must be *clean* for best results. Mag-

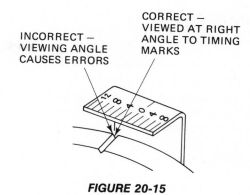

FIGURE 20-15

netic timing is an excellent tool. It permits accurate ignition timing without the possible visual error (called parallax) that often results from using a timing light. See Figure 20-15.

SET TIMING USING A VACUUM GAUGE?

Even though not as accurate as a timing light, *approximate* ignition timing can be set with a vacuum gauge in the event of a missing timing mark or slipped crankshaft pulley.

NOTE: This method is to be used only in an emergency and may not be accurate enough for proper exhaust emission levels.

To set the timing with a vacuum gauge, connect a vacuum gauge to a manifold vacuum source.

HINT: A manifold vacuum source has vacuum at idle. The most common source of manifold vacuum that is accessible is the connection to the vacuum-controlled air cleaner assembly.

Adjust the ignition timing by rotating the distributor until the highest vacuum reading is obtained and then retard the timing 2 in. Hg. This procedure is most accurate with the vacuum line(s) left connected to the distributor and the engine at normal operating temperature (upper radiator hose hot and pressurized).

For example, if by rotating the distributor, the highest possible engine vacuum obtained on the gauge is 23.5 in. Hg, retard the distributor until the vacuum gauge reads 21.5 in. Hg (23.5 − 2 = 21.5). Test drive the vehicle to be assured that engine-damaging ping or spark knock is not occurring.

AVERAGE TIMING

Some engines must be timed using an average timing method. The average timing method is used on 2.5-liter, 2.0-liter, and some 1.8-liter four-cylinder General Motors engines equipped with electronic fuel injection. Some engines require the timing light to be connected to the *coil wire*. The crankshaft pulley has two notches 180° apart. By using the coil wire as the signal to light the timing light,

a slight jiggling of the timing notches may be seen. The timing should be adjusted to the center of the notches' width as viewed with a timing light. Magnetic timing units *cannot* be used for this method. See Figures 20-16 and 20-17.

The other method of averaging the ignition timing is to use the spark plug wire for both cylinders 1 and 4 and "average" the two readings. Since exact timing procedures can vary according to engine, year, and type of fuel system, *always* consult the engine emission decal for the correct timing procedures.

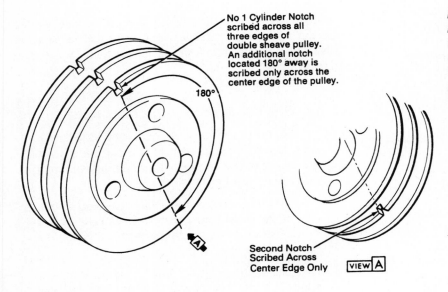

No 1 Cylinder Notch scribed across all three edges of double sheave pulley. An additional notch located 180° away is scribed only across the center edge of the pulley.

180°

Second Notch Scribed Across Center Edge Only   VIEW A

FIGURE 20-16 *Some GM four-cylinder engines must be timed using the averaging timing method. Notice that in this example, there are two timing notches on the crankshaft pulley, 180° apart. (Courtesy of Chevrolet Motor Division, GMC.)*

FIGURE 20-17 *Some GM four-cylinder engines must be timed with the timing light attached to the coil wire, while other engines must average the timing from two different cylinders. (Courtesy of Chevrolet Motor Division, GMC.)*

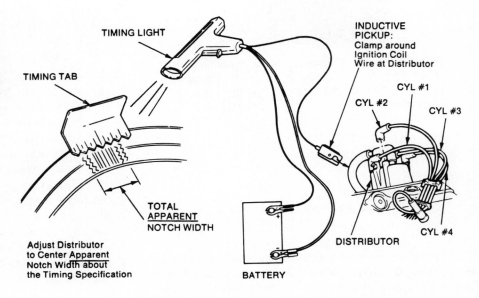

TIMING LIGHT

TIMING TAB

INDUCTIVE PICKUP: Clamp around Ignition Coil Wire at Distributor

CYL #2   CYL #1   CYL #3

TOTAL APPARENT NOTCH WIDTH

Adjust Distributor to Center Apparent Notch Width about the Timing Specification

BATTERY

DISTRIBUTOR

CYL #4

IGNITION TIMING AND DIESELING

Since the early 1970s, many engines continue to run after the ignition switch is turned "off." This is called "dieseling" or "run on." Many technicians blame the ignition timing and they retard the timing to prevent the engine from dieseling. While ignition timing may be a determining factor, too high an idle speed is the greatest cause of dieseling. Whenever the timing is changed, on most engines, the idle speed is also affected. More advanced ignition timing causes the engine to idle faster and retarded timing results in lower idle speed. Therefore, if the ignition timing is advanced and the idle speed is not lowered to its original setting, dieseling often results. When a technician retards ignition timing, the idle speed is lowered and the dieseling may be corrected. However, an engine running with retarded timing often does not perform as well nor give as good a fuel economy as the same engine with correctly set ignition timing. Always check and adjust the idle speed of any engine before attempting to "cure" a dieseling problem by retarding the ignition timing. After advancing the ignition timing, a technician should check and adjust the idle speed to the correct specification to prevent possible dieseling problems.

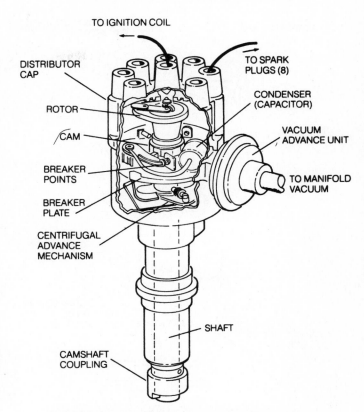

FIGURE 20-18 *The direction of rotation of the distributor shown is counterclockwise as viewed from the top. (Courtesy of Texas Instruments Incorporated, copyright 1982.)*

DISTRIBUTOR ROTATION DIRECTION

Distributors can rotate either clockwise (CW) or counter clockwise (CCW) as viewed from the top of the distributor. To quickly determine the direction of rotation, follow either of the following two methods for non-computer-controlled distributors:

1. Align your hand with the vacuum advance and curl your fingers around the distributor. The direction your fingers are pointing is the direction of rotation. See Figure 20-18.
2. With the distributor cap removed, attempt to rotate the rotor. The direction that the rotor moves is the direction of rotation.

NEED FOR TIMING ADVANCE

For a given fuel mixture and engine design, the time required for the mixture to burn remains constant [approximately 2–3 milliseconds (0.002 to 0.003 second)]. As engine speed increases, the spark must occur earlier to assure complete burning of the fuel mixture. Depending on engine design, the fuel-mixture burning should be completed between 10 to 23° ATDC (after top dead center) of the piston on the power stroke. This constant burning speed requires that the spark occur earlier with increasing engine speed (rpm). See Figure 20-19.

CENTRIFUGAL (MECHANICAL) ADVANCE

Most distributors in engines that are *not* computer controlled are equipped with a mechanical mechanism which advances the spark timing (up to a point) with increasing engine speed. These mechanical advance units use centrifugal force to advance the ignition timing by rotating the primary circuit control (breaker cam or reluctor). Centrifugal force is the outward force exerted on a weight from a center of rotation during rotation. The faster the rotation, the greater the force on the weight(s) to move outward. See Figures 20-20 and 20-21.

Centrifugal (mechanical) advance greatly affects engine *performance*. If the mechanical advance mechanism

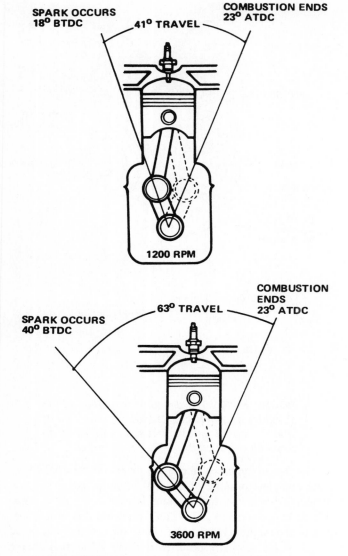

SPARK OCCURS 18° BTDC — 41° TRAVEL — COMBUSTION ENDS 23° ATDC

1200 RPM

SPARK OCCURS 40° BTDC — 63° TRAVEL — COMBUSTION ENDS 23° ATDC

3600 RPM

FIGURE 20-19   *The ignition timing must be advanced as the engine speed increases (up to a point) for maximum power and economy. (Courtesy of Ford Motor Company.)*

is rusted or prevented from working due to varnish or sludge buildup on the distributor shaft, lack of engine performance is the most noticeable symptom. If the retaining springs are too light to adequately control the centrifugal advance weights, severe engine "ping" (detonation) can result by allowing excessive spark advance. A correctly operating centrifugal advance mechanism should produce a steady spark advance with increasing engine speed. Most factory-installed advance mechanisms provide between 7 and 15 *distributor* degrees (14 and 30 crankshaft degrees) and do not advance beyond about 4500 engine rpm.

NOTE: Most distributor specifications are given in distributor degrees and distributor rpm. The dis-

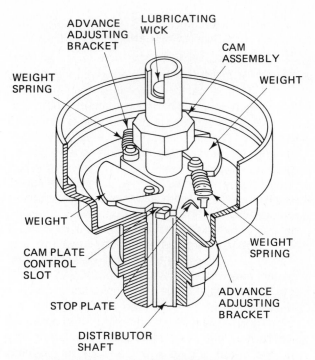

ADVANCE ADJUSTING BRACKET · LUBRICATING WICK · CAM ASSEMBLY · WEIGHT SPRING · WEIGHT · WEIGHT · CAM PLATE CONTROL SLOT · STOP PLATE · DISTRIBUTOR SHAFT · WEIGHT SPRING · ADVANCE ADJUSTING BRACKET

FIGURE 20-20   *Typical centrifugal (mechanical) advance mechanism used on most non-computer-equipped engines. The advance mechanism is the same for both point-type and electronic ignition systems. (Courtesy of Sun Electric Corporation.)*

FIGURE 20-21   *As the engine speed increases, centrifugal force rotates the breaker cam (point-type) or reluctor (electronic ignition). This movement advances the timing by breaking the primary ignition sooner in relation to the piston position. (Courtesy of Chrysler Corporation.)*

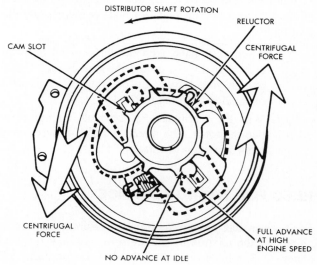

DISTRIBUTOR SHAFT ROTATION · RELUCTOR · CENTRIFUGAL FORCE · CAM SLOT · CENTRIFUGAL FORCE · NO ADVANCE AT IDLE · FULL ADVANCE AT HIGH ENGINE SPEED

tributor is turned by a gear on the engine camshaft. The camshaft rotates at one-half the speed of the engine crankshaft. Therefore, to convert distributor degrees or rpm into crankshaft degrees and crankshaft rpm, multiply by 2. For example, 12 distributor degrees at 1000 distributor rpm equals 24 engine crankshaft degrees at 2000 engine crankshaft rpm.

Distributors for use in high-performance engines can be modified to provide maximum centrifugal advance by 3000 engine rpm (1500 distributor rpm) for automatic-transmission-equipped vehicles. Manual-transmission-equipped vehicles can have the distributor modified to provide total centrifugal advance by 2500 engine rpm.

> **CAUTION:** Federal, state, province, or local laws may prohibit modifying engine spark advance mechanisms for vehicles operating on public streets and highways.

The proper amount of centrifugal advance varies among engine manufacturers and engine design. Typical centrifugal advance, when combined with base (initial) timing, generally equals 34 to 38 crankshaft degrees.

For example, if the initial (base) timing is 8° BTDC, the proper amount of centrifugal advance should be 26 to 30 crankshaft degrees or 14 to 15 distributor degrees.

8 (base) + 26 (maximum centrifugal) = 34° (total)

8 (base) + 30 (maximum centrifugal) = 38° (total)

This total advance of 34 to 38° is a general rule of thumb that applies to almost all engines. Be certain to check the manufacturer's specifications for the *exact* specifications for the engine being tested. Most factory specifications, however, will adhere closely to the 34 to 38° total for combined initial plus maximum centrifugal advance.

MANIFOLD VERSUS PORTED VACUUM

Vacuum-advance distributor units are used on most non-computer-controlled and on some computer-controlled ignition distributors. The purpose of vacuum advance units is to provide additional spark advance during light engine load conditions for maximum fuel economy. The vacuum advance units may be connected to either *ported* or *manifold* vacuum sources, depending on engine design. Vacuum is measured in inches or millimeters of mercury, abbreviated in. Hg (or mm Hg). Hg is the chemical symbol for the liquid metal, mercury. Manifold vacuum is vacuum (low pressure) that is available in the intake manifold or a port on the carburetor which is connected to an area *below* the throttle plates. The manifold vacuum is highest under

light-engine-load conditions such as during idle and deceleration. Manifold vacuum is lowest under heavy engine load. Ported vacuum is actually manifold vacuum that is controlled by the throttle plates of a carburetor or single-point fuel injection unit. Ported vacuum is also load sensitive, but ported vacuum is not available when the throttle is closed. Therefore, ported vacuum is zero at idle and deceleration, and as the throttle is opened, gradually increases to the same value as manifold vacuum. Ported vacuum and manifold vacuum are both almost zero (0 to 2 in. Hg) at wide-open throttle (WOT). Vacuum is affected by engine *load* only, not by engine speed.

VACUUM ADVANCE OPERATION

The distributor vacuum advance unit contains a rubber diaphragm, which in turn moves the breaker plate in the distributor depending on the vacuum applied to the vacuum advance unit. A spring inside the vacuum advance unit keeps the advance unit off (no advance) if no vacuum is applied. See Figure 20-22. When engine vacuum reaches a value high enough to overcome the internal spring pressure, the advance unit "pulls" on and rotates the breaker plate, which advances the ignition timing between 14 and 30 crankshaft degrees or 7 to 15 distributor degrees. This additional spark advance greatly improves fuel economy (4 or more mpg) under light-engine-load conditions. See Figure 20-23.

If the vacuum advance unit is connected to a manifold vacuum source, the ignition timing will be advanced

FIGURE 20-22 *Vacuum advance units rotate the pickup coil on the breaker plate in relation to the reluctor. This rotation causes the spark to occur earlier in relation to the piston position. (Courtesy of Chrysler Corporation.)*

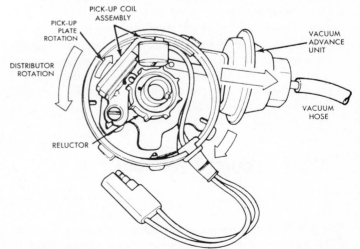

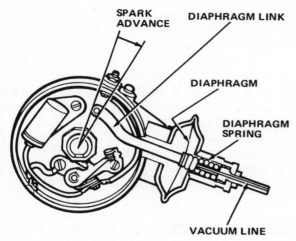

FIGURE 20-23 *The amount of vacuum, the strength of the internal spring, and the amount of vacuum advance link travel determine the rate and the amount of vacuum advance.*

(maximum vacuum advance) at idle, cruise, and deceleration. If ported vacuum is used, the amount of spark advance will be the same as manifold vacuum advance at cruise (throttle opened past the port), but does not provide any advance at idle or deceleration (throttle closed) because there is no vacuum when the throttle plate is closed.

Manifold vacuum most accurately reflects the load on the engine. The maximum possible advance is present at idle if manifold vacuum is used. However, this vacuum advance (up to 30° as indicated on the crankshaft) does not remain when the load on the engine increases. As the car is driven (engine load increases), the manifold vacuum drops and the vacuum advance decreases. Also, as the load on any engine increases, the combustion pressures inside the engine increase and the engine *requires* less spark advance, to avoid possible engine damage due to the increased pressures that fully advanced ignition timing could create.

Many engines are designed to operate correctly using only ported vacuum to the vacuum advance unit. Ported vacuum is zero at idle and gradually increases as the throttle is opened. Ported vacuum has the same value as manifold vacuum once the throttle plate has exposed the "port" or opening to the low pressure inside the intake manifold. Therefore, at heavy loads, ported vacuum is also almost zero, the same as manifold vacuum.

TOTAL ADVANCE OPERATION

A properly operating engine requires more spark advance for maximum fuel economy, but less spark advance during high-load conditions when combustion pressures are high. To prevent serious engine damage, it is important not to exceed 38° of total advance (base timing plus *maximum* centrifugal). Since additional spark advance is possible under light-load conditions (high vacuum), an additional 20° advance on the crankshaft is possible.

total advance (initial + maximum centrifugal)
= 34 to 38°

total advance (initial + maximum centrifugal
+ maximum vacuum) = 55 to 60°

Analysis of most manufacturers' specifications indicates that total maximum of both advance units plus initial timing should be between 55 and 60° BTDC. Remember that this total may never actually be seen by the engine because the combination of maximum centrifugal (high rpm) *and* maximum vacuum advance (light engine loads) may not occur at the same time.

SYMPTOMS OF DEFECTIVE ADVANCE MECHANISMS

If the centrifugal advance is not advancing the engine timing properly, the following situations may result:

1. Lack of power due to lack of spark advance during acceleration.
2. Possible constant overadvanced ignition timing caused by a stuck (varnished) mechanical advance shaft. This could cause:
   (a) Constant engine "ping" at low rpm.
   (b) Slow, jerky cranking, especially when the engine is warm.
3. Reduced fuel economy.

If the vacuum advance is not working correctly, it may *not* be noticeable by the driver except for reduced gasoline mileage (up to 4 to 5 mpg) if the vacuum advance is connected to *ported* vacuum. If the vacuum advance is connected to *manifold* vacuum, the following symptoms may be noticeable:

1. Poor or reduced gasoline mileage
2. Low and possible rough idle with possible stalling
3. Part-throttle ping [no ping at heavy throttle (WOT) where vacuum is almost zero]

Since all vacuum advance units use engine vacuum (ported or manifold), if there is a defective vacuum advance diaphragm, there may be a slight vacuum leak. Any vacuum leak can cause a lean fuel mixture, resulting in a high or rough idle, engine overheating, or other engine operating problems.

CHECKING ADVANCE UNITS ON THE CAR

Centrifugal advance can easily be checked for normal operation by removing the distributor cap. Attempt to rotate the rotor. The rotor should rotate and quickly spring back to its original location. If the rotor returns slowly or does not return at all, the advance weights or distributor shaft must be cleaned and lubricated. The distributor usually has to be removed from the engine to clean and service the centrifugal advance mechanism thoroughly.

The amount of *centrifugal* advance can be checked using a timing light or magnetic timing meter and a tachometer with the engine running.

1. Start and run the engine until normal operating temperature is reached.
2. Check and adjust the ignition timing, if necessary, to the factory specifications at the specified rpm.
3. Remove and plug the vacuum hose(s) from the vacuum advance (if equipped) and slowly increase the engine speed.
4. As the engine speed increases, the timing mark should gradually increase.
5. If an adjustable advance timing light or a magnetic timing meter is used, the amount of spark advance can be compared with the specifications to be certain that the amount and rate of centrifugal advance is correct. See Figure 20-24.
6. If the centrifugal advance does not advance the ignition timing at all or less than the specifications, the distributor should be removed from the engine for service.

Vacuum-advance units can easily be checked using a hand-operated vacuum pump or other suitable unit capable of applying 21 in. Hg of vacuum to the vacuum advance unit. The vacuum advance unit should hold vacuum. The operating arm of the vacuum advance unit should also smoothly rotate the breaker plate of the distributor.

REMOVING THE DISTRIBUTOR FROM THE CAR

The distributor must be removed if the centrifugal-advance mechanism is not advancing properly. The distributor of an engine can be easily removed and replaced if certain methods are used:

1. With the engine "off," remove the distributor cap.
2. Mark the direction that the rotor is pointing on the firewall (cowl panel) or other location near the

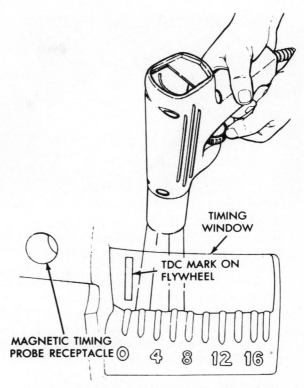

FIGURE 20-24 *The timing mark should advance smoothly while gently accelerating the engine. (Courtesy of Chrysler Corporation.)*

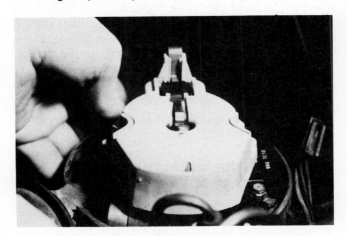

FIGURE 20-25 *Before removing the distributor, the direction that the rotor is pointing should be marked.*

distributor with a single chalk or grease pencil mark. See Figure 20-25.
3. Mark the location of the vacuum advance unit (if equipped) with a "V."
4. Remove the hold-down bolt or nut and hold-down clamp (if equipped) and the primary wiring from the coil.

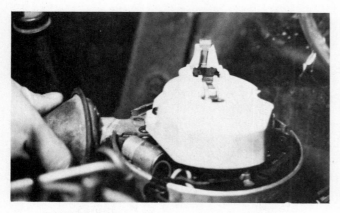

FIGURE 20-26 *As soon as the distributor stops rotating as it is being lifted from the engine, mark the direction the rotor is pointing.*

5. Slowly lift up on the distributor. The rotor will turn slightly during removal. As soon as the distributor has been lifted up enough so that the rotor stops to rotate, make *two* marks indicating the position of the rotor. This will assist in replacing the distributor back to its original location. See Figure 20-26.

6. To reinstall the distributor, line up the rotor with the two marks. As the distributor is installed, the drive gear will rotate the distributor shaft. When the distributor is fully installed, the rotor should line up with the single mark. This single mark is used to double check that the distributor has been reinstalled correctly. Rotate the distributor until the vacuum advance unit is pointing toward the "V." This helps assure that the distributor is installed very close to its original location.

7. Start the engine and check or adjust the base timing as required.

SERVICING THE MECHANICAL (CENTRIFUGAL) ADVANCE

With the distributor out of the engine, remove the distributor drive gear using a punch and hammer. The distributor shaft can now be pulled up and out of the distributor housing.

> **CAUTION:** If the distributor shaft will not pull easily out of the distributor housing, do not use excessive force. Use penetrating oil or spray carburetor cleaner to dissolve or loosen the varnish that can accumulate on the shaft. If the distributor shaft is forced out of the housing, the shaft may also force the distributor bushing out of the housing. The bushing location in most distributors is critical to the proper operation of the distributor.

After the distributor shaft is removed from the distributor, remove the centrifugal advance weights and springs. Remove clip(s) as necessary to separate the distributor drive shaft from the breaker cam (reluctor) shaft. It is this connection between the distributor drive shaft and the breaker cam (reluctor) shaft that often becomes stuck together due to the varnish from the engine oil. This varnish buildup prevents the two distributor shafts from moving freely and prevents proper operation of the centrifugal (mechanical) advance.

The varnish can usually be removed by using emery (polishing) cloth and/or a cleaning solvent. The advance-weight pivot pins should be inspected carefully for wear and lubricated lightly with engine oil. GM mechanical advance weights are often found to be excessively corroded (rusted) and pivot pins or advance weights worn. See Figure 20-27.

Special repair kits are often available at auto-parts suppliers for redrilling GM advance weights and replacement pivot pins. Rust on centrifugal advance weights on GM cars is often traced to a possible clogged PCV system, which could permit crankcase moisture to accumulate in

FIGURE 20-27 *Check for corroded and/or worn distributor advance weights.*

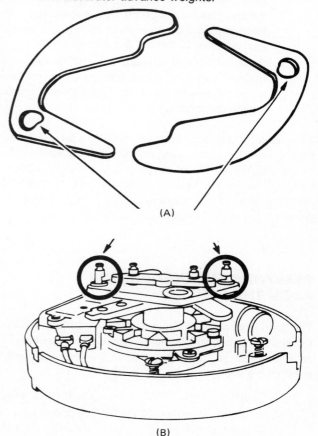

(A)

(B)

the distributor area. High-voltage corona effects, which form a concentrated form of oxygen called ozone, in the distributor cap area also contribute to rapid rusting and wearing of advance weights in GM vehicles.

SERVICING THE VACUUM ADVANCE

As stated earlier, the vacuum advance (and retard unit, if equipped) can easily be checked using a hand vacuum pump. A good vacuum unit should be able to hold 21 in. Hg of vacuum and should move the adjusting lever. Manufacturers' specifications indicates when the vacuum advance should *start* to move (in in. Hg) and when the maximum advance should occur (in in. Hg). The specifications also indicate the *number of degrees* of advance that the unit should provide. This specification may be indicated in either distributor degrees or crankshaft degrees. GM factory vacuum advance units are stamped with two numbers. The three-digit number indicates the last three numbers of the part number and the two-digit number indicates the maximum advance in *crankshaft degrees*. See Figure 20-28.

If the vacuum-advance unit will not hold vacuum or the correct amount of advance, the advance unit must be replaced.

> **NOTE:** Some vacuum advance units are adjustable. Most Ford advance units, for example, can be adjusted using a hex wrench through the vacuum connection hole. Turning the internal adjusting screw counterclockwise delays the starting advance point. Turning the screw clockwise reduces the tension on the internal return spring and permits the advance unit to start to advance at a lower vacuum reading. Two turns of the adjusting screw changes the beginning vacuum setting approxi-

FIGURE 20-28 *Code numbers on GM vacuum advance units indicate the number of crankshaft degrees of advance. The first three numbers represent the last three numbers of the factory part number for the vacuum advance unit.*

mately 1 in. Hg. Older-model Ford vacuum advance units used calibrating spacers (washers) to control the tension on the internal spring. These vacuum advance units can be adjusted to provide a wide range of beginning vacuum settings and total vacuum advance by varying the thickness and number of calibrating washers.

COMPUTER-CONTROLLED TIMING

Most computer-controlled engine systems use the computer to vary the ignition time according to the engine's operating conditions. Sensors supply information regarding engine temperature, speed, and load, and the computer directs the ignition timing. Most computers operate so quickly that one cylinder can be 20° advanced and the very next cylinder in the firing order could be advanced 25°, depending on sensor information.

WHAT IS SPARK SCATTER?

Spark scatter is the variation of spark timing among the cylinders of the engine. Because of engine design, a certain amount of ignition timing variation is commonly detected. For example, a typical engine with electronic ignition may have a variation among cylinders of 12 crankshaft degrees! This much spark timing variation is due to the stack-up of tolerances and torsional forces of the engine crankshaft, timing chain (gear), camshaft, and distributor shaft. The use of direct-fire (distributorless) ignition systems is a major step toward reducing cylinder-to-cylinder timing variation. However, because some direct-fire ignition systems use a crank sensor on the front of the crankshaft and a cam sensor run off of the camshaft, a variation of 6° on the crankshaft is typical. Newer direct-fire ignition systems use a crankshaft sensor off the *center* of the crankshaft and timing variation among cylinders (spark scatter) is reduced to less than 1° throughout all engine rpm ranges. With more accurate ignition timing, fuel delivery and ignition timing can be controlled more accurately than was previously possible.

Some manufacturers purposely vary the ignition timing among cylinders. For example, some engine designs utilize timing retard or advance to quickly and precisely control idle speed. Some manufacturers also retard ignition timing on a cylinder-by-cylinder basis depending on knock sensor activity. For example, the automobile's computer "knows" that if a spark knock occurs immediately following the firing of a spark plug, the computer will only retard the spark timing for the cylinder causing the spark knock. This permits the engine to produce maximum power and fuel economy without engine damage. This timing retard of individual cylinders based on knock sensor information is also referred to as spark scatter.

Most variation in ignition timing occurs during periods of engine speed variation such as occurs at idle. Timing variation at idle is most frequent in four-cylinder engines where crankshaft torsional forces are more severe due to the spacing between cylinder firings. Some manufacturers recommend setting ignition tim-

ing using an averaging method to more accurately achieve precise timing.

Therefore, spark scatter can be undesirable wherever cylinder-to-cylinder timing variation occurs randomly and without prediction or control. Spark scatter can be beneficial to precise engine control if cylinder-to-cylinder timing variations can be controlled.

Most computer-equipped engines use various engine sensors to permit the computer to advance and retard timing as required. Most computer systems can provide as much as 70° of spark advance during certain light-load conditions for maximum fuel economy. If engine knock occurs, many computer systems can provide up to 30° of ignition retard to prevent possible engine damage. See Chapters 25 and 26 for computer engine details.

SUMMARY

1. The proper ignition timing is very important for proper operation of any engine. If the ignition timing is retarded from specifications, low power, performance, and reduced gasoline mileage result. If the ignition timing is advanced from specifications, the engine may ping (spark knock), and slow, jerky starting may occur whenever the engine is warm.

2. For maximum power and performance, the spark timing must be advanced as engine speed increases. Mechanical (centrifugal) advance units advance the timing with increased engine speed. The maximum amount of spark advance due to base timing plus maximum mechanical advance usually ranges from 34 to 38° as measured on the crankshaft.

3. For maximum fuel economy, the spark should be advanced beyond that supplied by the mechanical advance during low-load engine conditions. The vacuum advance unit is designed to provide approximately 20° of spark advance, as measured on the crankshaft, during light engine conditions. The vacuum advance unit can be tested using a vacuum pump to check if the vacuum advance unit can hold a vacuum. Some vacuum advance units are adjustable. Magnetic timing can be used on most newer engines (except Fords) for exact and accurate ignition timing. Timing averaging is used on some engines. On most computer-equipped engines, the ignition timing is automatically advanced as necessary for maximum power and economy with lowest possible exhaust emission. Many computer-equipped engines use a spark knock control to retard the timing as necessary to prevent possible engine damaging detonation.

STUDY QUESTIONS

20-1. What is ignition timing? What are characteristics of retarded and advanced ignition timing?

20-2. Explain why engine detonation (ping) can be harmful to an engine.

20-3. Describe the proper timing procedure using a timing light.

20-4. What is magnetic timing? What is average timing?

20-5. Explain the purpose and operation of vacuum and mechanical advance mechanisms.

MULTIPLE-CHOICE QUESTIONS

20-1. When should the ignition timing be adjusted?
  (a) before adjusting the ignition contact points.
  (b) after adjusting the ignition contact points.
  (c) either before or after adjusting the contact points.
  (d) none of the above.

20-2. Technician A says that the vacuum advance advances the ignition timing with increasing rpm. Technician B says that the vacuum and mechanical advance can be tested using a timing light. Which technician is correct?
  (a) A only.
  (b) B only.
  (c) both A and B.
  (d) neither A nor B.

20-3. Technician A says that retarded timing can cause ping. Technician B says that point gap increases with wear. Which technician is correct?
  (a) A only.
  (b) B only.
  (c) both A and B.
  (d) neither A nor B.

20-4. Ignition timing can be adjusted using:
  (a) a timing light.
  (b) a magnetic probe.
  (c) a vacuum gauge.
  (d) all of the above.

20-5. Technician A says that a mechanical advance mechanism advances the timing by changing the dwell (point gap). Technician B says that a defective vacuum advance unit usually causes a decrease in fuel mileage. Which technician is correct?
  (a) A only.
  (b) B only.
  (c) both A and B.
  (d) neither A nor B.

20-6. Technician A says that increasing dwell 1° advances the timing 1°. Technician B says that increasing point gap advances the timing. Which technician is correct?
  (a) A only.
  (b) B only.
  (c) both A and B.
  (d) neither A nor B.

20-7. Technician A says that over advanced ignition timing can cause slow, jerky cranking. Technician B says that timing changes dwell a degree for a degree. Which technician is correct?
  (a) A only.
  (b) B only.
  (c) both A and B.
  (d) neither A nor B.

20-8. Technician A says that timing should be adjusted at the specified rpm. Technician B says that a defective vacuum advance reduces power and performance. Which technician is correct?
  (a) A only.
  (b) B only.
  (c) both A and B.
  (d) neither A nor B.

20-9. Magnetic offset timing can be used:
  (a) on all engines.
  (b) on many engines.
  (c) for average timing.
  (d) on all Ford-built engines.

20-10. Technician A says that the direction the rotor moves when rotated by hand while in the engine is the direction of rotation. Technician B says that a defective vacuum advance unit is most noticeable by the reduction of fuel economy. Which technician is correct?
  (a) A only.
  (b) B only.
  (c) both A and B.
  (d) neither A nor B.

21

Spark Plug Wires, Distributor Caps, and Rotors

SPARK PLUG WIRES, DISTRIBUTOR CAPS, AND DISTRIBUTOR ROTORS ARE all very important components for the proper operation of the ignition system. Most ignition systems, including electronic ignition systems, use spark plug wires. All cars except those built without distributors use distributor caps and rotors. These components should be carefully inspected at least once a year to ensure proper engine operation. The topics covered in this chapter include:

1. Electromagnetic radiation standards
2. Spark plug wires
3. Symptoms of defective spark plug wires
4. Purchasing considerations for spark plug wires
5. Spark plug wire testing
6. Distributor cap construction and inspection
7. Symptoms of defective distributor caps and rotors
8. Purchasing considerations for distributor caps and rotors

ELECTROMAGNETIC RADIATION STANDARDS

The high voltages produced in an automotive ignition system can create radio and television interference. Before 1958, there were no standards for reducing electromagnetic interference (EMI), also known as radio-frequency interference (RFI).

In 1958, SAE (then an abbreviation for Society of Automotive Engineers and now known as just SAE) established Standard J551, limiting interference in the frequency range 30 to 400 MHz (millions of cycles per second). The SAE standard was based on regulations first established by the Federal Communication Commission (FCC). To meet this standard, the automobile manufacturers installed one or more of the following controls:

1. Resistance spark plug wire
2. Resistance spark plugs
3. Resistance built in the distributor cap center tower
4. Resistance built in the distributor rotor
5. Capacitors (condensers) on the positive (+) side of the coil, voltage regulator, and generator (alternator)

The current SAE J551 Standard specifies that total electromagnetic interference be controlled over the wider frequency range 20 to 1000 MHz from *any* engine or electric motor.

HOW RESISTANCE STOPS RADIO AND TELEVISION INTERFERENCE

The high voltage produced by the coil oscillates back and forth before and after the spark plug fires. This oscillating voltage produces electromagnetic interference throughout a broad frequency range. The resistance built into various secondary ignition components dampens and eliminates electromagnetic radiation.

SPARK PLUG WIRES

Before the late 1950s, most spark plug wires were constructed of metallic wire. This type of wire was excellent and lasted a long time, but it allowed electromagnetic energy to escape and caused static (noise) in all radios and televisions. Current standards now require all ignition systems to control radio-frequency interference (RFI), also known as electromagnetic interference (EMI). Most auto manufacturers use resistance spark plug wires to be able to meet these standards.

TVRS. "TV and radio suppression" wire is manufactured using a nylon, rayon, or aramid thread impregnated with carbon (same substance as charcoal) and surrounded with rubber. TVRS wire has been used as original equipment (OE) on all new cars since the early 1960s. The electricity flows through the carbon, which resists the flow and effectively eliminates unwanted TV and radio interference without affecting engine operation. See Figure 21-1.

With age and temperature changes, the resistance value can and does change. Resistance is measured in ohms. The best and most effective method of determining the condition of TVRS wire is measurement with an ohmmeter in addition to a visual inspection. See Figure 21-2.

SAE currently lists two types of spark plug wire. HR (high resistance) wire, which is seldom used at the present time, is specified at between 6000 and 12,000 Ω per foot (30cm) of length. LR (low resistance) wire, which is generally used at present, is 3000 to 6000 Ω per foot (30cm) of length. For convenience, it is a generally accepted practice to use 10,000 Ω or *less* per foot of wire length as a general testing specification.

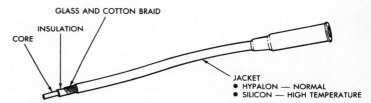

FIGURE 21-1 *Typical television and radio suppression (TVRS) spark plug wire. (Courtesy of Chrysler Corporation.)*

FIGURE 21-2 *Checking a spark plug wire with an ohmmeter. Either ohmmeter test lead can be used on either end of the wire (300 to 1500 Ω/ in. is approximately 4000 to 18,000 Ω/ft). (Courtesy of Ford Motor Company.)*

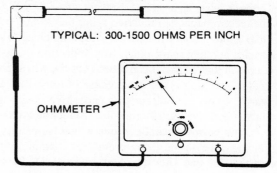

If resistance is *double* this amount per foot, it should be replaced. Dirty or wet spark plug wires should be cleaned with a soft rag. Dirt and moisture increase the voltage required to fire the spark plug by increasing the capacitive load on the ignition system.

GM HEI Wire.   HEI stands for "High Energy Ignition" and is a term used for General Motors' electronic ignition. Due to higher voltage (electrical pressure), HEI wire is 8 mm OD (outside diameter) versus standard 7-mm-OD spark plug wire. HEI wire is also surrounded in silicon rubber, which is more resistant to heat. The internal construction of HEI wire is the same as TVRS wire and has the same specifications (10,000 Ω or less per foot) as TVRS wire. HEI wire is not interchangeable with standard wire because HEI wire has female connections at both ends. Other manufacturers, besides General Motors, also use 8-mm-OD wires.

Other Spark Plug Wire Types.   Aftermarket high-performance spark plug wire can be yellow, orange, or blue in color and either 7 or 8 mm OD. These special wires are either TVRS type or metallic-core type. Another type of non-original-equipment wire uses a Monel wire core. This wire uses thousands of turns of real wire over an iron-impregnated core, and has enough resistance to prevent static in televisions and radios. The specification for testing this type is 400 to 700 Ω or less per foot. An advantage of this type of wire is that the internal resistance does not change with age. Many spark plug wires are covered with Hypalon, a brand name of the DuPont Company for a good-quality insulating rubber compound. Some newer spark plug wires use ethylene vynathene acetate (EVA), which is more abrasion and moisture resistant than silicone.

SPARK PLUG WIRE ROUTING AND CROSS-FIRE

All spark plug wires must be in the correct location in the distributor cap and the spark plug. The wires must also be in the correct wire separator to prevent cross-fire. Cross-fire is the electromagnetic induction spark that can be transferred into another spark plug wire if running close to *and* parallel with another spark plug wire. Induction cross-fire does *not* mean that a spark occurs between the wires. Electromagnetic induction is the usual cause of cross-fire. Latest research seems to indicate that capacitive coupling is also a factor in the cause of cross-fire. Induction cross-fire usually occurs between parallel wires which happen to fire one after the other in the firing order. An example of two cylinders that fire one after the other, and are side by side, are the two left-rear cylinders on many V-8 engines. See Figure 21-3.

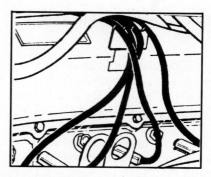

FIGURE 21-3   *Typical spark plug wire routing on the left side (driver's side) of a V-8 engine. Notice that the right two spark plug wires are kept as far apart as possible.*

SYMPTOMS OF DEFECTIVE SPARK PLUG WIRES

The following is a list of symptoms (conditions that would be noticeable) of possibly defective spark plug wires.

> **NOTE:**   There may be other reasons for these symptoms, but spark plug wires should be one of the first items checked.

1. Rough idle (the engine rocks or shakes when running).
2. Misses on acceleration (the engine does not run smoothly and is usually most noticeable at slower engine speeds).
3. Hard to start (the engine cranks normally)—The engine cranks a long time before starting (this may be most noticeable when the weather is damp).

> **NOTE:**   Dirt and/or moisture on spark plug wires adds to the capacitive resistance of the secondary circuit. If the secondary voltage is not high enough to overcome this increased resistance, no spark will occur at the spark plugs.

4. Lack of power (the engine does not seem to have normal power).
5. Poor gas mileage—Because a spark may not be occurring inside the engine, the unburned gasoline simply goes out the exhaust system unburned.

SPARK PLUG WIRE TESTING PROCEDURE

Using an ohmmeter set on the $\times 1000$ or 1K scale (K means "kilo" = 1000):

1. Calibrate the ohmmeter. (Touch both test leads together and adjust the ohmmeter to zero.)

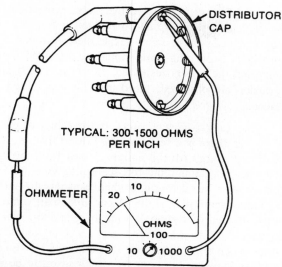

TYPICAL: 300-1500 OHMS
PER INCH

OHMMETER

Measuring Wire Resistance Through Distributor Cap

FIGURE 21-4 Some manufacturers recommend testing spark plug wires through the distributor cap inserts. (Courtesy of Ford Motor Company.)

2. Attach one lead to each end of the spark plug wire. (It does not matter which lead is attached to which end.)

NOTE: Some manufacturers recommend checking spark plug wires at the inside distributor cap inserts and the spark plug boot. If the resistance is excessive, remove and test the wire separately. This helps assure a good spark plug wire-to-distributor cap connection. See Figure 21-4.

3. Carefully wiggle the wire. Watch the meter for a change in the resistance. If the resistance changes more than a slight amount, replace the spark plug wire.
4. If the meter reads infinity, the wire is defective.
5. If the meter reads low ohms (within specifications), the wire is usable.

Results:

Metallic wire = 0 Ω

TVRS type = 10,000 Ω or less per foot

Monel type = 400 to 700 Ω or less per foot

HEI type = 10,000 Ω or less per foot

NOTE: Ford permits spark plug wire resistance up to 5000 Ω/in. (60,000 Ω/ft). New Ford replacement wires, however, should measure 7000 Ω or less per foot. See Figure 21-5.

FIGURE 21-5 Shown is a length of new vacuum hose being used for a coil wire. The vacuum hose conducts the high voltage. An engine will start and run. The carbon content of the vacuum hose is high enough to provide about 100,000 Ω of resistance. Notice the spark from the hose's surface to a grounded screwdriver.

PURCHASING CONSIDERATIONS FOR SPARK PLUG WIRE

Spark plug wires can be purchased individually or in complete sets. If only one or two wires are needed, it would be best to purchase only the wires required. When asking for replacement spark plug wires *individually*, take the wire with you *or* you will need to know the following:

1. Length of spark plug wire needed
2. Type of wire [HEI (8 mm) or standard 7 mm OD]
3. The angle of the boot or connection that fits over the spark plug
   (a) Straight
   (b) Angled (45°)
   (c) 90° (right angle)

If buying a *complete set* of ignition wires, you need to know the following:

1. Make, model, and year of car.
2. Engine size and number of cylinders. (This is very important because many manufacturers use engines similar in displacement, but which are different engines.)

DISTRIBUTOR CAPS

Distributor caps are the plastic replaceable top part of the distributor where all the spark plug wires attach. Distributor caps are constructed of Bakelite, alkyd plastic, phenol resin, or fiberglass-reinforced polyester resin plastic.

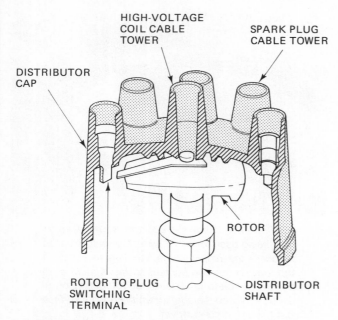

FIGURE 21-6 *Typical distributor cap and rotor. (Courtesy of Sun Electric Corporation.)*

NOTE: Bakelite is a brand name of the Union Carbide Company for phenol-formaldehyde resin plastic.

All distributor caps are made from a high-dielectric-strength material which is the resistance to electrical penetration. See Figure 21-6.

DISTRIBUTOR CAP INSPECTION

When inspecting the distributor cap, always check the following items:

1. *Dirt on the distributor cap.* Dirt and/or moisture on the distributor cap increases the voltage required to fire the spark plugs. If the available voltage is not high enough to overcome this increased load, no spark will occur at the spark plug. Wipe off all accumulated dirt with a soft rag.

2. *Locating tab.* Every distributor cap has to be correctly located on the distributor. Therefore, every cap has a locating tab or notch located on the bottom of the cap which matches up with a corresponding area on the distributor.

3. *Hold-down methods.* Distributor caps are held in place by clips, spring-loaded clamps, or screws. If screws are used (usually two), the screws may have slightly different diameter threads, assuring that the cap will be correctly installed.

4. *Towers.* These are the raised projections where the spark plug wires are attached. Moisture or loose-fitting plug wires can cause arcing inside these towers and cause corrosion. Each plug wire should be removed (one at a time) and the towers inspected. If any darkened areas or corrosion are observed, the distributor cap should be replaced.

5. *Center carbon insert.* In the center of the inside of the cap there is a small rounded insert that contacts the distributor rotor. All the current comes from the ignition coil and goes through the coil spark plug wire and into the center tower. A rotating rotor then transmits this current to the proper spark plug at the proper time. This center insert is constructed of carbon (a form of graphite), which is a conductor of electricity and acts as a lubricant for the spring-metal arm of the rotor. See Figure 21-7. Inspection involves looking closely at the center insert to be assured that the brittle carbon has not chipped or eroded away. If found to be cracked or missing, a replacement cap is required.

6. *Side inserts.* Each spark plug tower insert (segment) is made from aluminum or brass (copper and zinc). If made of a white metal (aluminum), look for excessive "dusting" around the side inserts. See Figure 21-8. (This dusting is aluminum oxide, an abrasive that could cause wear in the distributor.) If dusting is excessive, a replacement cap would be required. If the side inserts are copper color or brass, a slight amount of a green color on the insert is normal and replacement is not needed due to this condition. It is highly recommended that a brass insert distributor cap be purchased for replacement, if possible.

7. *Cracks and/or carbon tracks.* Constructed inside of many distributor caps are ribs, between the side inserts

FIGURE 21-7 *Notice that the center carbon insert has been chipped away. This distributor cap should be replaced because of this problem.*

FIGURE 21-8 *Good example of excessive aluminum oxide dusting of the side inserts of this HEI distributor cap. Also notice that the area around the center electrode is completely destroyed.*

FIGURE 21-9 *Notice the carbon track from the center carbon insert across the anti-flash-over ribs to a side insert. A defective spark plug wire often causes the distributor cap to carbon track.*

FIGURE 21-10 *All GM HEI distributors with internal coils use a rubber washer between the coil and the center carbon insert to help seal out moisture.*

FIGURE 21-11 *Before installation, this rubber seal should be coated with a thin layer of silicone grease on both sides.*

and around the center insert, for the purpose of preventing sparks from arcing to the wrong insert. If moisture is present, due to humid atmospheric conditions, it is possible for a spark to arc between side inserts or from the center insert to a side insert. These ribs are called *antiflashover ribs*. Arcing would prevent the spark from reaching the proper spark plug and could cause a "no-start" situation, especially when raining. However, if you look very closely, you may find a carbon track, which is a dark cracklike mark around the inserts, indicating that a spark has occurred. If any dark lines are observed, the distributor cap must be replaced. See Figure 21-9.

HEI DISTRIBUTOR CAPS

The larger distributor cap also provides a greater distance between the spark plug wires plug space for the ignition coil. The secondary of the coil discharges through a replaceable center carbon insert (included with replacement HEI distributor caps). Between the coil and the center insert is a rubber seal which must be coated (on both sides) with silicone dielectric grease to seal out moisture. See Figures 21-10 and 21-11.

SYMPTOMS OF A DEFECTIVE DISTRIBUTOR CAP OR ROTOR

A defective distributor cap or rotor could have one or more of the following symptoms. There may be other items that could cause these same symptoms, but the distributor cap and rotor should be one of the first items checked.

FIGURE 21-12 Notice the extreme electrical wear from arcing that has occurred at the tip of this rotor.

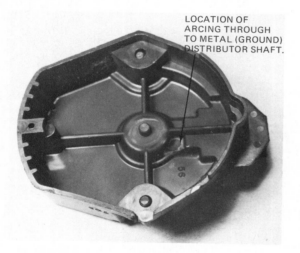

LOCATION OF ARCING THROUGH TO METAL (GROUND) DISTRIBUTOR SHAFT.

FIGURE 21-13 Notice the white ring near the center of this rotor. This is how "punch through" or arcing through the rotor can be identified. This rotor caused missing during part-throttle acceleration.

1. Hard to start during high-moisture conditions. For example, "long cranking before starting" whenever it is raining. See Figure 21-12.
2. Missing on acceleration (especially during damp weather or during the engine warm-up period). See Figure 21-13.
3. Reduced gas mileage.
4. Reduced performance (engine does not run as smoothly as it should).
5. "No start" situation with normal cranking.

FIGURE 21-14 Typical locating notch on a rotor.

ROTOR INSTALLATION

All distributor rotors are designed to fit in only one position. Every rotor has a notch or locating tab which must fit snugly down on the top of the distributor shaft or over the mechanical advance mechanism. See Figure 21-14. Always use care when installing a rotor to make certain that it fits snugly and in the correct locating notch. Some Fords use a double-level (bilevel) rotor and the firing order is not the same as the spark plug wire sequence. Chrysler rotors used on Hall-effect distributors *must* be grounded to the distributor shaft by a narrow metal strip.

> **NOTE:** It is *possible* for rotors to be installed incorrectly. Since the rotor distributes the electrical current from the ignition coil to the correct spark plug, if the rotor is not installed correctly, the engine will not start and may cause a dangerous backfire and possible engine fire.

PURCHASING CONSIDERATIONS FOR DISTRIBUTOR CAPS AND ROTORS

1. Purchase a distributor cap that fits your car with brass inserts, if possible. (If the inserts are brass, the cap itself is usually of premium quality.) See Figures 21-15 and 21-16.

> **NOTE:** If solid metallic core spark plug wire is used, a distributor cap with brass inserts is *required*. If an aluminum insert distributor cap is used with solid core wire, the inserts will rapidly erode, due to the resulting increased current flow and possible galvanic corrosion caused by the junction of two dissimilar metals.

2. Always inspect new caps and rotors to be certain that they are the correct parts and check to see if they were damaged. See Figure 21-17.

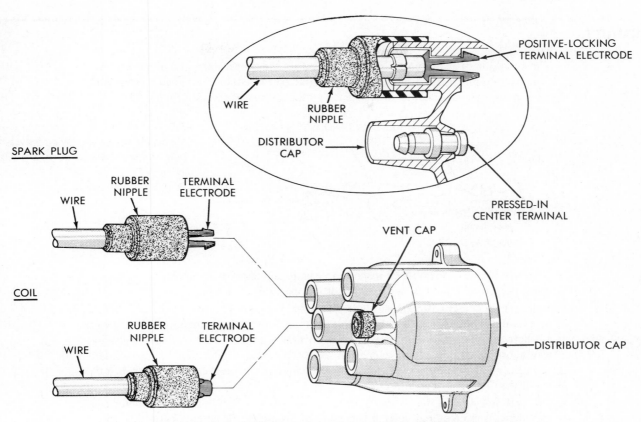

FIGURE 21-15 *Some Chrysler four-cylinder engine distributor caps use the terminal of the spark plug wire as the side inserts for the cap. (Courtesy of Chrysler Corporation.)*

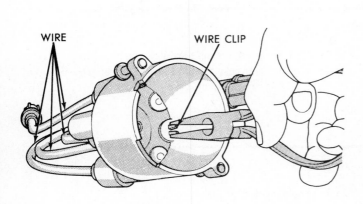

FIGURE 21-16 *Needle-nose pliers should be used to remove the spark plug wire from the distributor cap on some Chrysler four-cylinder engines. (Courtesy of Chrysler Corporation.)*

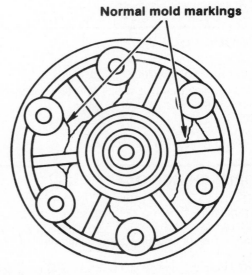

FIGURE 21-17 *Some normal mold marks can be mistaken for carbon tracks.*

SUMMARY

1. The condition of spark plug wires is important for the proper operation of any engine. Original-equipment spark plug wire is constructed of carbon-impregnated fabric and does deteriorate and change in resistance with time and usage. See Figure 21-18.

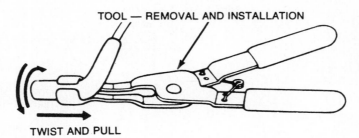

TOOL — REMOVAL AND INSTALLATION

TWIST AND PULL

FIGURE 21-18 *All spark plug wire should be twisted slightly on the spark plug end and then gently removed by grasping only the boot of the spark plug wire.*

2. All spark plug wires should be inspected visually and tested with an ohmmeter. Good spark plug wire should have 10,000 Ω or less per foot of length.

3. Whenever a defective spark plug wire is found, the distributor cap and rotor should be examined carefully for defects such as carbon tracks. High-voltage electricity can arc between distributor cap inserts trying to find a ground. High-voltage electricity also causes a corona effect, which ionizes the air near an electrical spark or high-voltage coil. This corona often causes deterioration of the plastic or fiberglass materials of the distributor cap and rotor.

3. Dirt and moisture on the outside surfaces of the spark plug wires and distributor cap can increase the required voltage necessary to fire the spark plugs. Spark plug wires should be tested whenever poor engine operation is discovered. Distributor caps should be inspected every year and replaced as necessary. Since *all* of the secondary ignition current must flow through the rotor, it is best to replace it every year. Distributor caps should be examined closely whenever poor engine operation or a "no start" situation occurs during damp or rainy weather. Distributor caps should be replaced approximately every 3 or 4 years of normal service.

STUDY QUESTIONS

21-1. What is RFI or EMI?

21-2. A good spark plug wire should have less than _____ ohms of resistance per foot of length.

21-3. Describe cross-fire and how it can be prevented.

21-4. List five items that should be checked when inspecting the condition of a distributor cap.

MULTIPLE-CHOICE QUESTIONS

21-1. Spark plug wire:
   (a) increases EMI.
   (b) is made of metal if original equipment.
   (c) increases RFI.
   (d) should measure 10,000 Ω or less per foot of length.

21-2. 8-mm spark plug wire:
   (a) uses 8-mm deep boots.
   (b) fits 8-mm spark plugs.
   (c) uses 8-mm-diameter metallic wire.
   (d) is 8 mm in outside diameter.

21-3. Cross-fire means that:
 (a) a spark occurs between two spark plug wires.
 (b) current from one spark plug wire is induced into another spark plug wire.
 (c) two sparks occur at the same instant from the ignition coil.
 (d) the distributor cap must be replaced.

21-4. Spark plug wires should be tested:
 (a) with an ohmmeter.
 (b) with a voltmeter.
 (c) visually.
 (d) both (a) and (c).

21-5. Distributor caps should be checked for:
 (a) chipped or cracked center carbon inserts.
 (b) cracks or carbon tracks.
 (c) a corroded tower(s).
 (d) all of the above.

21-6. Dirt on the distributor cap and spark plug wires can cause:
 (a) rapid spark plug wear.
 (b) a higher voltage to be required to fire the spark plugs.
 (c) a lower voltage to be required to fire the spark plugs.
 (d) rapid center carbon insert wear.

21-7. A ''no start'' or poor engine operation during damp or wet weather conditions could be an indication of:
 (a) a defective distributor cap.
 (b) a defective spark plug wire.
 (c) dirty spark plug wires and/or distributor cap.
 (d) all of the above.

21-8. Silicone grease should be installed on both sides of the rubber seal inside a GM HEI distributor cap with integral coil to:
 (a) lubricate the rotor.
 (b) lubricate and cool the coil.
 (c) help seal out moisture.
 (d) help heat escape from coil.

22

Spark
Plugs

THE CONDITION OF SPARK PLUGS IS IMPORTANT FOR CORRECT OPERATION of the engine. Worn spark plugs can cause hard starting, poor gas mileage, and engine missing. Spark plug inspection can also determine possible engine, carburetor, or ignition timing problems. Spark plugs should be inspected whenever there is an engine performance problem and replaced every year for best fuel economy. The topics covered in this chapter include:

1. Spark plug construction
2. Spark plug operation and heat range
3. Resistor spark plugs
4. Spark plug code numbers
5. Spark plug purchasing considerations
6. Spark plug installation considerations

SPARK PLUG PARTS AND FEATURES

A spark plug must be able to transfer high-voltage electricity to the combustion chamber, where the electrical spark ignites the air/fuel mixture in the engine's cylinder. The high-voltage secondary current must be insulated from the metal components of the engine. The spark plug is constructed to insulate this high voltage and must be able to survive in the heat of thousands of miles of continuous usage.

Parts of a typical automotive spark plug include:

1. *Shell.* Large spark plugs require a $^{13}/_{16}$-in. socket, small spark plugs require a $^5/_8$-in. socket.

NOTE: Most automotive spark plugs are 14-mm-diameter *threads.* Older-model Ford engines often used 18-mm-diameter thread spark plugs. Farm equipment and lawn and garden equipment often use various sizes of spark plugs. See Figure 22-1.

2. *Ceramic insulator.* The ceramic insulator has ribs on the outside portion for two reasons:
   (a) Anti-flash-over
   (b) Helps retain plug wire boot
3. *Center electrode.* This electrode is constructed of nickel steel alloy.
4. *Side electrode.* The side electrode is welded to the shell.
5. *Gap*—0.020 to 0.080 in. (0.5 to 2.0 mm). A spark occurs across this gap between the center and side electrodes.
6. *Terminal.* The terminal is the top electrical connector where the spark plug wire connects. See Figure 22-2.

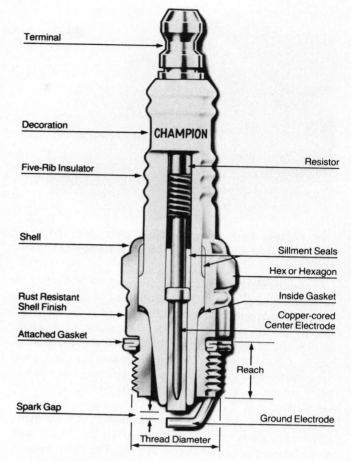

FIGURE 22-2 *(Courtesy of Champion Spark Plug Company.)*

Whenever removing any spark plug for any reason, check its condition for wear and color. The condition of spark plugs can help diagnose possible engine operating conditions.

Normal Wear and Color. When removing and inspecting used spark plugs, it is often difficult to determine if replacement spark plugs should be installed. It is important to know what is generally considered "normal wear"; therefore, the technician should learn how to determine if and when to replace spark plugs. The first noticeable wear occurs as the center electrode becomes rounded. See Figure 22-3. The more rounded the center electrode, the greater the voltage required to fire the spark plug. The gap between the center and side electrodes increases approximately 0.001 in. per 1000 miles (2.5 mm per 1600 km) for regular spark plugs. Resistor spark plugs wear at a rate of approximately 0.001 in. per 5000 miles (2.5 mm per 8000 km). The normal color of the inside insulator around the center electrode should be *tan* or *light gray.*

FIGURE 22-1 *(Courtesy of Champion Spark Plug Company.)*

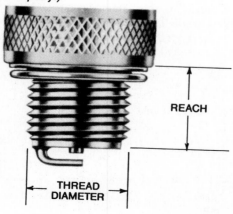

FIGURE 22-3 *Worn-out spark plug. Note the excessive rounding of the center electrode.*

Spark Plug Heat Range.   The correct heat range of any spark plug should be between fouling temperature and the preignition temperature. Fouling occurs if the temperature of the tip of the plug does not allow for burnoff of deposits at approximately 850°F (450°C). Preignition occurs if the temperature (the temperature of the spark plug tip, which could cause ignition by itself without a spark being present) is approximately 1700°F (900°C). See Figure 22-4.

The spark plug number as specified by the car manufacturer is always to be considered to be the correct heat range for each particular engine and car. The "heat range" of any spark plug is determined by how far the center insulator goes up into the shell around the center electrode. The greater this distance, the longer it takes heat to reach the shell. The shell is in contact with the engine. The heat travels from the hot spark plug to the cooler engine (where it is finally cooled by the engine), and therefore controls the temperature of the tip. See Figure 22-5.

Abnormal Color(s) and Possible Causes.   If the center insulator is *dark* or *black* on all spark plugs, this means (see Figure 22-6):

1. A possible stuck choke in the carburetor, causing an overly rich (too much gas) mixture. Gas fouling is usually identified by dry, fluffy deposits.

2. A possible engine problem which causes oil burning due to defective valve seals (most common) or piston rings. Oil fouling is usually identified by wet, sludgelike deposits.

3. Possibility of all city driving or slow driving for the heat range of the spark plugs being used.

4. If one or more, but not all the spark plugs are dark, the problem could be defective valve seals or piston rings on those particular cylinders.

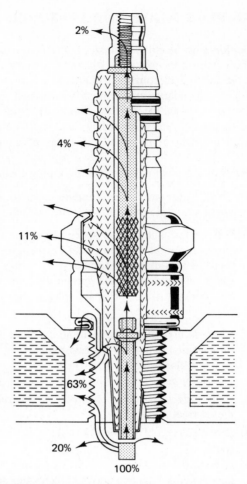

FIGURE 22-4 *Typical heat flow from the spark plug. How rapidly the heat escapes from the center electrode determines the spark plug heat range.*

FIGURE 22-5 *The shorter the distance from the tip to the shell, the cooler the spark plug tip. This temperature of a spark plug tip is called the "heat range" of the plug. (Courtesy of Chrysler Corporation.)*

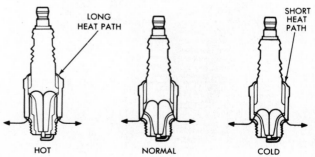

FIGURE 22-6 *Spark plug that has been running too cold for conditions. Excessive deposits have built up around the electrodes.*

FIGURE 22-7 *Spark plug that has been running too hot. Notice that the center insulator is snow white and the center electrode has started to become rounded.*

FIGURE 22-8 *Normal spark plug.*

If the center insulator is *very light* or *white*, this means (see Figure 22-7):

1. Possibly too far advanced ignition timing
2. Possibly a lean mixture, most likely due to a vacuum leak (a vacuum hose disconnected or defective)
3. Too hot a heat range for the type of driving

A normal spark plug is shown in Figure 22-8. Figures 22-9 through 22-19 show worn and damaged spark plugs.

FIGURE 22-9 *Worn-out spark plug. Notice the greatly rounded off center electrode.*

FIGURE 22-10 *Ash-fouled spark plug.*

FIGURE 22-11 *Carbon-fouled spark plug.*

FIGURE 22-12   *Oil-fouled spark plug.*

FIGURE 22-13   *Gap bridging spark plug.*

FIGURE 22-14   *Overheated spark plug.*

FIGURE 22-15   *Splash-fouled spark plug. The material splashed onto the spark plug is usually aluminum from melting engine components.*

FIGURE 22-16   *Spark plug glazing.*

FIGURE 22-17   *Spark plug detonation damage.*

FIGURE 22-18   *Spark plug preignition damage.*

FIGURE 22-19   *Spark plug mechanical damage.*

RESISTOR SPARK PLUGS

Resistor spark plugs are spark plugs that have a built-in resistor inside the spark plug in the center electrode core. The amount of resistance can vary according to the spark plug manufacturer. If measured with a standard ohmmeter, the resistance values for the following manufacturers may be considered acceptable:

AC	2,500 to 7,500 Ω
Champion	2,500 to 50,000 Ω
Autolite	2,500 to 22,000 Ω

The resistor, used in spark plugs, operates with high voltages and temperatures. The resistance of the spark plug resistor is different during its operation in an engine than it is at room temperature without high-voltage current flowing through the resistor. Many spark plug resistors are made of semiconductor materials which may show higher resistance when tested with an ohmmeter than is actually present during spark plug usage. Resistors for resistor spark plugs are constructed of carbon or SAC (strontium, alumina, and copper oxide). Therefore, an ohmmeter should only be used to test a resistor spark plug at room temperature to determine if the spark plug is open (infinity ohms) or for the absence of a resistor (zero ohms).

The purpose of the resistor in the spark plug is to reduce electromagnetic radiation created in the ignition system. The closer the resistance is to the actual spark, the more effective the resistor is to reducing radiation. Many automotive manufacturers combine resistance spark plug wire with resistor spark plugs to meet the radiation standards. (See Chapter 21 for details.) Resistor spark plugs also reduce electrode erosion (wear) by dampening (removing) lower-voltage and secondary voltage pulse frequencies. Resistor spark plugs, if used with unleaded fuel, can achieve a service life of over 20,000 miles (32,000 km).

SPARK PLUG CODE NUMBERS (AC, CHAMPION, AND OTHERS)

The code numbers used on spark plugs are used for identification and should be used by the technician to be sure that the spark plug is correct for the vehicle. The various factors included in spark plug code numbers are explained in detail below using examples of AC and Champion spark plugs. Other spark plug brands may use various letters or number codes to represent different factors.

AC Spark Plug Codes (see Figure 22-20) (AC is the brand name originally named for its founder, Albert Champion)

AC 45. The "4" means that it has a 14-mm-diameter thread. The "5" is the heat range (the higher this number, the hotter the plug). The normal range of AC spark plug heat is from 2 (the coldest) to 7 (the hottest).

AC R45. The "R" means "resistor" (approximately 2500 to 7500 Ω).

NOTE: Resistance spark plug wire usually has enough resistance to eliminate radio and television interference. However, the closer the resistance is to the actual spark, the more effective it is. Resistor spark plugs are used *with* resistance spark plug wire (LR) to achieve the specified RFI limits.

AC R45S. The "S" means "extended tip" (also known as projected nose). The purpose of the extended tip is to place the center electrode farther into the combustion chamber. Because the insulator around the center electrode is also extended, this expands the heat range of the plug because the incoming fuel/air mixture tends to cool the spark plug tip, reducing the possibility of preignition. Preignition is ignition of the fuel mixture by something hot in the combustion chamber *before* the spark occurs.

AC R45T. The "T" means "tapered seat." This taper on the shell of the spark plug matches a similar taper of the engine and will seal without the need of a gasket.

"T" FOR TINY?

If the letter "T" is part of an AC code for a 14-mm plug, it is also a "tiny" spark plug requiring a ⅝-in. socket.

Many manufacturers started using a small-shell (⅝ in.) spark plug in the early 1970s. After 1971, all engines had to be manufactured to be *able* to use unleaded fuel. Using fuel without tetraethyllead increased valve temperatures. Therefore, the cylinder heads were redesigned on many engines to provide for larger cooling jackets around the valves. This reduced the space around the spark plugs. With the cylinder head redesign, smaller-shell spark plugs were required. The tapered seat design was included on most small spark plugs to eliminate the need for a separate sealing gasket.

AC R45TS. This spark plug combines all codes above: R = resistor, 4 = 14-mm-diameter thread, 5 = heat range, T = tapered seat, and S = extended tip.

SPARK PLUG IDENTIFICATION CHART

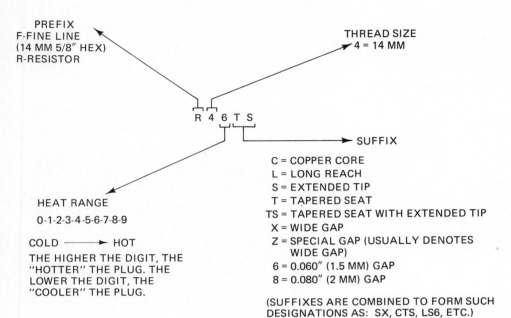

PREFIX
F-FINE LINE
(14 MM 5/8" HEX)
R-RESISTOR

THREAD SIZE
4 = 14 MM

R 4 6 T S

SUFFIX

C = COPPER CORE
L = LONG REACH
S = EXTENDED TIP
T = TAPERED SEAT
TS = TAPERED SEAT WITH EXTENDED TIP
X = WIDE GAP
Z = SPECIAL GAP (USUALLY DENOTES
      WIDE GAP)
6 = 0.060" (1.5 MM) GAP
8 = 0.080" (2 MM) GAP

(SUFFIXES ARE COMBINED TO FORM SUCH
DESIGNATIONS AS: SX, CTS, LS6, ETC.)

HEAT RANGE

0-1-2-3-4-5-6-7-8-9

COLD ——————→ HOT
THE HIGHER THE DIGIT, THE
"HOTTER" THE PLUG. THE
LOWER THE DIGIT, THE
"COOLER" THE PLUG.

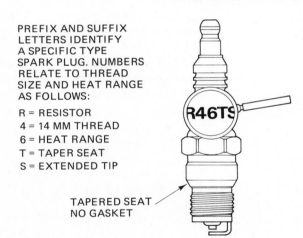

PREFIX AND SUFFIX
LETTERS IDENTIFY
A SPECIFIC TYPE
SPARK PLUG. NUMBERS
RELATE TO THREAD
SIZE AND HEAT RANGE
AS FOLLOWS:

R = RESISTOR
4 = 14 MM THREAD
6 = HEAT RANGE
T = TAPER SEAT
S = EXTENDED TIP

R46TS

TAPERED SEAT
NO GASKET

FIGURE 22-20   *(Courtesy of Oldsmobile Division.)*

AC R45CTS. The "C" used behind the heat range number indicates that the center electrode has a copper core. The copper conducts heat away from the tip, quickly reducing preignition conditions when using a hot plug to control carbon deposits on the plug.

AC R45TSX. The "X" (can also be a "Z") indicates "extra wide gap." An "X" means a 0.060- or 0.080-in. gap (check specifications for correct gap) and the "Z" indicates a slightly narrower (0.045 and 0.060-in.) extra-wide gap.

AC R45TS8. The number at the end of the spark plug code indicates the plug gap in hundredths of an inch. The "8" means gapped at 0.080 in.

COPPER CORE SPARK PLUGS

Most spark plug manufacturers are manufacturing copper-core plugs to be able to meet or exceed the temperature ranges required by all types of driving. The spark plug manufacturers can also decrease the number of heat ranges that need to be built and stocked, since a copper-core plug will function correctly throughout a broad heat range. Many spark plugs that have copper cores may or *may not* have the letter "C" in their code number.

Other commonly used letter codes include:

N   ¾-in. reach, threaded ⅜ in.

XL   ¾-in. reach, threaded the entire ¾-in. length

The purpose of becoming familiar with these basic code numbers is to be certain that the spark plug found in the engine or purchased for the engine is correct. For example, if an AC R45TSX is substituted for an AC R45TS6 and the recommended gap is 0.060 in., there is no problem. Therefore, some variation in spark plug code numbers is okay, but others cannot be changed.

> **CAUTION:** Serious mechanical engine damage can result if the incorrect spark plug reach is used. Some engines may also be damaged if extended-tip (projected nose) spark plugs are installed where non-extended-tip plugs should be used.

Champion Spark Plug Codes (see Figure 22-21):

J-11Y. The "J" indicates 14-mm-diameter thread using a gasket. The "11" is the heat range (the higher the number, the hotter). The "Y" indicates an extended tip (projected nose).

RJ-11Y. The resistor equal to the spark plug above.

F-11Y. The "F" can best be remembered as fitting older-model Ford products which require an 18-mm-diameter thread. The heat range (11) and extended tip (Y) are the same.

BL-11Y. The "BL" indicates a 14-mm-diameter thread, tapered seat, tiny spark plug.

> **HINT:** The "BL" can be thought of as meaning "became little," because it is a small-shell (⅝-in), 14-mm-thread-diameter spark plug.

RBL-11Y6. Same "tiny" spark plug as above, except that it is a resistor (R) and gapped at 0.060 in. (6).

RV-11YC6. The "V" replaced the "BL" plug and can be remembered by thinking that it means "very small." The "C" indicates copper-core construction.

RN-9Y. This commonly used spark plug is a resistor (R) with a ¾-in. reach threaded the entire ¾ in. The "9" is the heat range number; the "Y" indicates the extended tip.

There are many other brand names of spark plugs and most use codes that mean different things. It is important that a professional automotive technician uses the recommended spark plug brand name and code that the factory recommended for the car.

THE CORRECT, BUT THE WRONG SPARK PLUG!

A technician at an independent service center replaced the spark plugs in a Pontiac with new Champion brand spark plugs of the correct size, reach, and heat range. When the customer returned to pay his bill, he inquired as to the brand name of the replacement parts used during the tune-up. When told that Champion spark plugs were used, he stopped signing his name on the check he was writing. He said that he owned a 1,000 shares of General Motors stock and he owned two General Motors cars and he expected to have General Motors parts used in his General Motors cars. The service manager had the technician replace the spark plugs with AC brand spark plugs because this brand was used in the engine when the car was new. Even though most spark plug manufacturers produce spark plugs that are correct for use in almost any engine, many customers prefer that original brand spark plugs are used in their engines.

However, spark plug codes and features can change; therefore, always consult a current spark plug application chart or booklet to be assured of installing the correct spark plugs. See Figures 22-22 through 22-25.

> **CAUTION:** State, federal, provincial, or local laws may not permit changing from the factory-specified heat range and spark plug type. If the spark plug does not perform satisfactorily and the engine condition has been checked, consult factory service bulletins for possible spark plug code number changes.

PURCHASING CONSIDERATIONS FOR SPARK PLUGS

Whenever a new set of spark plugs is purchased, always check the following on each plug before installing:

1. Make certain that the spark plug purchased is the correct spark plug for the car.
2. Make certain that *all* of the spark plugs in the box purchased are the same.
3. Make certain that the side electrode is centered on the center electrode.
4. Always check for the correct spark plug gap.
5. Check by trying to twist the plug to see if the insulator is loose in the shell.

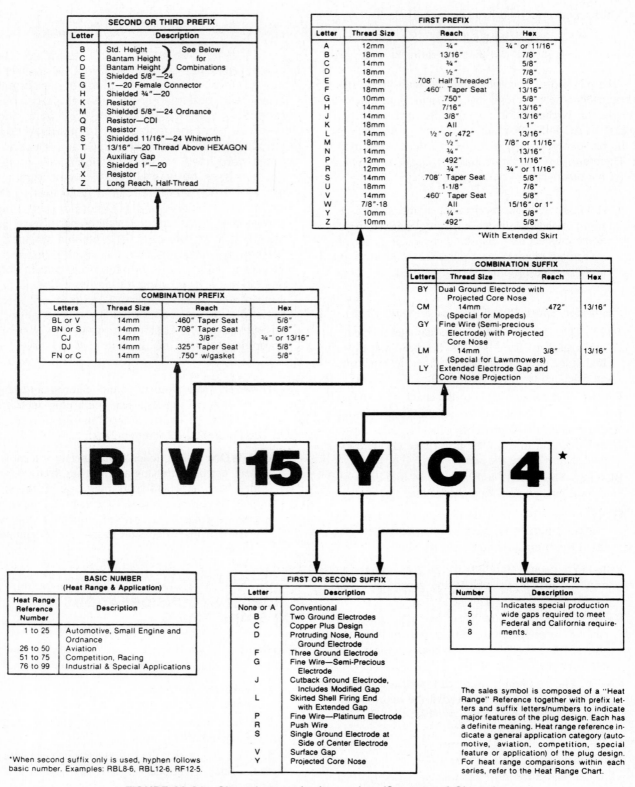

SECOND OR THIRD PREFIX	
Letter	**Description**
B	Std. Height See Below
C	Bantam Height } for
D	Bantam Height } Combinations
E	Shielded 5/8″—24
G	1″—20 Female Connector
H	Shielded 3/4″—20
K	Resistor
M	Shielded 5/8″—24 Ordnance
Q	Resistor—CDI
R	Resistor
S	Shielded 11/16″—24 Whitworth
T	13/16″—20 Thread Above HEXAGON
U	Auxiliary Gap
V	Shielded 1″—20
X	Resistor
Z	Long Reach, Half-Thread

FIRST PREFIX			
Letter	**Thread Size**	**Reach**	**Hex**
A	12mm	3/4″	3/4″ or 11/16″
B	18mm	13/16″	7/8″
C	14mm	3/4″	5/8″
D	18mm	1/2″	7/8″
E	14mm	.708′ Half Threaded*	5/8″
F	18mm	.460′ Taper Seat	13/16″
G	10mm	.750′	5/8″
H	14mm	7/16″	13/16″
J	14mm	3/8″	13/16″
K	18mm	All	1″
L	14mm	1/2″ or .472″	13/16″
M	18mm	1/2″	7/8″ or 11/16″
N	14mm	3/4″	13/16″
P	12mm	.492″	11/16″
R	12mm	3/4″	3/4″ or 11/16″
S	14mm	.708′ Taper Seat	5/8″
U	18mm	1-1/8″	7/8″
V	14mm	.460′ Taper Seat	5/8″
W	7/8″-18	All	15/16″ or 1″
Y	10mm	1/4″	5/8″
Z	10mm	.492″	5/8″

*With Extended Skirt

COMBINATION PREFIX			
Letters	**Thread Size**	**Reach**	**Hex**
BL or V	14mm	.460″ Taper Seat	5/8″
BN or S	14mm	.708″ Taper Seat	5/8″
CJ	14mm	3/8″	3/4″ or 13/16″
DJ	14mm	.325″ Taper Seat	5/8″
FN or C	14mm	.750″ w/gasket	5/8″

COMBINATION SUFFIX			
Letters	**Thread Size**	**Reach**	**Hex**
BY	Dual Ground Electrode with Projected Core Nose		
CM	14mm (Special for Mopeds)	.472″	13/16″
GY	Fine Wire (Semi-precious Electrode) with Projected Core Nose		
LM	14mm (Special for Lawnmowers)	3/8″	13/16″
LY	Extended Electrode Gap and Core Nose Projection		

R V 15 Y C 4 *

BASIC NUMBER (Heat Range & Application)	
Heat Range Reference Number	**Description**
1 to 25	Automotive, Small Engine and Ordnance
26 to 50	Aviation
51 to 75	Competition, Racing
76 to 99	Industrial & Special Applications

*When second suffix only is used, hyphen follows basic number. Examples: RBL8-6, RBL12-6, RF12-5.

FIRST OR SECOND SUFFIX	
Letter	**Description**
None or A	Conventional
B	Two Ground Electrodes
C	Copper Plus Design
D	Protruding Nose, Round Ground Electrode
F	Three Ground Electrode
G	Fine Wire—Semi-Precious Electrode
J	Cutback Ground Electrode, Includes Modified Gap
L	Skirted Shell Firing End with Extended Gap
P	Fine Wire—Platinum Electrode
R	Push Wire
S	Single Ground Electrode at Side of Center Electrode
V	Surface Gap
Y	Projected Core Nose

NUMERIC SUFFIX	
Number	**Description**
4	Indicates special production
5	wide gaps required to meet
6	Federal and California require-
8	ments.

The sales symbol is composed of a "Heat Range" Reference together with prefix letters and suffix letters/numbers to indicate major features of the plug design. Each has a definite meaning. Heat range reference indicate a general application category (automotive, aviation, competition, special feature or application) of the plug design. For heat range comparisons within each series, refer to the Heat Range Chart.

FIGURE 22-21 *Champion spark plug codes. (Courtesy of Champion Spark Plug Company.)*

Design symbols used in NGK spark plugs

—PREFIX— —SUFFIX— —WIDE GAP—

B P 5 E S - 15

WIDE GAP
10 : 1.0 mm (.040'')
11 : 1.1 mm (.044'')
13 : 1.3 mm (.050'')
14 : 1.4 mm (.055'')
15 : 1.5 mm (.060'')
20 : 2.0 mm (.080'')

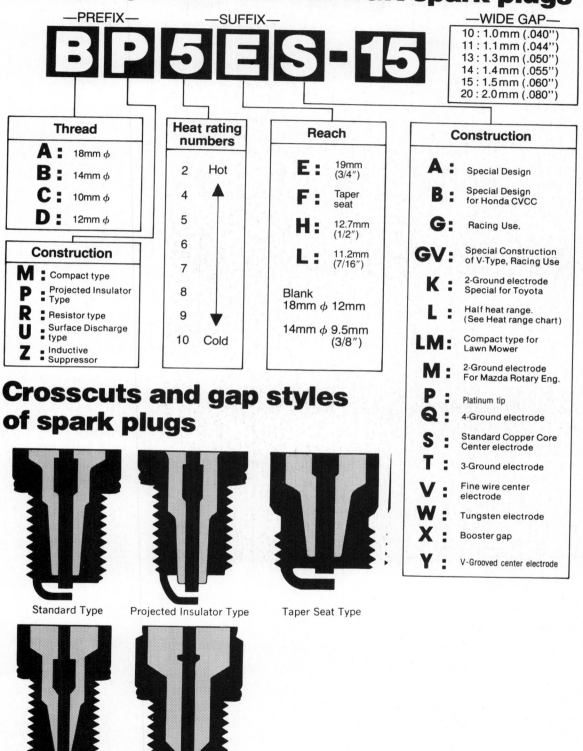

Thread
A : 18mm φ
B : 14mm φ
C : 10mm φ
D : 12mm φ

Construction
M : Compact type
P : Projected Insulator Type
R : Resistor type
U : Surface Discharge type
Z : Inductive Suppressor

Heat rating numbers
2 — Hot
4
5
6
7
8
9
10 — Cold

Reach
E : 19mm (3/4'')
F : Taper seat
H : 12.7mm (1/2'')
L : 11.2mm (7/16'')

Blank
18mm φ 12mm
14mm φ 9.5mm (3/8'')

Construction
A : Special Design
B : Special Design for Honda CVCC
G : Racing Use.
GV : Special Construction of V-Type, Racing Use
K : 2-Ground electrode Special for Toyota
L : Half heat range. (See Heat range chart)
LM : Compact type for Lawn Mower
M : 2-Ground electrode For Mazda Rotary Eng.
P : Platinum tip
Q : 4-Ground electrode
S : Standard Copper Core Center electrode
T : 3-Ground electrode
V : Fine wire center electrode
W : Tungsten electrode
X : Booster gap
Y : V-Grooved center electrode

Crosscuts and gap styles of spark plugs

Standard Type   Projected Insulator Type   Taper Seat Type

V Type   Surface Discharge Type

FIGURE 22-22 *NGK spark plug codes. (Courtesy of NGK Spark Plugs.)*

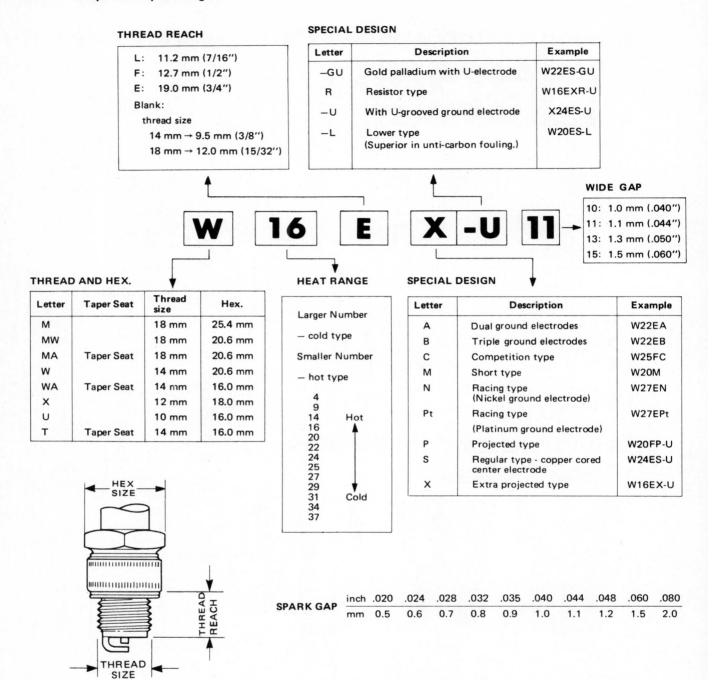

THREAD REACH

L:	11.2 mm (7/16")
F:	12.7 mm (1/2")
E:	19.0 mm (3/4")

Blank:
  thread size
  14 mm → 9.5 mm (3/8")
  18 mm → 12.0 mm (15/32")

SPECIAL DESIGN

Letter	Description	Example
−GU	Gold palladium with U-electrode	W22ES-GU
R	Resistor type	W16EXR-U
−U	With U-grooved ground electrode	X24ES-U
−L	Lower type (Superior in unti-carbon fouling.)	W20ES-L

WIDE GAP

10:	1.0 mm (.040")
11:	1.1 mm (.044")
13:	1.3 mm (.050")
15:	1.5 mm (.060")

W 16 E X -U 11

THREAD AND HEX.

Letter	Taper Seat	Thread size	Hex.
M		18 mm	25.4 mm
MW		18 mm	20.6 mm
MA	Taper Seat	18 mm	20.6 mm
W		14 mm	20.6 mm
WA	Taper Seat	14 mm	16.0 mm
X		12 mm	18.0 mm
U		10 mm	16.0 mm
T	Taper Seat	14 mm	16.0 mm

HEAT RANGE

Larger Number
— cold type

Smaller Number
— hot type

4
9
14
16
20
22
24
25
27
29
31
34
37

Hot ↑
Cold ↓

SPECIAL DESIGN

Letter	Description	Example
A	Dual ground electrodes	W22EA
B	Triple ground electrodes	W22EB
C	Competition type	W25FC
M	Short type	W20M
N	Racing type (Nickel ground electrode)	W27EN
Pt	Racing type (Platinum ground electrode)	W27EPt
P	Projected type	W20FP-U
S	Regular type - copper cored center electrode	W24ES-U
X	Extra projected type	W16EX-U

HEX SIZE

THREAD REACH

THREAD SIZE

SPARK GAP

inch	.020	.024	.028	.032	.035	.040	.044	.048	.060	.080
mm	0.5	0.6	0.7	0.8	0.9	1.0	1.1	1.2	1.5	2.0

FIGURE 22-23 *Nippondenso (Denso) spark plug codes. (Courtesy of Nippondenso of Los Angeles, Inc.)*

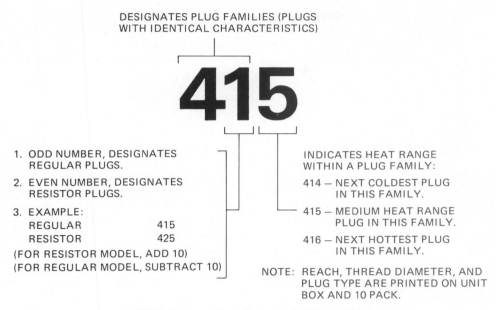

DESIGNATES PLUG FAMILIES (PLUGS WITH IDENTICAL CHARACTERISTICS)

415

1. ODD NUMBER, DESIGNATES REGULAR PLUGS.

2. EVEN NUMBER, DESIGNATES RESISTOR PLUGS.

3. EXAMPLE:
   REGULAR   415
   RESISTOR   425

(FOR RESISTOR MODEL, ADD 10)
(FOR REGULAR MODEL, SUBTRACT 10)

INDICATES HEAT RANGE WITHIN A PLUG FAMILY:

414 – NEXT COLDEST PLUG IN THIS FAMILY.

415 – MEDIUM HEAT RANGE PLUG IN THIS FAMILY.

416 – NEXT HOTTEST PLUG IN THIS FAMILY.

NOTE: REACH, THREAD DIAMETER, AND PLUG TYPE ARE PRINTED ON UNIT BOX AND 10 PACK.

FIGURE 22-24 *Autolite spark plug codes.*
(Courtesy of Autolite.)

FIGURE 22-25 *Bosch spark plug codes.*
(Courtesy of Robert Bosch Corporation.)

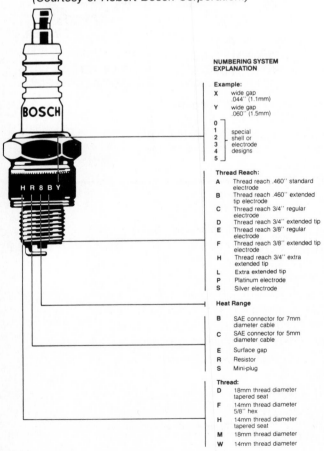

NUMBERING SYSTEM EXPLANATION

Example:

X   wide gap .044" (1.1mm)
Y   wide gap .060" (1.5mm)

0
1
2   special
3   shell or
4   electrode
5   designs

Thread Reach:

A   Thread reach .460" standard electrode
B   Thread reach .460" extended tip electrode
C   Thread reach 3/4" regular electrode
D   Thread reach 3/4" extended tip
E   Thread reach 3/8" regular electrode
F   Thread reach 3/8" extended tip electrode
H   Thread reach 3/4" extra extended tip
L   Extra extended tip
P   Platinum electrode
S   Silver electrode

Heat Range

B   SAE connector for 7mm diameter cable
C   SAE connector for 5mm diameter cable
E   Surface gap
R   Resistor
S   Mini-plug

Thread:

D   18mm thread diameter tapered seat
F   14mm thread diameter 5/8" hex
H   14mm thread diameter tapered seat
M   18mm thread diameter
W   14mm thread diameter

6. Check for a cracked insulator.
7. Check for a possible chipped center insulator around the center electrode.

According to the spark plug manufacturers, there are no "seconds." The price for a spark plug varies because of convenience, delivery, or sales promotion. Therefore, the *new* brand name spark plug for sale at the discount store is the same plug that you can buy at an auto-parts store.

SPARK PLUG REMOVAL AND INSTALLATION

Most manufacturers recommend that the engine be allowed to cool before attempting to remove spark plugs, especially engines with aluminum cylinder heads. All spark plugs should be properly gapped before installation. Do not depend on the factory-set gap to be correct after shipping and handling. Every spark plug is designed for a specific gap. Do not exceed 0.010 in. wider or narrower than the normal design gap for the spark plug. Spark plugs should be installed being certain that the spark plug threads are not crossed. Use a rubber hose or an old spark plug wire boot on the spark plug to start the threads. If cross-threaded, the

Spark Plug	Torque Wrench		Without Torque Wrench	
	Cast-Iron Head	Aluminum Head	Cast-Iron Head	Aluminum Head
Gasket				
14 mm	26–30 lb/ft	18–22 lb/ft	1/4 turn	1/4 turn
18 mm	32–38 lb/ft	28–34 lb/ft	1/4 turn	1/4 turn
Tapered seat				
14 mm	7–15 lb/ft	7–15 lb/ft	1/16 turn	1/16 turn
18 mm	15–20 lb/ft	15–20 lb/ft	Snug	Snug

hose or boot will simply slip and no damage will be done to the threads. Install spark plugs to the proper tightness (torque) to be assured that the spark plug shell makes proper contact with the cylinder head for maximum heat transfer. See Figure 22-26.

Some manufacturers recommend the use of a high-temperature antiseize compound on the threads of spark plugs being installed in aluminum heads. Other manufacturers do not recommend the use of any lubricants on spark plug threads because the lubricant can cause torque-reading errors. If a lubricant is used, the torque specifications (as listed above) should be reduced by 40%.

GASKET TYPE PLUGS

THREAD PLUG INTO CYLINDER HEAD BY HAND

FINGER TIGHT

TIGHTENING WITH SOCKET WRENCH

1/4 TURN   **FINGER TIGHT**

TAPERED SEAT PLUGS

FINGER TIGHT

THREAD PLUG INTO CYLINDER HEAD BY HAND

1/16 TURN   **FINGER TIGHT**

TIGHTENING WITH SOCKET WRENCH

FIGURE 22-26 *Installation of a spark plug with a gasket. For new spark plugs, tighten 90° (¼ turn) after contacting the gaskets. For used spark plugs, tighten only 30° after contacting the gaskets. For a tapered seat spark plug, tighten only 15° (¹⁄₁₆ turn) after contact with the tapered seat. (Courtesy of Champion Spark Plug Company.)*

SUMMARY

1. The condition of spark plugs is an important part of the ignition system. Spark plugs should be carefully inspected for wear and color whenever engine operation is poor or during once-a-year inspection. The rounding of the center electrode is the first section of the spark plug that wears. See Figure 22-27.

2. Normal spark plug color is tan or light gray. Dark or black center insulator color indicates too cold a spark plug or faulty engine condition. Too white an insulator indicates too hot a spark plug or an engine problem such as a vacuum leak. Resistor spark plugs are used to help RFI interference and result in longer spark plug life.

3. Spark plug code numbers identify the heat range, reach, thread diameter, electrode type, and sometimes spark plug gap.

4. Copper-core spark plugs are used to allow the use of longer center insulators for foul resistance while keeping the temperature of the tip low to prevent preignition.

5. The proper tightness of any spark plug is important for proper heat transfer from the spark plug to the engine cylinder heat.

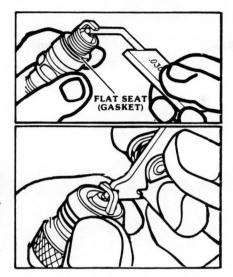

FLAT SEAT (GASKET)

FIGURE 22-27 Whenever inspecting or installing new spark plugs, the gap should be checked and corrected, if required, using a special spark plug gap tool to prevent possible damage to the spark plug. (Courtesy of Sun Electric Corporation.)

STUDY QUESTIONS

22-1. Describe the normal wear and color of a spark plug.

22-2. Describe the possible problem if a spark plug center insulator is black or dark. What possible conditions exist if the spark plug is white?

22-3. Explain why resistor spark plugs are used in most car engines.

22-4. List four items that can be identified about a spark plug by knowing its code number.

MULTIPLE-CHOICE QUESTIONS

22-1. Spark plug numbering and lettering identifies:
(a) the heat range.
(b) reach.
(c) electrode type.
(d) all of the above.

22-2. For American-made spark plugs and many import brands, the heat range code number indicates:
(a) the higher the number, the hotter the plug.
(b) the higher the number, the colder the plug.
(c) that resistor plugs are colder than nonresistor types.
(d) that resistor plugs are hotter than nonresistor types.

22-3. A spark plug that could correctly replace an AC R46TSX is:
(a) AC 45XLS.
(b) AC R45TX.
(c) AC R46TS6.
(d) AC R44NSX.

22-4. A spark plug that has a white center insulator indicates that:
(a) the engine may have a vacuum leak.
(b) the ignition timing may be overadvanced.
(c) the ignition timing may be retarded.
(d) both (a) and (b).

22-5. A spark plug that has a black center insulator indicates that:
(a) the engine may have a vacuum leak.
(b) the ignition timing may be overadvanced.
(c) the choke in the carburetor may be stuck closed.
(d) both (a) and (b).

22-6. Resistor spark plugs are used:
(a) to help control RFI.
(b) to reduce electrode wear.
(c) with unleaded fuel only.
(d) (a) and (b) only.

22-7. Most spark plug brands use the same code letter for:
(a) reach.
(b) resistor.
(c) diameter thread.
(d) electrode type.

22-8. Copper-core spark plugs:
(a) should be used only in engines run on unleaded fuel.
(b) will not last as long as standard plugs.
(c) are more likely to foul during rich-mixture conditions.
(d) none of the above.

23

Ignition System Testing

THE IGNITION MUST WORK CORRECTLY FOR ANY ENGINE TO PERFORM properly. The topics covered in this chapter include:

1. How to check for spark
2. Basic ignition system troubleshooting procedure
3. Testing ignition coils
4. Testing and troubleshooting point-type ignition
5. Testing and troubleshooting electronic ignition pickup coils
6. Testing and troubleshooting electronic ignition modules

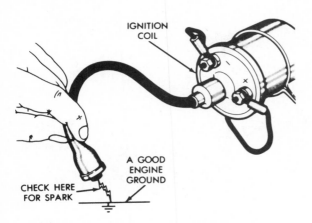

FIGURE 23-1 *Checking for spark. (Courtesy of Chrysler Corporation.)*

IGNITION SYSTEM TROUBLESHOOTING

The ignition system must operate correctly for the engine to start and run correctly. If the starter motor is operating correctly, yet the engine will not start, the ignition system must be checked for correct operation.

How do you check for spark? Remove the coil wire from the center of the distributor cap and hold the end of the wire ¼ in. away from the engine. Crank the engine. There should be a blue spark. See Figure 23-1. If there is a yellow spark, the spark is weak. Check:

1. Points and condenser (if equipped)
2. Coil wire (if equipped)
3. Coil (see below)
4. Connections and wires to the coil

NOTE: Many HEI ignitions do not use a coil wire; therefore, a slightly different procedure is required. If testing a 1975 or newer General Motors car equipped with HEI electronic ignition with integral coil, remove one spark plug and place the steel shell of the plug against the engine block with the spark plug wire attached. Crank the engine. A spark should be heard and seen at the spark plug. An occasional spark should be considered a "no spark" situation.

TROUBLESHOOTING A "NO SPARK" SITUATION

All ignition systems (point-type and all types of electronic ignitions) operate by opening and closing the ground return path circuit of the ignition coil primary. A simple two-step test procedure can be used to determine ignition system problems.

Step 1. With the ignition switch "on" ("run"), use a voltmeter or a 12-V test light to determine if voltage is available at the positive (+) of the coil (could be labeled "BAT"). The voltage reading should be 5-7 V for point-type ignition systems with the points closed, or above 8 V while cranking the engine, regardless of ignition type.

NOTE: The voltmeter may indicate less than battery voltage or the test light may be dim because of the voltage drop caused by the ballast resistor or the resistance wire (if equipped).

If no voltage is present at the positive (+) terminal of the ignition coil, the problem could be due to:

(a) A defective ballast resistor (if equipped).
(b) A defective or misadjusted ignition switch. Most steering-column-mounted ignition switches are actually located on an adjustable bracket attached to the steering column. The attaching bracket is adjustable by loosening two or three bolts or screws and moving the switch up or down the steering column. The key switch on the steering column operates the ignition switch by moving a metal rod that goes into the ignition switch. See Figure 23-2.
(c) Other electrical problems which could prevent voltage from getting to the coil primary.

Step 2. Use a voltmeter or test light and check for a *pulsing* current at the negative (−) side of the coil while *cranking the engine.* [The negative (−) side of the coil could be labeled "DEC" or "TACH."] If current is pulsing at the negative (−) side of the coil and there is no spark from the coil, replace the ignition coil. If current is

FIGURE 23-2 *Adjustment location for typical steering column-mounted ignition switch. This style of ignition switch is mounted on top of the steering wheel behind the dash panel and operated by a rod from the key switch.*

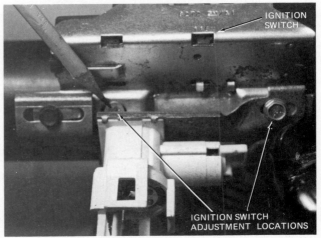

not pulsing, check for correct operation of points and condenser if point-type ignition. On electronic ignition, if the current is not pulsing, check the pickup coil (see p. 309). If the pickup coil tests okay, check the wiring between the pickup coil and the module. If the wiring is okay, replace the module.

IGNITION COIL TESTING

The ignition coil can be tested with an ohmmeter. For accurate test results, all wires should be disconnected from the ignition coil before testing. With the ohmmeter set on the low scale, test the coil's primary windings by touching one lead of the meter to the primary positive (+) terminal of the coil and the other end to the negative (−) terminal. See Figure 23-3.

Specifications vary for each model of car, but the usual specification is between 1 and 3 Ω. If not between 1 and 3 Ω, check the manufacturer's exact specification before replacing the ignition coil because some coils have very low primary winding resistance, less that 1 Ω. The secondary coil resistance usually ranges from 6000 to 30,000 Ω. See Figures 23-4 and 23-5.

> **NOTE:** Many Chrysler products use a ceramic ballast resistor in the ignition circuit. If this ballast resistor is defective, no spark will occur from the coil. This resistor is ceramic and gets hot to the touch during normal operation. Rainwater often cracks these resistors, which can cause the engine to quit suddenly. The ballast resistor is usually bolted to the firewall. Many technicians recommend keeping a spare ballast resistor in the event of failure.

FIGURE 23-3 *(Courtesy of Chrysler Corporation.)*

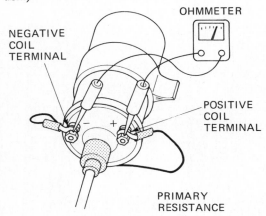

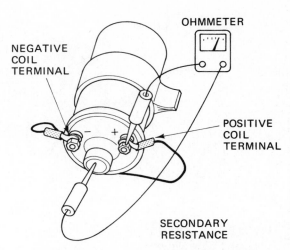

FIGURE 23-4 *(Courtesy of Chrysler Corporation.)*

FIGURE 23-5 *Testing the primary and secondary coil winding of an internal coil GM HEI. Step 1: Connect the ohmmeter as shown in figure 1. The reading should be zero or almost zero (very low ohms). If not, replace the coil. Step 2: Connect the ohmmeter both ways as shown in figure 2. Set the ohmmeter on the high scale. Replace the coil only if both readings are infinite. (Courtesy of Oldsmobile Division, GMC.)*

TESTING IGNITION COIL

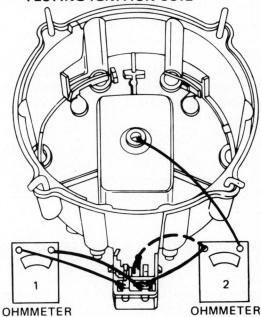

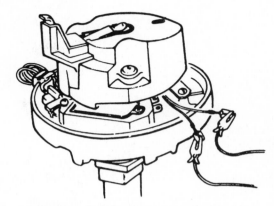

FIGURE 23-6 *Testing the pickup coil resistance on a typical GM HEI system. The ohmmeter reading should be 500 to 1500 Ω.*

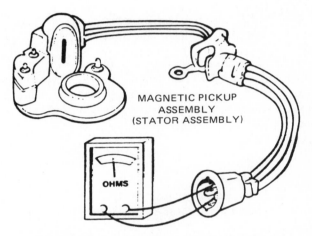

FIGURE 23-7 *Testing a typical Ford-type pickup coil. The ohmmeter reading should be 400 to 1000 Ω.*

IGNITION COIL DAMAGE

The ignition coil can be damaged if the ignition switch is left in the "on" position without the engine running. With point-type ignition systems, no harm is done if the ignition points happen to be "open," because this opens the circuit through the coil. However, Chrysler electronic ignition systems (and perhaps others) provide a ground return path through the module (electronic ignition control unit) whenever the ignition is "on" ("run"). The signal (pulse) from the pickup coil triggers the module to "open" the coil primary circuit to produce a spark whenever the engine is rotated. Therefore, to protect the ignition coil from possible overheating damage, always turn the ignition to the "accessory" position whenever accessory power is needed with the engine not running.

PICKUP COIL TESTING

The pickup coil replaced the points as the switch that controls the module that controls the operation of the ignition coil on most electronic ignition systems. If the pickup coil is defective, the coil will not be triggered to produce the high voltage necessary to fire the spark plugs. Each manufacturer has its own names for the pickup coil assembly. The most commonly found assembly names and testing specifications are summarized below.

GM. HEI electronic ignition identifies the pickup coil as *pole piece* and *timer core*. The acceptable resistance between the two wires leading to the pole piece from the module is 500 to 1500 Ω. The green and white wires must be disconnected from the module before testing. See Figure 23-6.

NOTE: The entire distributor has to be removed from the engine and disassembled to replace the pole piece.

Ford. Ford's electronic ignition pickup coil assembly is called the *stator* and *armature*. The acceptable resistance between the two colored leads (orange and purple) is 400 to 1000 Ω. See Figure 23-7.

NOTE: The stator can be replaced without removing the distributor from the engine.

Chrysler. The pickup coil and reluctor are the proper names for the pulse generator unit in most Chrys-

ler models. The proper resistance between the two leads of the pickup coil (orange and black) is between 150 and 900 Ω. See Figure 23-8. The pickup coil can be replaced without removing the distributor from the engine.

MODULE TESTING

Various car and test equipment manufacturers offer electronic ignition module (control unit) testers. These testers are usually electrical pulsing units. Most electronic ignition modules (control units) can be tested using an electric soldering gun to "pulse" the module. An electric soldering gun (or similar electric unit) contains electrical coils that produce a strong alternating (moving) magnetic field. This

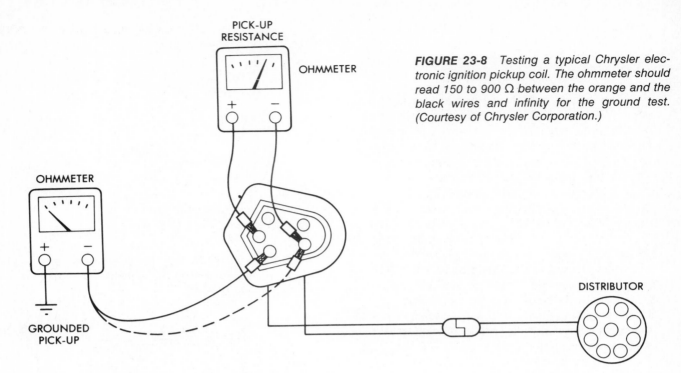

FIGURE 23-8 *Testing a typical Chrysler electronic ignition pickup coil. The ohmmeter should read 150 to 900 Ω between the orange and the black wires and infinity for the ground test. (Courtesy of Chrysler Corporation.)*

moving magnetic field can produce a small current in the pickup coil assembly of an electronic ignition system. *If the pickup coil has been tested to be okay, a soldering gun may work to test many ignition modules using the following procedure:*

1. Remove the coil wire from the center of the distributor cap.

2. Install a spark plug at the end of the coil wire and ground the shell of the spark plug.

   NOTE: If testing a GM HEI unit without a separate coil wire, temporarily remove the distributor cap and make certain that the distributor rotor is pointing *directly* toward *any* spark plug terminal of the distributor cap. Rotate the engine, if necessary, to align the rotor to a plug terminal. Reinstall the distributor cap and remove the spark plug that corresponds to the spark plug wire lined up with the rotor.

3. Turn the ignition switch to the "on" ("run") position (do not crank or start the engine!).

4. Turn on an electric soldering gun and hold the soldering gun as close to the distributor cap as possible.

5. The rapidly changing (alternating 60 times per second because of the alternating ac 110-V current) creates a rapidly changing magnetic field. This changing magnetic field cuts across the coil windings of the pulse generator (pickup) coil. This small current generated in the pickup coil windings "triggers" the module by turning the main power transistor "on" and "off." See Figure 23-9.

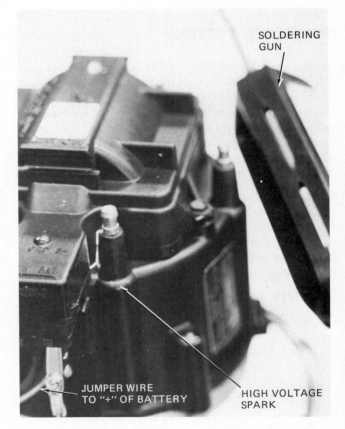

FIGURE 23-9 *Testing an HEI distributor with a soldering gun. Jumper wires are attached from a battery to the "BAT" terminal of the HEI and ground. Turn the soldering gun on and hold it as close to the distributor as possible. The high-voltage spark indicates that everything in the system is functioning correctly.*

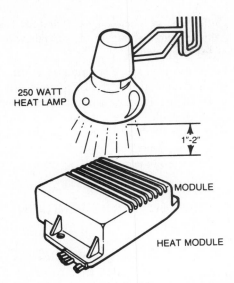

HEAT LAMP

1"-2"

MODULE

HEAT MODULE

FIGURE 23-10 *Ford modules can be heated with a heat lamp to check for intermittent problems. Do not permit the module temperature to exceed 212°F (100°C). (Courtesy of Ford Motor Company.)*

Results: If there is a spark at the test spark plug, the ignition module (control unit) is working (at least able to create a spark). If no spark occurs at the test spark plug, the pickup coil (pulse generator), module (control unit), or interconnecting wires are defective. Recheck the pickup coil and wires before replacing the electronic ignition module.

NOTE: Ford electronic ignition modules (control units) and pickup coils should be tested using a heat lamp to heat the module to ensure proper hot-temperature operation. The module and pickup coil should also be *lightly* tapped with a screwdriver handle during testing to check for possible loose or intermittent module problems. See Figures 23-10 and 23-11.

GM HEI MODULE TEST

Tools Needed

Test lamp with leads

Three jumper wires

Test Procedure (See Figure 23-12)

1. The module must be removed from the distributor.

2. Connect a 12-V dc test lamp between the B and C module terminals.

3. Connect a jumper wire from a 12-V dc source to the B module terminal.

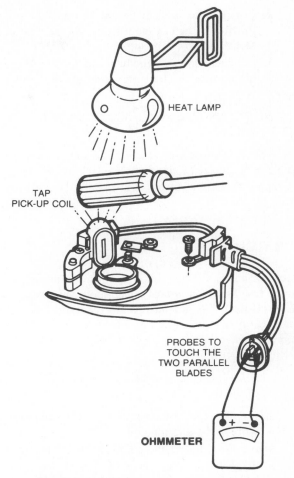

HEAT LAMP

TAP
PICK-UP COIL

PROBES TO
TOUCH THE
TWO PARALLEL
BLADES

OHMMETER

FIGURE 23-11 *Ford pickup coils can be heated with a heat lamp and tapped lightly to check for intermittent ignition problems. Replace any pickup coil that does not measure within specifications. (Courtesy of Ford Motor Company.)*

FIGURE 23-12 *Hookup for testing a GM module using only a test light and three jumper wires.*

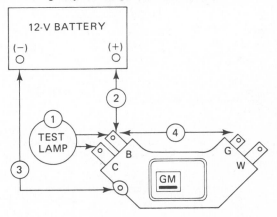

12-V BATTERY

(−)   (+)

TEST
LAMP

B

C

GM

G

W

4. Connect the module ground terminal to a good ground. If the test lamp lights, the module is defective and must be replaced.

5. Connect a jumper wire between the B and G module terminals. The lamp will light if the module is okay.

NOTE: For a seven-pin module, use terminal P instead of G.

The lack of heat-conducting silicone grease underneath the module can cause complete lack of ignition when the engine is warm, yet test and operate normally when cool.

SCRATCH TEST

A *scratch test* is a simple ignition coil operation test. The scratch test works on almost all ignition coils except GM HEI and direct-fire systems, where additional test leads would be necessary. The procedure can easily be performed on the car and includes the following steps.

NOTE: For this test to work, voltage from the ignition must be available at the positive (+) side of the coil.

Step 1. Disconnect the negative (−) (distributor side) wire on the ignition coil.

Step 2. Disconnect the coil wire from the center of the distributor cap and hold it ¼ in. from a good ground.

Step 3. Turn the ignition "on" ("run") (engine off).

Step 4. Attach a short jumper wire connected to a condenser, as shown in Figure 23-13, to the negative (−) side of the coil. By "scratching" this jumper wire or by turning the switch "on" and "off," the coil should be "charged" and then "discharged" whenever the jumper wire is lifted away from ground.

Results: If a good spark is produced at the coil wire, the ignition coil is okay. (If there is still no spark after all the ignition wires are reconnected, the problem could be the pickup coil, module, or the wiring.) If no spark is produced, check for proper voltage at the positive (+) side of the coil. If okay, replace the ignition coil.

CHRYSLER ELECTRONIC IGNITION SCRATCH TEST

Chrysler electronic ignition systems (EIS) can be tested using a variation of the scratch test. A simple three-step procedure tests the entire ignition circuit *except* the pickup coil.

Step 1. Disconnect the two-wire connector near the distributor (orange and black wires). See Figure 23-14.

Step 2. Disconnect the coil wire from the center of the distributor cap and hold the wire approximately ¼ in. from the engine block.

Step 3. With the ignition switch "on" ("run"), "scratch" the exposed wire end of the disconnected two-wire connector from the control unit to the engine block (ground).

Results: If a spark occurs at the coil wire whenever the exposed wire is touched to ground, the ignition switch, ignition coil, and electronic control unit are functioning. If after reconnecting the two-wire connector, no spark occurs, the problem is the pickup coil. If no spark occurs during the scratch test, check for voltage at the ignition coil and test the ignition coil. If current is available at the coil and the coil is good, the problem is a defective electronic control unit or a poor ground connection on the control unit.

FIGURE 23-13 *(Courtesy of Ford Motor Company.)*

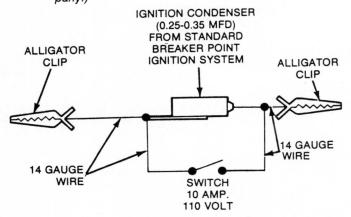

FIGURE 23-14 *Unplug the two-wire connector to scratch test a Chrysler electronic ignition system.*

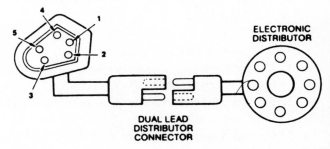

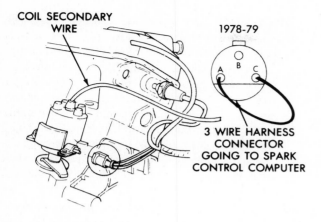

COIL SECONDARY WIRE

1978-79

3 WIRE HARNESS CONNECTOR GOING TO SPARK CONTROL COMPUTER

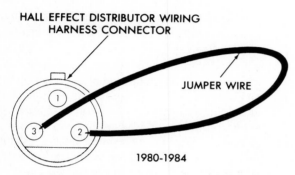

HALL EFFECT DISTRIBUTOR WIRING HARNESS CONNECTOR

JUMPER WIRE

1980-1984

FIGURE 23-15 *Use a jumper wire as shown to scratch test a Chrysler Hall-effect ignition system. (Courtesy of Chrysler Corporation.)*

HALL-EFFECT SCRATCH TEST

A similar test to the scratch test above is the scratch test that Chrysler recommends to check their Hall-effect switches and control circuits. See Figure 23-15.

POINT-TYPE IGNITION PROBLEMS

The primary ignition circuit must be complete to charge and discharge the coil. If the points are not closing or not opening, there will be *no* spark from the coil. A proper spark should be able to jump a ¼ in. gap and the spark should be blue. If the spark is yellow, it may be too weak to ignite the fuel mixture in the cylinders. Common causes of a weak or no spark for a point-type ignition include:

1. Incorrect dwell (point gap). If the points are adjusted too far apart, there may not be enough time to charge the coil fully. If the points are adjusted too close, excessive current flow through the points can cause burning and pitting of the points. This pitting can prevent the coil from becoming

fully charged due to the lack of current flow through the coil because of the voltage drop of the pitted points.

2. Defective or wrong capacity of the ignition condenser (capacitor).

3. Lack of a good ground connection between the stationary ignition point and the distributor housing. Most distributors provide a ground wire connection between the movable breaker plate and the distributor housing. This ground wire must be continuous to provide for the proper ground return path circuit for the primary ignition circuit.

NOTE: A loose condenser hold-down bracket or loose contact point mounting screws can cause intermittent ignition system failure. A common symptom is an engine miss or backfire due to lack of a proper and consistent ground connection.

TROUBLESHOOTING INTERMITTENT IGNITION PROBLEMS

Engine "missing" or intermittent ignition problems can be difficult to find. Below are listed several possible causes of occasional or intermittent ignition problems. These problems could result in a miss at a certain speed range, missing at all speed ranges, or could cause a "no start" situation.

1. Broken or defective pickup coil wires or pickup coil assembly. This fault is commonly found on electronic ignition systems equipped with vacuum advance units. The vacuum advance moves the pickup coil and its wires. The pickup coil wires can fray or break, which can cause occasional engine cutout or missing. This missing usually occurs only at a particular vehicle speed and/or load condition. A broken pickup coil wire is often first noticeable as a missing or engine cutout between 35 and 55 mph (56 and 88 kph). Idle, low-speed, and high-speed engine operation is often unaffected because of the operation of the vacuum advance unit during these periods of operation.

2. A coil with *shorted* primary or secondary windings could result in a weak spark or no spark from the coil secondary tower.

3. Open or high-resistance coil wire will cause a weak or reduced voltage spark.

4. A dirty distributor cap or coil cap can cause a reduction of high voltage out of the coil due to increased secondary circuit capacitance. The top of the coil (*cap*) must be kept *clean* and dry to ensure maximum coil output. Many manufactur-

FIGURE 23-16 *A dirty, moist, or cracked distributor cap can cause high-voltage current to seek the path of least resistance to ground. This photo was created by connecting a battery to the BAT terminal and the housing of the HEI distributor and then rotating the distributor shaft while it was being held in a vise.*

ers (especially of import cars) install a protective cover to help keep moisture and dirt form accumulating on coils and distributor caps.

5. A cracked coil secondary tower can cause the high-voltage spark to travel to one of the primary terminals. See Figure 23-16.

QUICK-AND-EASY SECONDARY IGNITION TESTS

Most engine running problems are caused by defective or out-of-adjustment ignition components. Many ignition problems involve the high-voltage secondary ignition circuit.

Test 1. If there is a crack in a distributor cap, coil, spark plug, or a defective spark plug wire, a spark may be visible at night. Since the highest voltage is required during part-throttle acceleration, the technician's assistant should accelerate the engine slightly with the gear selector in "drive" or second gear (if manual transmission) the brake firmly applied. If *any* spark is visible, the location should be closely inspected and the defective parts replaced. A blue glow or "corona" around the shell of the spark plug is normal and *not* an indication of a defective spark plug.

Test 2. For intermittent problems, use a spray bottle to apply a water mist to the spark plugs, distributor cap, and spark plug wires. With the engine running, the water may cause an arc through any weak insulating materials and cause the engine to miss or stall.

Test 3. To determine if the rough engine operation is due to secondary ignition problems, connect a 6- to 12-V test light to the negative (−) side (tach) of the coil. Connect the other lead of the test light to the positive (+) lead of the battery (not to ground!). With the engine running, the test light should be dim and steady in brightness. If there is high resistance in the secondary circuit (such as a defective spark plug wire), the test light will pulse bright at times. If the test light varies noticeably, this indicates that the secondary voltage cannot "find" ground easily and is feeding back through the primary windings of the coil. This feedback causes the test light to become brighter.

SUMMARY

1. Most automotive engine running problems are due to defects in the ignition system.

2. All ignition systems must supply current to the positive (+) side of the ignition coil and turn the ground circuit of the negative (−) side of the coil "on" and "off." The coil can be tested with an ohmmeter. Current at the positive (+) side of the coil can be measured with a voltmeter (above 8 V). The negative (−) side of the coil should be pulsing whenever the engine is cranked or running. If the test light on the voltmeter does not indicate pulsing current at the negative (−) side of the coil, the points, condenser (if equipped), pickup coil, or module is (are) defective.

3. Pickup coils can be tested with an ohmmeter. If the resistance valves differ from the allowable range, the pickup coil must be replaced. Modules can be tested by pulsing the module with a signal created in a known-good pickup coil by an electric soldering gun or similar unit. Modules can also be tested with special module testers. GM modules can be tested using a simple jumper wire and test light hookup.

4. Intermittent ignition problems are often found by close inspection of all ignition components and connecting wires.

STUDY QUESTIONS

23-1. Describe how to check for spark on an engine equipped with a General Motors HEI ignition system.

23-2. Explain how to check an ignition coil using an ohmmeter.

23-3. Describe the procedure and specifications for testing a pickup coil using an ohmmeter.

23-4. Explain how a soldering gun can be used to test electronic ignition modules.

23-5. Describe the scratch test.

MULTIPLE-CHOICE QUESTIONS

23-1. Technician A says that a pickup coil (pulse generator) can be tested with an ohmmeter. Technician B says that ignition coils can be tested with an ohmmeter. Which technician is correct?
- **(a)** A only.
- **(b)** B only.
- **(c)** both A and B.
- **(d)** neither A nor B.

23-2. Technician A says that a defective spark plug wire can cause an engine miss. Technician B says that a defective pickup coil wire can cause an engine miss. Which technician is correct?
- **(a)** A only.
- **(b)** B only.
- **(c)** both A and B.
- **(d)** neither A nor B.

23-3. The _____ sends a pulse signal to an electronic ignition module.
- **(a)** ballast resistor.
- **(b)** pickup coil.
- **(c)** ignition coil.
- **(d)** condenser.

23-4. Typical primary coil resistance specifications usually range from:
- **(a)** 100 to 450 Ω.
- **(b)** 500 to 1500 Ω.
- **(c)** 1 to 3 Ω.
- **(d)** 6000 to 30,000 Ω.

23-5. Typical secondary coil resistance specifications usually range from:
- **(a)** 100 to 450 Ω.
- **(b)** 500 to 1500 Ω.
- **(c)** 1 to 3 Ω.
- **(d)** 6000 to 30,000 Ω.

23-6. Technician A says that an engine will not start and run without a good ground on the points and condenser (breaker plate). Technician B says that one wire of any pickup coil must be grounded. Which technician is correct?
- **(a)** A only.
- **(b)** B only.
- **(c)** both A and B.
- **(d)** neither A nor B.

23-7. Technician A says that a GM HEI distributor rotor can burn through and cause an engine miss during acceleration. Technician B says that a defective pickup coil can cause an engine miss during acceleration. Which technician is correct?
- **(a)** A only.
- **(b)** B only.
- **(c)** both A and B.
- **(d)** neither A nor B.

23-8. The secondary ignition circuit can be tested using:
- **(a)** an ohmmeter.
- **(b)** a test light.
- **(c)** an ammeter.
- **(d)** both a and b.

23-9. Electronic ignition modules can be tested using:
- **(a)** special electronic ignition module testers.
- **(b)** a soldering gun for some modules.
- **(c)** a test light and jumper wires for some modules.
- **(d)** all of the above.

23-10. The scratch test tests operation of:
- **(a)** the pickup coil.
- **(b)** ignition coil.
- **(c)** ignition switch and ignition coil.
- **(d)** spark plug wires.

24

Oscilloscope Testing

OSCILLOSCOPE (COMMONLY SHORTENED TO "SCOPE") TESTING IS A VALuable engine diagnostic tool. An oscilloscope is an electronic optical device that indicates changes in electrical voltage by means of a cathode ray tube or CRT. Scope testing is used to display ignition system and alternator patterns. These patterns are compared to the shape and contours of a normally operating system. The area of the malfunction of a system can be identified by the section of the scope pattern that does *not* look like the normal pattern. The topics covered in this chapter include:

1. Scope testing procedures
2. Scope patterns for point-type ignition systems
3. Scope patterns for electronic ignition systems
4. Examples of defective scope patterns
5. Alternator scope patterns

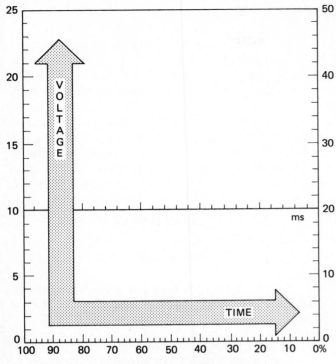

FIGURE 24-1 *(Courtesy of Sun Electric Corporation.)*

SCOPE TESTING THE IGNITION SYSTEM

Any automotive scope will show an ignition system pattern. All ignition systems, whether electronic, Hall-effect, or point-type, must charge and discharge an ignition coil. With the engine "off," most scopes will display a straight horizontal line. With the engine running, this horizontal line is changed to a pattern which will indicate sections both above and below this zero line. The section of this pattern which is *above* the zero line indicates that the ignition coil is *discharging*. Sections of the scope pattern *below* the zero line indicate *charging* of the ignition coil. The height of the scope pattern indicates *voltage*. The length (from left to right) of the scope pattern indicates *time*. See Figure 24-1.

CONNECTING THE SCOPE LEADS

All scopes must be connected at four points to function correctly (see Figure 24-2):

1. *Connection to the coil wire.* This connection or inductive pickup around the coil wire (or on the

FIGURE 24-2 *Typical oscilloscope hookup. (Courtesy of Sun Electric Corporation.)*

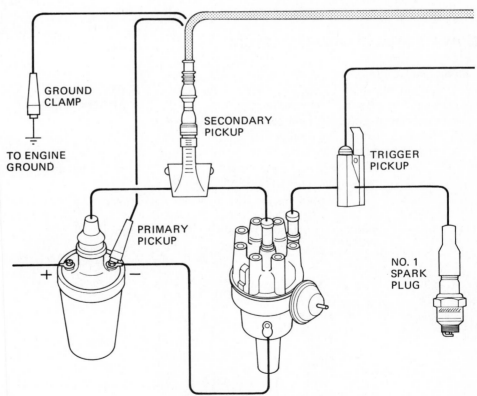

distributor cap for HEI systems) signals the electronics of the scope, the intensity, and polarity of the discharging and charging of the coil. This connection primarily determines the scope pattern.

2. *Connection to the number 1 spark plug wire.* The connection or inductive pickup around the number 1 spark plug wire signals the scope to start the sequence of coil discharges so that the patterns will be displayed correctly on the screen. The number 1 spark plug connection is commonly used to operate the timing light and tachometer units of the scope (if equipped).

3. *Connection to the negative (−) side of the coil.* The connection on the negative (−) side of the coil triggers the dwell meter section on the scope (if equipped), but *not* the dwell section of the scope pattern. The connection to the negative (−) side of the coil also allows the ignition system to ''ground out'' a single cylinder one at a time to check the power balance of the engine.

4. *Connection to ground.* This last connection must be to a clean metal connection on the engine block or a noninsulated engine bracket to ensure a proper ground connection for power balance testing and accurate dwell meter readings.

HOW TO READ A SCOPE PATTERN

Automotive oscilloscopes provide controls that permit viewing a pattern for the primary ignition circuit and the secondary ignition circuit. Many technicians view just the secondary pattern because the ignition primary is reflected in the secondary pattern. Each section of a scope pattern has a name. The name of the scope pattern section also describes the section of the ignition operation.

FIRING LINE

The leftmost vertical (upward) line is called the *firing line*. The height of the firing line should be between 5000 and 15,000 V (5 and 15 kV) with not more than a 3-kV difference between the highest and the lowest cylinder's firing line. See Figure 24-3.

The height of the firing line indicates the *voltage* required to fire the spark plug. It requires a high voltage to make the air inside the cylinder electrically conductive (ionize the air). A higher-than-normal height (or higher than other cylinders) can be caused by one or more of the following:

1. A spark plug gapped too wide
2. A lean fuel mixture
3. A defective spark plug wire

If the firing lines are higher than normal for *all* cylinders, then some possible causes include one or more of the following:

1. A worn distributor cap and/or rotor (if equipped)
2. All spark plugs excessively worn
3. A defective coil wire (the high voltage could still ''jump'' across the open section of the wire to fire the spark plugs)

See page 320 for further information on firing line diagnosis.

SPARK LINE

The spark line is a short horizontal line connected to the firing line. The ''height'' of the spark line represents the voltage required to maintain the spark across the spark plug

CONVENTIONAL SECONDARY PATTERN

FIGURE 24-3 (Courtesy of Sun Electric Corporation.)

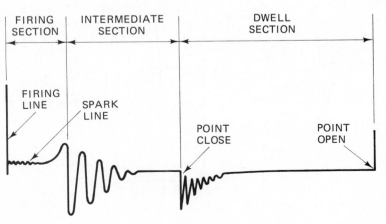

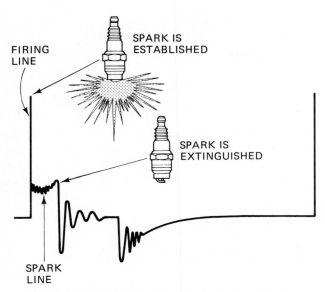

FIRING LINE

SPARK IS ESTABLISHED

SPARK IS EXTINGUISHED

SPARK LINE

FIGURE 24-4 *(Courtesy of Sun Electric Corporation.)*

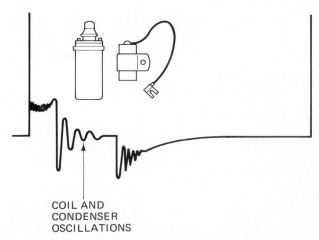

COIL AND CONDENSER OSCILLATIONS

FIGURE 24-5 *(Courtesy of Sun Electric Corporation.)*

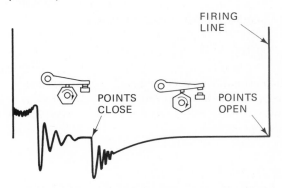

FIRING LINE

POINTS CLOSE

POINTS OPEN

FIGURE 24-6 *(Courtesy of Sun Electric Corporation.)*

after the spark has started. The height of the spark line should be one-fourth of the height of the firing line (between 1.5–2.5 kV). The "length" (from left to right) of the line represents the length of time (duration) of the spark. The spark stops at the end (right side) of the spark line as shown in Figure 24-4. See page 322 for a complete detailed explanation of spark lines.

INTERMEDIATE OSCILLATIONS

After the spark has stopped, there is still some remaining energy left in the coil.

This remaining energy dissipates in the coil windings and the entire secondary circuit. The oscillations are also called the "ringing" of the coil as it is pulsed.

In a point-type ignition system, the condenser, which is in the primary ignition circuit, also affects the oscillations of the secondary coil. The secondary pattern amplifies any voltage variation occurring in the primary circuit because of the turns ratio between the primary and secondary windings of the ignition coil.

A correctly operating ignition system should display three or more "bumps" (oscillations). See Figure 24-5.

POINT-CLOSE POINT

After the intermediate oscillations, the coil is empty (not charged), as indicated by the scope pattern being on the zero line for a short period. When the points close (on a point-type system) or the transistor turns "on" (electronic system), the coil is being "charged." Note that the charg-

ing of the coil occurs slowly (coil charging oscillations) due to the inductive reactance of the coil. See Figure 24-6.

DWELL SECTION

Dwell is the number of degrees of distributor shaft rotation that the points are closed. During this time, the scope pattern (from point-close point to the right edge) indicates the time that the current is charging the coil. At the end of the dwell section is the beginning of the next firing line. This point is called the *point-open point* (transistor "off") and indicates that the primary current of the coil is stopped, resulting in a high-voltage spark out of the coil.

PRIMARY PATTERN

The scope pattern of the primary ignition circuit shown in Figure 24-7 is similar to the secondary because the secondary reflects the operation of the primary circuit.

CONVENTIONAL PRIMARY PATTERN

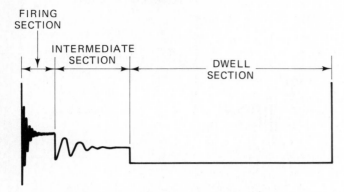

FIGURE 24-7 *(Courtesy of Sun Electric Corporation.)*

SCOPE PATTERN DIAGNOSIS

To get accurate problem-solving results using a scope, it is best to have the engine at normal operating temperature. Extremely cold or overheated engines may not be operating correctly due to the temperature rather than to a mechanical problem. The engine is warm when the upper radiator hose is hot and pressurized.

PATTERN SELECTION

The basic pattern illustrated in Figure 24-7 is not seen totally on a scope. Ignition oscilloscopes use three positions to view certain sections of the basic pattern more closely. These three positions are called:

1. *Superimposed.* This position is used to look at differences in patterns between cylinders in all areas except the firing line. There are no firing lines illustrated in superimposed positions. See Figure 24-8.

2. *Raster (stacked).* The number 1 cylinder is at the bottom on most scopes. Use the raster pattern to look at the spark line length and point-close point. The raster (stacked) pattern shows all areas of the scope pattern except the firing lines. See Figure 24-9.

3. *Display (parade).* Display (parade) is the only position where firing lines are visible. The firing line section for the number 1 cylinder is on the far-right side of the screen, with the remaining portions of the pattern on the left side. This selection is used to compare the height of firing lines among all cylinders. See Figure 24-10.

READING THE SCOPE ON DISPLAY (PARADE)

Start the engine and operate at approximately 1000 rpm to ensure a smooth and accurate scope pattern. Firing lines are visible only on the display (parade) position. The firing lines should all be 5 to 15 kV in height and be within 3 kV of each other. If one or more cylinders have high firing lines, this could indicate a defective (open) spark plug wire, a spark plug gapped too far, or a lean fuel mixture affecting just those cylinders.

A lean mixture (not enough fuel) requires a higher voltage to ignite because there are not as many droplets of fuel in the cylinder for the spark to use as "stepping stones" for the voltage to jump across. Therefore, a lean mixture is "less conductive" than a rich mixture.

EXAMPLE OF FIRING LINE PROBLEMS

If the height of the two firing lines that are side by side is lower than the rest of the firing lines, the distributor cap is defective (carbon tracked between the two inserts) or the spark plug wires are crossed. The firing lines are low be-

FIGURE 24-8 *(Courtesy of Sun Electric Corporation.)*

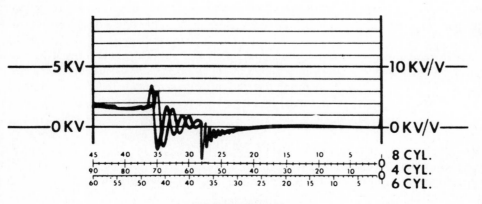

SUPERIMPOSED

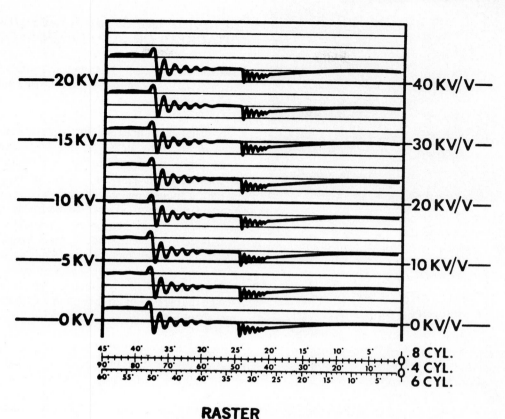

RASTER

FIGURE 24-9   *(Courtesy of Sun Electric Corporation.)*

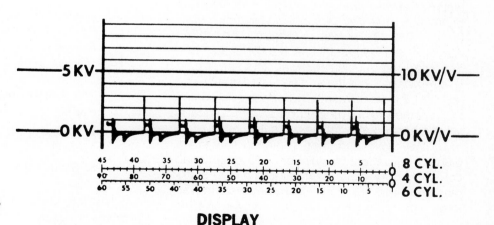

FIGURE 24-10   *(Courtesy of Sun Electric Corporation.)*

DISPLAY

cause the spark is being sent to a cylinder *not* under the high pressure of the compression stroke. It requires a lower voltage (lower firing line) to fire a spark plug under low compression. If the firing lines are high, lower, high, lower (like a roller coaster), the fuel mixture (usually the idle mixture) is unequal, with the high firing lines representing a lean fuel mixture. See Figure 24-11.

A high firing line that also extends below the zero line is an indication of a defective (open) spark plug wire. If the firing line is high, yet the firing does *not* drop below the zero line, the high-voltage spike has "found" a path to ground. This path to ground could be through a carbon track in the distributor cap, rotor or be caused by poor insulation on a plug wire.

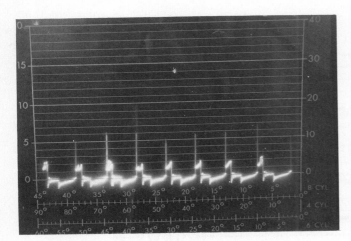

FIGURE 24-11 *Typical eight-cylinder display (parade) pattern. Notice that the firing line for the number 1 cylinder is at the far right of the screen. Also notice the differences in firing line heights and lengths and slope of the spark lines.*

A Technician's Toughie

A car on the scope ran poorly, yet the scope patterns were "perfect." Remembering that the scope indicates only that a spark has occurred (not necessarily inside the engine), the technician removed one spark plug wire at a time. Every time a plug wire was disconnected, the engine ran worse, until the last cylinder was checked. When the last spark plug wire was disconnected, the engine ran the same. The technician checked the spark plug wire with an ohmmeter; it tested within specifications (less than 10,000 Ω per foot of length). The technician also removed and inspected the spark plug. The spark plug looked normal. The spark plug was reinstalled and the engine tested again. The test had the same results as before—an engine that seemed to be running on seven cylinders, yet the scope pattern was perfect.

The technician then replaced the spark plug for the affected cylinder. The engine ran correctly. Very close examination of the spark plug showed a thin crack between the wire terminal and the shell of the plug. Why didn't the cracked plug show on the scope? The scope simply indicated that a spark had occurred. The scope cannot distinguish between a spark inside or outside the engine. In this case, the required voltage to travel through the spark plug crack to ground was about the same voltage required to jump the spark plug electrodes inside the engine. The spark that occurred across the cracked spark plug may, however, have been visible at night with the engine running.

READING THE SPARK LINES [RASTER (STACKED) OR SUPERIMPOSED]

Spark lines can easily be seen on either superimposed or raster (stacked) positions. On the raster (stacked) position, each individual spark line can be viewed. See Figure 24-12.

The *spark lines* should be level and one-fourth as high as the firing lines (1.5–2.5 kV). The *length* of the spark line is the critical factor for proper operation of the engine because it represents the spark duration time. There is only a limited amount of energy in an ignition coil. If most of the energy is used to ionize the air gaps of the rotor and the spark plug, there may be not enough energy remaining to create a spark duration long enough to completely burn the air/fuel mixture. Many scopes are equipped with a *millisecond (ms) sweep*. This means that the scope will sweep only that position of the pattern that can be shown during a 5- or 25-ms setting. The length of the spark line should be

0.8 ms	Too short
1.5 ms	Average
2.0 ms	Too long

If the spark line is too short, possible causes include

1. Spark plug(s) gapped too wide
2. Rotor to distributor cap gapped too wide (worn cap or rotor)
3. High-resistance spark plug wire
4. Air/fuel mixture too lean (vacuum leak, broken valve spring, etc.

If the spark line is too long, possible causes include

1. Fouled spark plug(s)
2. Spark plug(s) gapped too closely
3. Shorted spark plug or spark plug wire

Many scopes do not have a millisecond scale. Some scopes are labeled in degrees and/or percent (%) of dwell. See Figure 24-13. The chart below can be used to determine acceptable spark line length.

NORMAL SPARK LINE LENGTH (AT 700 to 1200 RPM)

Number of Cylinders	Milliseconds	Percent Dwell Scale	Degrees
4	1.0–1.5	3%–6%	3°–5°
6	1.0–1.5	4%–9%	2°–5°
8	1.0–1.5	6%–13%	3°–6°

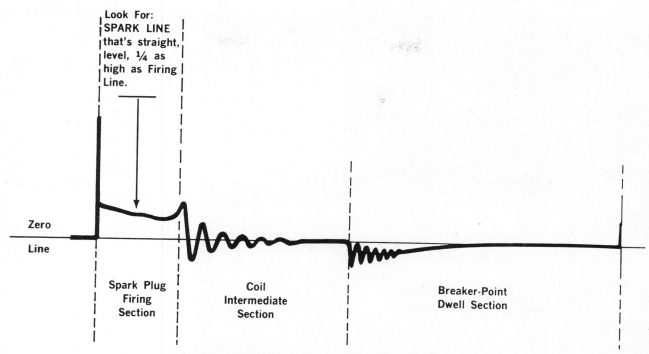

Look For:
SPARK LINE
that's straight,
level, ¼ as
high as Firing
Line.

Zero

Line

Spark Plug
Firing
Section

Coil
Intermediate
Section

Breaker-Point
Dwell Section

FIGURE 24-12 *(Courtesy of Sun Electric Corporation.)*

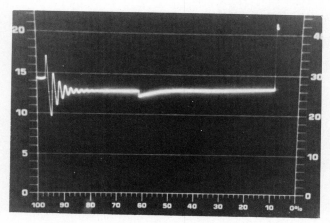

FIGURE 24-13 *Notice that the length of the spark line is about 4% for this four-cylinder engine. The pattern length control should be adjusted before reading the spark line length to make certain that the pattern is full length on the screen.*

SPARK LINE SLOPE

Downward-sloping spark lines indicate that the voltage required to maintain the spark duration is decreasing during the firing of the spark plug. This downward slope usually indicates that the spark energy is "finding ground" through spark plug deposits (the plug is "fouled") or other ignition problems. See Figures 24-14 and 24-15.

An *upward*-sloping spark line usually indicates an engine mechanical problem. A defective piston ring or valve would tend to seal better during the increasing pressures of combustion. As the spark plug fires, the effective increase in pressures increases the voltage required to maintain the spark, and the height of the spark line rises during the duration of the spark.

WHAT IS HASH?

Hash is a term used to describe a "fuzzy" or unclear section of a scope pattern. The word "hash" means to "make a mess or botch" something. Hash is commonly detected on spark lines. Hash on spark lines (fuzzy looking) means worn spark plugs or an engine condition affecting the firing of the spark plugs. Typical engine conditions that can cause hash on the spark lines include:

1. Burned valve(s)
2. Defective head gasket
3. Broken valve spring
4. Defective spark plug wire, distributor cap, or rotor
5. High-lift, long-duration racing-type camshaft *(cont'd on p. 325)*

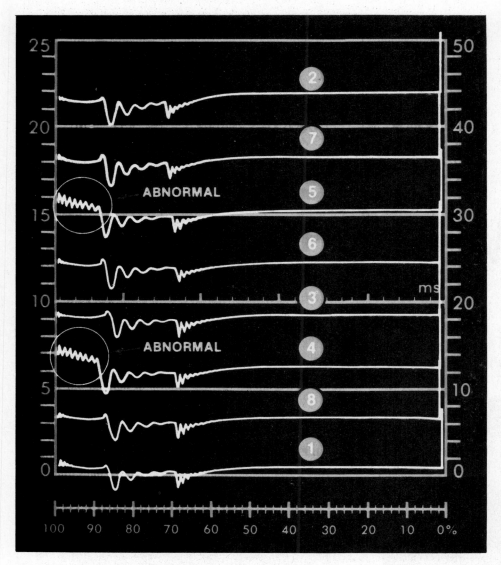

FIGURE 24-14 (Courtesy of Sun Electric Corporation.)

FIGURE 24-15   (Courtesy of Sun Electric Corporation.)

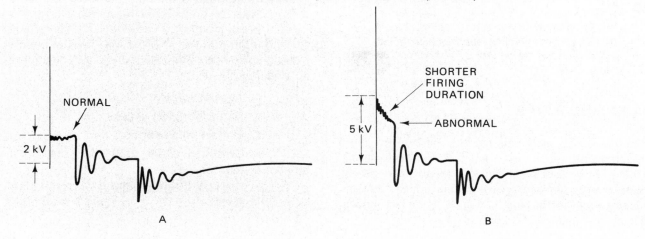

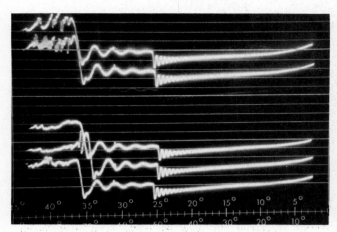

FIGURE 24-16 *Notice the hash on the spark lines of all cylinders shown. This eight-cylinder engine was only operating on five cylinders because of worn distributor bushings that prevented the points from opening at all on three cylinders.*

The usual cause of hash on the spark lines is worn or defective spark plugs. If the spark lines still have hash after installing new spark plugs, further testing may be necessary to determine the exact cause. Many engines operate perfectly, yet display hash on the spark lines. The reason may be minor, undetectible engine problems, or characteristic of a particular engine design. If hash occurs at the point-open point or point-close point on a point-type ignition system engine, this indicates poor condenser action and possible burned points. See Figure 24-16.

READING THE INTERMEDIATE SECTION

The intermediate section should have three *or more* oscillations (bumps) for a correctly operating ignition system. Since there are approximately 250 V in the primary ignition circuit when the spark stops flowing across the spark plugs, this voltage is reduced about 75 V per oscillation. Additional resistances in the primary circuit would decrease the number of oscillations. If there are fewer than three oscillations, possible problems include (See Figure 24-17):

1. Shorted ignition coil
2. Leaky condenser (if point-type ignition)
3. Loose or high-resistance primary connections on the ignition coil or primary ignition wiring

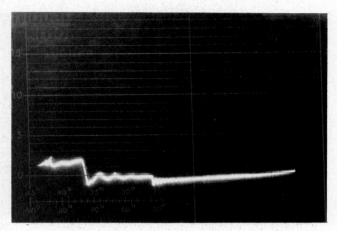

FIGURE 24-17 *Lack of oscillations in the intermediate section.*

READING THE DWELL SECTION

The point-close "point" is a short downward line below the zero line that indicates when the primary current for the coil is turned back "on." If testing a point-type ignition system, the point-closed point should display a sharp downward line. If the point-close point has a nonuniform point-close point, the points could be worn or pitted. See Figure 24-18.

Also, directly below the point-close point is where the dwell reading is indicated on the bottom of the scope screen. Normal dwell is 30° for eight-cylinder engines, 35° for six-cylinder engines, and 35 to 50° for four-cylinder engines. Check the car manufacturer's specifications for exact dwell. Using the percent dwell scale used on many scopes, normal dwell for most engines is approximately 66%.

FIGURE 24-18 *The dwell angle is indicated on a scope pattern at the point-close point. The dwell in this photo is 23°.*

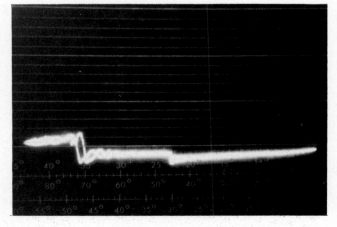

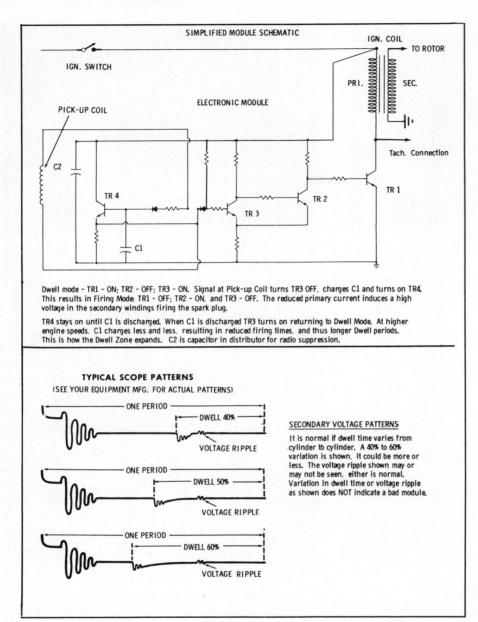

SIMPLIFIED MODULE SCHEMATIC

IGN. SWITCH

PICK-UP COIL

ELECTRONIC MODULE

IGN. COIL

TO ROTOR

PRI.   SEC.

Tach. Connection

C2

TR 4

C1

TR 3

TR 2

TR 1

Dwell mode - TR1 - ON; TR2 - OFF; TR3 - ON. Signal at Pick-up Coil turns TR3 OFF, charges C1 and turns on TR4. This results in Firing Mode: TR1 - OFF; TR2 - ON, and TR3 - OFF. The reduced primary current induces a high voltage in the secondary windings firing the spark plug.

TR4 stays on until C1 is discharged. When C1 is discharged TR3 turns on returning to Dwell Mode. At higher engine speeds, C1 charges less and less, resulting in reduced firing times, and thus longer Dwell periods. This is how the Dwell Zone expands. C2 is capacitor in distributor for radio suppression.

TYPICAL SCOPE PATTERNS
(SEE YOUR EQUIPMENT MFG. FOR ACTUAL PATTERNS)

ONE PERIOD
DWELL 40%
VOLTAGE RIPPLE

ONE PERIOD
DWELL 50%
VOLTAGE RIPPLE

ONE PERIOD
DWELL 60%
VOLTAGE RIPPLE

SECONDARY VOLTAGE PATTERNS

It is normal if dwell time varies from cylinder to cylinder. A 40% to 60% variation is shown, it could be more or less. The voltage ripple shown may or may not be seen, either is normal. Variation in dwell time or voltage ripple as shown does NOT indicate a bad module.

FIGURE 24-19 *Basic GM HEI operation and a typical scope pattern. (Courtesy of Oldsmobile Division, GMC.)*

ELECTRONIC IGNITION AND THE DWELL SECTION

Electronic ignitions also use a dwell period to ''charge'' the coil. Dwell is not adjustable with electronic ignition, but it does change with increasing rpm with many electronic ignition systems. This change in dwell with rpm should be considered normal. See Figure 24-19.

Many electronic ignition systems also produce a ''hump'' in the dwell section, which reflects a current-limiting circuit in the control module. This current-limiting hump may have a slightly different shape depending on the exact module used. For example, the humps produced by a white four-pin GM HEI module and the black four-pin module differ slightly even though the modules are interchangeable. See Figure 24-20.

COIL TESTING

With the scope connected and the engine running, observe the scope pattern in the ''superimposed'' mode. If the pattern is upside down, the primary wires on the coil may be reversed, causing the coil polarity to be reversed. See Figure 24-21.

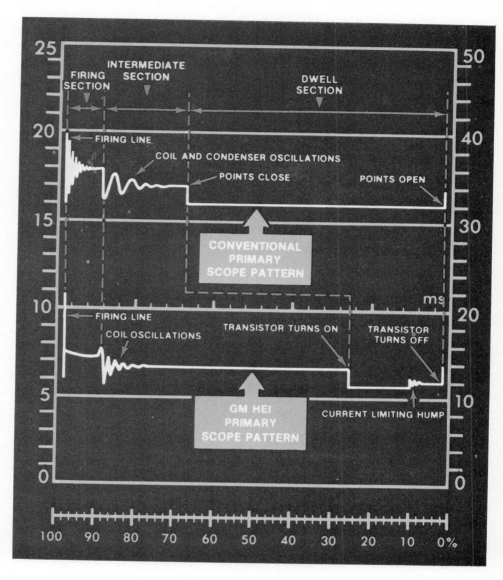

FIGURE 24-20 *(Courtesy of Sun Electric Corporation.)*

NOTE: Check the scope hookup and controls before deciding that the coil polarity is reversed.

The reversed polarity greatly increases the voltage required to fire the spark plugs and often causes the engine to miss on acceleration.

To test the output of a coil, it must be "forced" to produce its maximum output voltage. This is done by disconnecting any spark plug wire (except number 1) and holding the spark plug wire away from ground. With the engine running and the scope pattern selector on display (parade) and the high-voltage scale, observe the height of the firing line for the cylinder with the disconnected plug wire. The minimum acceptable coil output voltage is (See Figure 24-22):

Point-type ignition: 20,000 V (20 kV) *minimum*

Electronic ignition: 25,000 V (25 kV) *mimimum*

Replace the coil if the output voltage is low.

NOTE: Check the distributor cap and rotor before replacing the ignition coil. A worn or defective distributor cap or rotor could cause the coil test to indicate too low a voltage output.

CAUTION: Some manufacturers do not recommend coil testing by checking for maximum open-circuit voltage, especially on computer-equipped engines. Most coils can be accurately tested by observing the spark line duration. If the coil is producing normal spark line duration and the spark plugs are not fouled, the coil is usable. If the ignition coil is still suspected of being defective, remove the coil from the engine and test it separately on the scope following the test equipment manufacturer's recommended procedures.

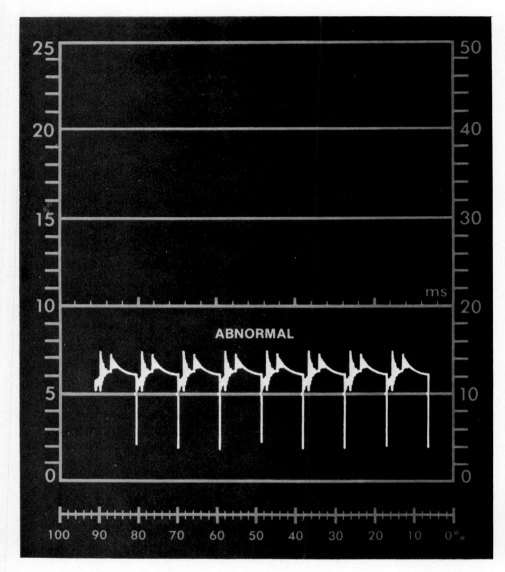

FIGURE 24-21 *The pattern is upside down if the ignition coil polarity is reversed (primary wires are on the wrong coil terminals). (Courtesy of Sun Electric Corporation.)*

ACCELERATION CHECK

With the scope selector set on the display (parade), snap accelerate the engine (gear selector in park or neutral with the parking brake ''on''). Results:

1. All the firing lines should raise evenly (not to exceed 75% of maximum coil output) for properly operating spark plugs.
2. If the firing lines on one or more cylinders *fail to rise*, this indicates fouled spark plugs.

ROTOR GAP VOLTAGE

This test measures the voltage required to jump the gap [0.030 to 0.050 in. (0.8 to 1.3 mm)] between the rotor and the inserts (segments) of the distributor cap. Select the display (parade) scope pattern and remove a spark plug wire using a jumper wire to provide a good ground connection. Start the engine and observe the height of the firing line for the cylinder being tested. Since the spark plug wire is connected directly to ground, the firing line height on the scope will indicate the voltage required to jump the air gap between the rotor and the distributor cap insert. The normal rotor gap voltage is 3 to 7 kV and should not exceed 8 kV. If the rotor gap indicated is near or above 8 kV, inspect and replace the distributor cap and/or rotor as required. See Figure 24-23.

POWER BALANCE TEST

The power balance test is used to determine if all cylinders are producing power equally. Each cylinder should have the same effect on engine speed. The power balance test

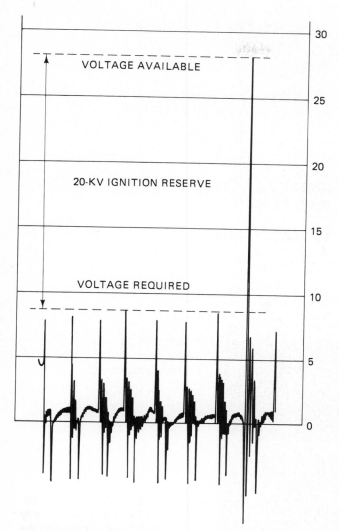

FIGURE 24-22 *Testing a coil's maximum voltage output using a scope on display (parade).*

FIGURE 24-23 *The voltage required to arc between the tip of the rotor and the distributor insert is 5 kV in this example.*

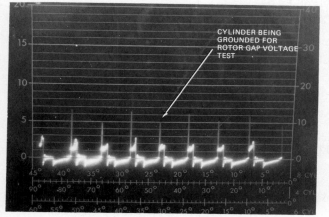

''grounds out'' the primary ignition circuit for only one cylinder at a time. If the cylinder was grounded out and no spark was sent to that cylinder, the idle speed should drop.

Most scopes have buttons, dials, or an automatic cylinder canceling section. With the engine at idle and the scope pattern selection on display (parade), cancel one cylinder at a time and note the engine rpm. The engine rpm should decrease *equally* for each cylinder. The amount of rpm drop is *not* important. All cylinders should, however, be within 50 rpm of each other. If the engine rpm does not drop enough (or drop at all), the cylinder is not ''working.'' If the rpm *increases* whenever a cylinder is canceled, this indicates:

1. Possible crossed spark plug wires
2. Defective (cracked) distributor cap
3. Spark occurring too early on the cylinder, caused by possible stray electrical impulses to the module from a spark plug wire, alternator, or other sources

If the power balance test indicates a ''dead'' or ''weak'' cylinder, further tests, including a compression test, should be performed to determine the exact cause.

POWER BALANCING A COMPUTER-EQUIPPED ENGINE

Whenever a balance test is being performed, a computer-equipped engine will automatically increase idle speed or otherwise compensate for the dead (canceled) cylinder. To prevent the computer controls from compensating for a canceled cylinder, disconnect the coolant sensor, oxygen sensor, and/or idle speed control device. With the sensor(s) disconnected, the computer will go closed loop and decrease the engine idle when a cylinder is canceled. After testing, the sensors should be reconnected and the computer fault code(s) erased. See Chapter 26 for the trouble-code clearing procedures.

NOTE: Check the vehicle service manual for the exact method to prevent possible computer or safety procedure errors.

SCOPE TESTING ALTERNATORS

Defective diodes and open or shorted stators can be detected on a scope. Connect the scope leads as normal, *except* connect the coil negative (−) connection, which attaches to the alternator output (BAT) terminal. With the pattern selection set to raster (stacked), start the engine and run to approximately 1000 rpm (slightly higher than normal idle speed). The scope pattern should indicate an even ripple pattern showing the slight alternating up-and-down level of the alternator output voltage.

If the alternator is controlled by an electronic voltage regulator, the rapid "on" and "off" cycling of the field current can create vertical spikes evenly throughout the pattern. These spikes are normal. If the ripple pattern is *jagged* or *uneven*, a defective diode (open or shorted) or a defective stator is indicated. See Figure 24-24. If the alternator scope pattern does not indicate even ripples, the alternator should be disassembled and all internal components tested. See Chapter 17 for alternator test procedures.

OTHER SCOPE TESTS

Many scopes are also capable of testing:

1. Electronic ignition pickup coil pulses

2. Fuel injection operating pulses
3. Coil testing (off the car)
4. Condenser testing (off the car)

See the scope manufacturer's instructions for exact hookups, procedures, and precautions.

SCOPE PATTERNS OF TYPICAL PROBLEMS

Figures 24-25 through 24-32 show scope patterns of some typical engine problems.

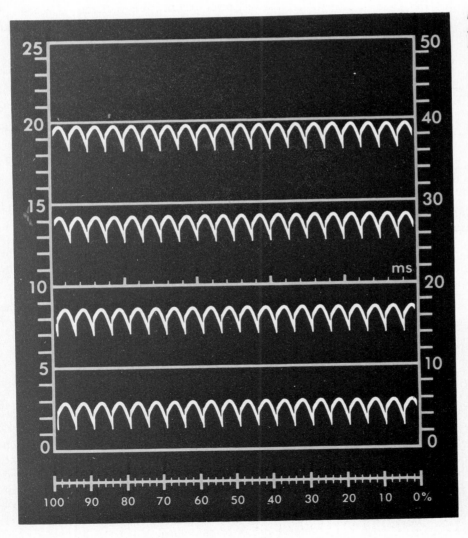

FIGURE 24-24 *Normal alternator scope pattern. (Courtesy of Sun Electric Corporation.)*

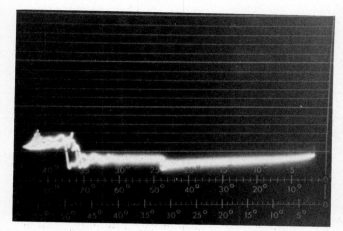

FIGURE 24-25 Scope set on "superimposed," showing one shorter-than-normal spark line and hash on almost all other cylinders. The short spark line indicates a high-resistance plug wire or a poor connection in the distributor cap or spark plug. Too long a spark line indicates a fouled plug or a shorted spark plug wire.

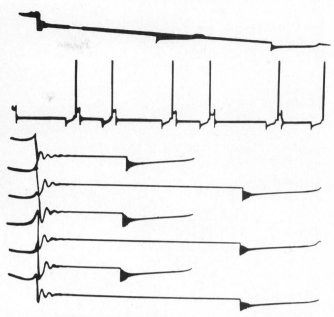

FIGURE 24-26 Typical normal scope patterns for an odd-firing V-6.

FIGURE 24-27 An abnormal point-close (and/or point open) point indicates defective points or defective condenser causing arcing of the points.

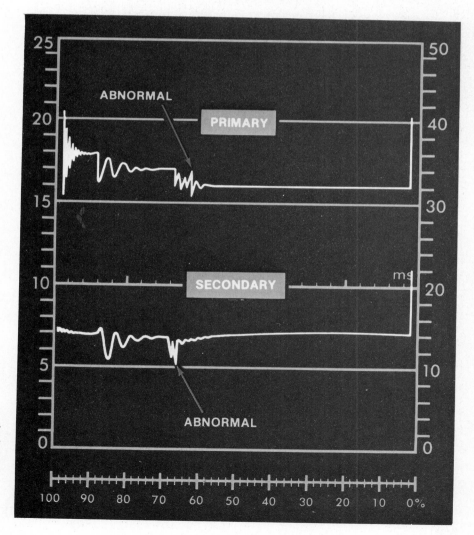

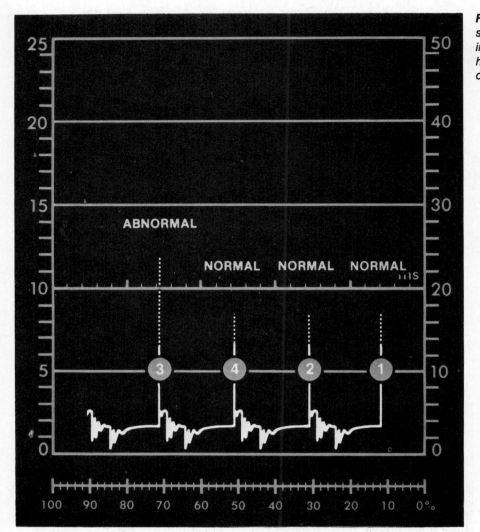

FIGURE 24-28 *All firing lines should be within 3 kV of each other in height. A high firing line indicates high resistance in the secondary circuit.*

FIGURE 24-29 *Low firing lines side by side usually indicate a carbon-tracked distributor cap because side by side on a scope pattern means side by side in the firing order.*

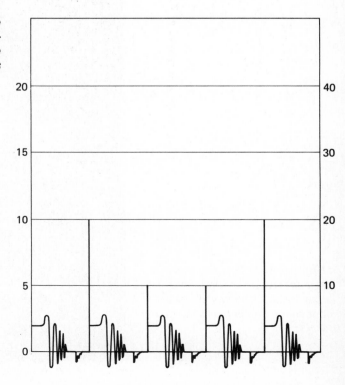

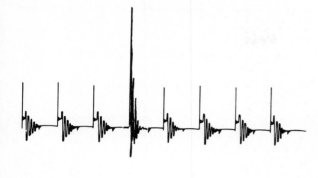

FIGURE 24-30 *Open spark plug wire (un-plugged or disconnected). (Courtesy of Allen Testproducts.)*

FIGURE 24-31 *Fouled spark plug. (Courtesy of Allen Testproducts.)*

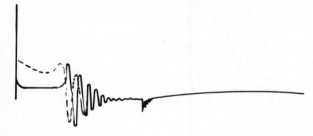

FIGURE 24-32 *Sticking valves or unequal fuel moisture. (Courtesy of Allen Testproducts.)*

SUMMARY

1. Oscilloscope testing is a popular engine testing procedure. An engine fault may be indicated on a scope by displaying a pattern that is not exactly the shape that it should display. Therefore, the technician must identify which section of the scope pattern is unlike the pattern for other cylinders and/or unlike that of a normal pattern. The basic secondary scope pattern sections and the engine components affecting the section include:

   (a) *Firing line.* The firing line should be 5 to 15 kV and within 3 kV of all the other cylinders. A higher-than-normal firing line indicates a lean mixture, wide spark plug gap, or high resistance in the secondary ignition circuit of the particular cylinder affected.

   (b) *Spark line.* The spark line is the horizontal line after the firing line and represents the duration of the spark. The length should be approximately 1.5 ms long. If the spark line is too long, a fouled spark plug is indicated. If the spark line is too short, a wide spark plug gap or high resistance in the secondary circuit of the particular cylinder affected is indicated. A downward-sloping spark line indicates an electrical problem and an upward-sloping spark line usually indicates a mechanical problem.

   (c) *Intermediate section.* The intermediate section should have five or more oscillations. If there are fewer than five oscillations, possible problem areas include a shorted ignition coil, a leaky condenser, or poor connections.

   (d) *Dwell section.* The point-closed (transistor ''on'') and point-open (transistor ''off'') points indicate the proper charging and discharging of the ignition coil. If this section of the scope is not normal, possible problems include: defective points and/or condenser (if point-type ignition) or a defective module (if electronic ignition).

   Oscilloscope testing is a useful troubleshooting tool and can often locate problems in all types of engines.

STUDY QUESTIONS

24-1. Draw and label the basic secondary ignition scope pattern and label each section.

24-2. Explain why the length of the spark line is critical to proper engine operation.

24-3. Explain why the slope of the spark line can be used to help distinguish between an electrical and a mechanical problem.

24-4. Describe how an ignition coil can be tested on an oscilloscope.

24-5. Describe how an alternator can be tested on an oscilloscope.

MULTIPLE-CHOICE QUESTIONS

24-1. The height of firing lines can be observed on:
(a) display (parade) only.
(b) raster (stacked) only.
(c) superimposed only.
(d) all settings.

24-2. The length of the spark line should be:
(a) 5 to 15 kV.
(b) 1.0 to 1.5 ms.
(c) 5 or more "bumps."
(d) 1.5 to 2.0 ms.

24-3. Too short a spark line on all cylinders could indicate:
(a) a worn distributor cap and/or rotor.
(b) an open coil wire (if equipped).
(c) too lean a fuel mixture.
(d) all of the above.

24-4. A high firing line without a downward spike below the zero line indicates:
(a) a defective spark plug wire with good insulation.
(b) a defective spark plug wire with poor insulation.
(c) a weak ignition coil.
(d) a rich fuel mixture in the affected cylinder.

24-5. The "roller coaster" height of the firing lines usually indicates:
(a) an engine mechanical problem.
(b) a distributor cap/rotor problem.
(c) a fuel mixture problem.
(d) a spark plug wire problem

24-6. Some electronic ignition systems show:
(a) no firing lines on the scope.
(b) a current-limiting hump.
(c) no point-close point.
(d) no point-open point.

24-7. To test the rotor gap voltage:
(a) a spark plug wire must be removed.
(b) a spark plug wire must be grounded.
(c) the rotor must be removed from the engine.
(d) the engine must not be running.

24-8. Dwell on a scope pattern is read at:
(a) the point-closed point.
(b) the point-open point.
(c) the coil section.
(d) the spark line section.

24-9. Technician A says that the dwell on the scope should increase as the engine speed is increased on some engines. Technician B says that all firing lines should be within 3 kV of each other. Which technician is correct?
(a) A only.
(b) B only.
(c) both A and B.
(d) neither A nor B.

24-10. Technician A says that if one cylinder drops in rpm much more than the remaining cylinders during a power balance test, that the cylinder is weak. Technician B says that rotor gap voltage must be tested with a spark plug wire grounded. Which technician is correct?
(a) A only.
(b) B only.
(c) both A and B.
(d) neither A nor B.

25

Computerized Engine Control System Operation

COMPUTERIZED ENGINE CONTROLS BECAME STANDARD EQUIPMENT ON most automotive engines in the early 1980s. The need to meet the federal corporate average fuel economy (CAFE) standards *and* exhaust standards has required that electronic engine controls be used to control ignition timing and fuel mixture. To reduce exhaust levels for oxides of nitrogen (NO_x), carbon monoxide (CO), and unburned hydrocarbons (HC), an air/fuel ratio of 14:7:1 is required for the correct operation of the three-way catalyst. In this chapter we present the operation of the sensors, controls, and circuits used by most manufacturers for proper engine operation. The topics covered in this chapter include:

1. Stoichiometric air/fuel ratios and three-way catalyst
2. Oxygen sensor operation
3. Coolant sensor operation
4. Manifold pressure sensors
5. Vehicle speed sensor/throttle position switch
6. Mixture control solenoids
7. Ignition timing control/detonation control
8. Air management solenoids
9. Normal modes of operation: open and closed loop
10. Precautions required when working with computer controls

COMPUTERS

A computer is an electronic unit that can turn other electrical units "on" and "off" very rapidly. A computer cannot "think," but it is capable of selecting what unit to turn "on" and "off" according to a predetermined set of guidelines. The guidelines are called a computer "program" and are stored in the computer in a type of memory. There are up to five types of memory used in automotive computer applications:

1. *Read-only memory (ROM)*. The ROM is programmed information that can be read only by the computer. The ROM program cannot be changed. If battery voltage is removed (battery disconnected), ROM information will be retained.

2. *Random access memory (RAM)*. This memory is part of the decision-making section for the computer. Information is stored temporarily and is used by the computer to make calculations similar to a pocket calculator. Sensor information, diagnostic codes, and results of calculations are stored temporarily in RAM. If battery voltage is removed, all information stored in RAM is lost.

3. *Programmable read-only memory (PROM)*. This memory is similar to ROM except that the programmed information may be slightly different for each car, due to variations in engine calibration, transmission, body style (vehicle weight), and options. A PROM is a "chip" that can be removed from the computer. A PROM permits one computer to function with many different engines.

4. *Erasable programmable read-only memory (EPROM)*. An EPROM is similar to a PROM except that the information (program) can be erased using ultraviolet light and replaced with another program. After the EPROM has been reprogrammed, a sticker must be installed over the chip to prevent normal daylight from erasing the program.

5. *Electrically erasable programmable read-only memory (EEPROM)*. EEPROM is commonly used for electronic odometers.

A typical automotive computer is capable of switching on and off many different engine controls many times per second. The computer is called by different names by different manufacturers:

Auto Manufacturer	Computer Name	Location
GM	ECM (Electronic Control Module)	Passenger compartment (usually kickpanel)
Ford	MCU (Microprocessor Control Unit)	Left fenderwell
Ford	EEC (Electronic Engine Control)	Dashboard
Chrysler	EFC (Electronic Fuel Control)	Kickpanel
Chrysler	SCC (Spark Control Computer)	Left fenderwell/air cleaner
American Motors	CEC (Computerized Emission Control)	Dashboard or kickpanel

Some computer systems control the fuel system, some the ignition timing only, but most control both fuel delivery and ignition timing, as well as other functions.

EMISSION REQUIREMENTS (GRAMS/MILE) (PASSENGER CARS)

FIGURE 25-1 (Courtesy of General Motors Corporation.)

MODEL YEAR	HYDROCARBON (HC) CALIFORNIA	FEDERAL	CARBON MONOXIDE (CO) CALIFORNIA	FEDERAL	OXIDES OF NITROGEN (NOx) CALIFORNIA	FEDERAL
1978	0.41	1.5	9.0	15.0	1.5	2.0
1979	0.41	0.41	9.0	15.0	1.5	2.0
1980	0.39	0.41	9.0	7.0	1.0	2.0
1981	0.39	0.41	7.0	3.4	0.7	1.0
1982	0.39	0.41	7.0	3.4	0.4	1.0
1983	0.39	0.41	7.0	3.4	0.4	1.0
1984	0.39	0.41	7.0	3.4	0.4	1.0
1985	0.39	0.41	7.0	3.4	0.7	1.0
1986	0.39	0.41	7.0	3.4	0.7	1.0
1960 (No Control)		10.6		84		4.1

THE NEED FOR COMPUTER ENGINE CONTROLS

Computer engine controls have been used on most cars since the early 1980s in order to meet emission, gas mileage, and performance standards. The U.S. federal emission standards are most easily met by using a three-way catalytic converter in the exhaust system. See Figure 25-1.

A three-way catalyst can control all three major pollutants; HC (hydrocarbons or unburned fuel), CO (carbon monoxide, a deadly gas), and NO_x (oxides of nitrogen, a smog and corrosion ingredient). The three-way catalyst, however, can operate efficiently only if the air/fuel ratio is closely controlled to 14.7:1. This specific air/fuel ratio is called *stoichiometric*. See Figure 25-2.

''Stoichiometric'' means that all of the fuel will be mixed with all of the air to create the most efficiently burning air/fuel mixture. Precise control of the fuel system is required to maintain the air/fuel ratio at 14.7:1 under all driving conditions. Most automobile manufacturers use a computer system to make adjustments to the fuel system to maintain the correct air/fuel ratio for all driving conditions. A computer-controlled engine can often provide the power, performance, and fuel economy of older non-emission-controlled engines. This increased power is possible because of the exact control of the air/fuel ratio and ignition timing possible with the computer operation. A computer must be able to ''sense'' engine temperature, vacuum, oxygen content of the exhaust, plus other operating conditions to be able to control the air/fuel ratio. The following sensors provide the *input* to the computer.

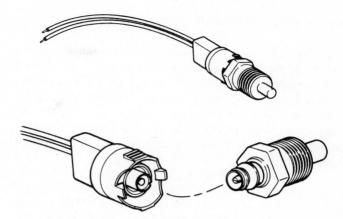

FIGURE 25-3 *Typical coolant sensor. Most coolant sensors are located near the thermostat. (Courtesy of General Motors Corporation.)*

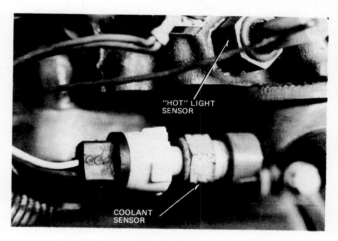

"HOT" LIGHT SENSOR

COOLANT SENSOR

FIGURE 25-4

COOLANT SENSORS

All computer systems must be able to sense the engine temperature. Engine temperature is best determined by the coolant temperature. A coolant sensor is a *thermistor*, which is a resistor that changes in resistance in relation to temperature. A typical coolant sensor has 100,000 Ω when the coolant temperature is cold [$-40°F$ ($-40°C$)] and less than 1000 Ω when the coolant temperature is warm [266°F (130°C)]. See Figures 25-3 and 25-4.

Coolant (engine) temperature is critical to proper engine operation. A cold engine requires a richer mixture than that required by a warm engine. A cold engine can also use more ignition spark advance than a warm engine.

NOTE: A separate sensor signals the coolant temperature gauge or light on the dash and does *not* use the coolant sensor.

FIGURE 25-2 *(Courtesy of General Motors Corporation.)*

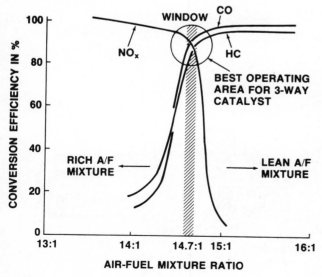

CONVERSION EFFICIENCY IN %

WINDOW — CO

NO_x — HC

BEST OPERATING AREA FOR 3-WAY CATALYST

RICH A/F MIXTURE →

← LEAN A/F MIXTURE

AIR-FUEL MIXTURE RATIO

OXYGEN SENSORS

Oxygen (O_2) sensors are used on most computer engine systems. An oxygen sensor compares the oxygen content in the exhaust to the oxygen content in the outside air. There is a small tube from the top to the inner chamber of the sensor. An oxygen sensor acts as a small battery and generates a low voltage when hot. The amount of voltage produced depends on the amount of oxygen in the exhaust gases. See Figures 25-5, 25-6, and 25-7.

A lean mixture creates exhaust which has more oxygen and a *lower* voltage (less than $1/2$ V). A rich mixture creates a higher voltage (more than $1/2$ V). The voltage output ranges from 0.1 V (100 mV) (lean) to 0.9 V (900 mV) (rich). To indicate exhaust oxygen content accurately, the temperature of the oxygen sensor must be hot [600°F (315° C)].

Oxygen sensor diagnosis includes connecting a jumper lead in series with the oxygen sensor lead(s). The voltage of the O_2 sensor can be safely read by connecting the positive (+) lead of an *high-impedance* voltmeter (20-V scale) to the junction of the jumper wire of the hot lead of the O_2 sensor and the black lead of the voltmeter to a good engine ground. Reading the output voltage of the O_2 sensor with a digital voltmeter will indicate if the sensor is working. With the engine warm, if the choke was closed (or air intake almost closed), the resulting rich mixture should cause the O_2 sensor to create almost 1.0 V. A large vacuum leak can be created by disconnecting a large vacu-

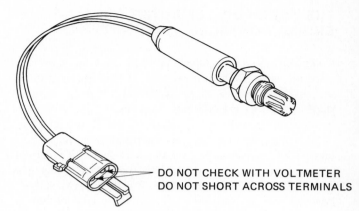

DO NOT CHECK WITH VOLTMETER
DO NOT SHORT ACROSS TERMINALS

FIGURE 25-5 *Early-design GM oxygen (O_2) sensors. The newer design uses only one wire. (Courtesy of General Motors Corporation.)*

FIGURE 25-7 *A newer-style one-wire GM oxygen sensor. (Courtesy of General Motors Corporation.)*

FIGURE 25-6 *Interior of an oxygen sensor. (Courtesy of General Motors Corporation.)*

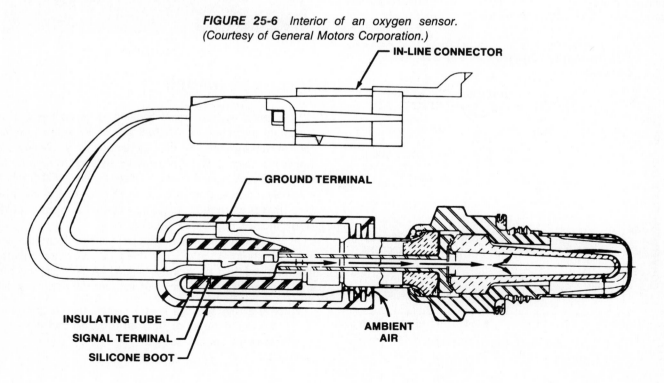

IN-LINE CONNECTOR

GROUND TERMINAL

INSULATING TUBE
SIGNAL TERMINAL
SILICONE BOOT

AMBIENT
AIR

um hose such as the power brake booster line. With this large vacuum leak, the O$_2$ sensor voltage should be very low. If the O$_2$ sensor does not respond to changing fuel mixtures, it must be replaced.

PRESSURE SENSORS

Pressure sensors are needed for accurate operation of the automotive engine computer. Engine air/fuel requirements change with changes in barometric (atmospheric) pressure and engine vacuum. Engine vacuum changes with engine *load*. With a load on the engine (engine "working" hard), the engine vacuum is low. If the engine load is light, the engine vacuum is high.

There are several types of pressure sensors:

1. *MAP*. MAP means manifold absolute pressure, which is atmospheric pressure minus intake manifold pressure. A MAP sensor measures the changes in the intake manifold which change with engine load. It produces a low voltage (approximately 1 V) when the manifold pressure is low (high manifold vacuum) and a voltage of about 4.5 V when the manifold pressure is high (low manifold vacuum). See Figures 25-8 and 25-9.

2. *BARO* (barometric pressure). A barometric pressure sensor measures changes in the barometric pressure due to changes in the weather and altitude. The BARO sensor produces a voltage of 3 to 4.5 V depending on the barometric pressure. A BARO sensor may be located inside the passenger compartment near the computer or in the engine compartment. See Figure 25-10.

3. *VAC* vacuum sensor (differential pressure sensor). A VAC sensor compares manifold vacuum to atmospheric pressure. The VAC sensor sends the

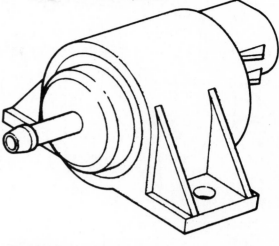

MANIFOLD ABSOLUTE PRESSURE (MAP) SENSOR W/CONNECTOR

FIGURE 25-9 *(Courtesy of Ford Motor Company.)*

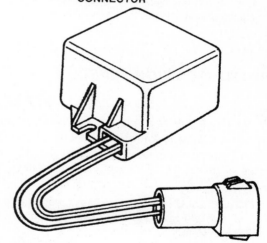

BAROMETRIC PRESSURE SENSOR & CONNECTOR

FIGURE 25-10 *(Courtesy of Ford Motor Company.)*

computer a signal (voltage) based on this differential in pressure. It produces a low voltage (about 1 V) when manifold vacuum is low and a high voltage (about 4.5 V) when the manifold vacuum is high. See Figure 25-11.

Computer engine control systems may have one or more pressure sensors.

FIGURE 25-8 *(Courtesy of General Motors Corporation.)*

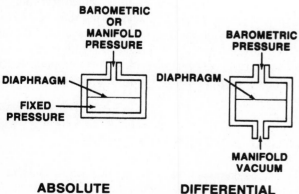

BAROMETRIC OR MANIFOLD PRESSURE

DIAPHRAGM

FIXED PRESSURE

ABSOLUTE

BAROMETRIC PRESSURE

DIAPHRAGM

MANIFOLD VACUUM

DIFFERENTIAL

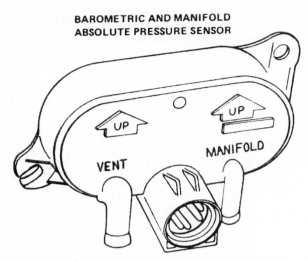

FIGURE 25-11 *(Courtesy of Ford Motor Company.)*

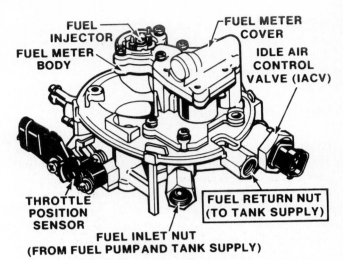

FIGURE 25-13 *Note the throttle position sensor (TPS) on the side of a throttle body injection (TBI) unit. (Courtesy of General Motors Corporation.)*

VEHICLE SPEED SENSORS

The speed of the vehicle is necessary for the correct operation of the torque converter clutch (lock-up) used on automatic transmission and electric cooling fans (if used). The vehicle speed is determined by a sensor (VSS) located in the speedometer or on the transmission. See Figure 25-12.

THROTTLE POSITION SENSORS

Most computer-equipped cars use a throttle position sensor (TPS, also known as the *throttle angle position sensor*). The computer needs to know where the accelerator pedal is and the *rate* that the accelerator pedal is being moved either up or down. The TPS sends a high voltage (about

FIGURE 25-12 *Typical vehicle speed sensor built into the speedometer head. (Courtesy of General Motors Corporation.)*

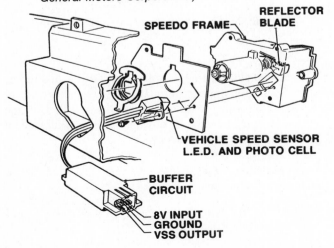

5.0 V) when the accelerator is depressed wide open (WOT) and a lower voltage (about 0.5 V) at idle. The computer uses this voltage reading from the TPS to control spark advance plus other systems. Most throttle position sensors are actually variable-resistor units that change in resistance with the throttle movement. Most computers send a fixed voltage (usually 5 V) to the sensor and receive a lower voltage back to the computer as the result of the varying resistors in the throttle sensor. Therefore, most sensors can also be tested with an ohmmeter or a voltmeter. See Figure 25-13.

ENGINE SPEED

Engine speed input is usually "sensed" by the tach signal from the negative (−) side of the ignition coil. Engine rpm is critical to proper functioning of the computer system. For example, if the engine speed is below 400 rpm, the computer "knows" that the engine is not running and will continue to provide the necessary amount of fuel (correct air/fuel mixture) and spark timing for starting. After starting, the engine speed is needed for the computer to calculate the proper spark timing for the most efficient operation. See Figure 25-14.

COMPUTER-OPERATED UNITS

The engine computer takes the various sensor input information and compares it against the program (instructions in ROM). Based on the various inputs, the computer op-

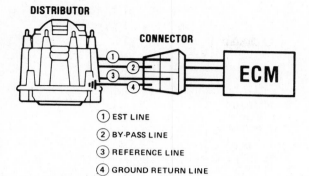

(1) EST LINE
(2) BY-PASS LINE
(3) REFERENCE LINE
(4) GROUND RETURN LINE

FIGURE 25-14 *The engine speed (RPM) is sensed by the computer (ECM) and used to calculate the proper air/fuel ratio and ignition timing for maximum power and fuel mileage with the lowest possible emissions. (Courtesy of General Motors Corporation.)*

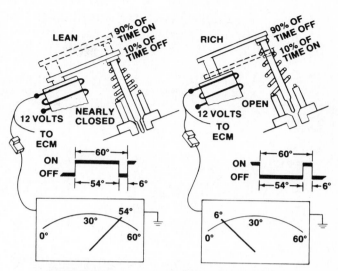

FIGURE 25-15 *A dwell meter connected to a test lead will indicate how long (in degrees) the mixture control solenoid is being actuated. (Courtesy of General Motors Corporation.)*

erates various units to keep the engine operating within the standards established in the computer memory (PROM). Most computers control the following units (computer outputs):

1. Mixture control (carburetor solenoid or fuel injectors).
2. Idle speed control.
3. Ignition timing control.
4. Air management (air pump).
5. Related functions may include: electric cooling fan operation, air-conditioner clutch operation, torque converter clutch operation, EGR valve operation, plus other functions that depend on exact application.

MIXTURE CONTROL

The computer provides the ground for a 12-V circuit to a solenoid or fuel injector coil. The amount of fuel can be controlled by grounding or ungrounding the circuit for the mixture control (M/C) solenoid. A typical mixture control solenoid opens and closes the fuel solenoid circuit 10 times per second. The computer determines the length of time the fuel solenoid is delivering fuel. The solenoid may be turned "on" and "off" 10 times per second, but the "on" time can vary. See Figure 25-15.

The "on" time is determined by the computer to deliver the exact amount of fuel necessary for the engine speed, throttle position, engine load, and oxygen content of the exhaust. The "on" time may be called "duty cycle" or "pulse width." See Figures 25-16 and 25-17.

FIGURE 25-16 *(Courtesy of General Motors Corporation.)*

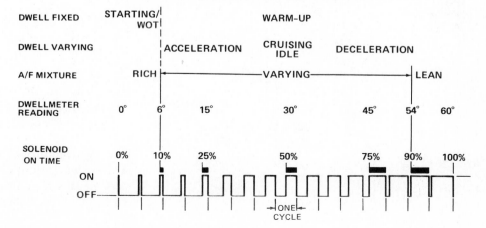

COMPUTER COMMAND CONTROL

FIGURE 25-17   *(Courtesy of General Motors Corporation.)*

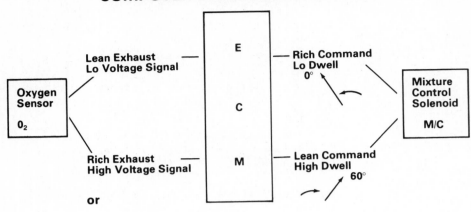

Command Corrects Condition

IDLE SPEED CONTROL

Most computer-equipped engines include some type of engine idle speed control. Carburetor-equipped engines often use an idle speed control (ISC) stepper motor operated by the computer. A stepper motor is a type of permanent-magnet motor that can be controlled precisely using the electronics of the computer. If the engine speed sensor detects a drop in engine speed (such as can happen when turning the steering wheel if equipped with power steering), the ISC increases the throttle opening to maintain the proper idle speed. See Figure 25-18.

On an engine equipped with fuel injection (TBI or port injection), the idle speed is controlled by increasing or decreasing the amount of air bypassing the throttle plate. Again, an electronic stepper motor is used to maintain the

correct idle speed. This control is often called the idle air control (IAC).

Many computer-controlled throttle linkage units use manifold vacuum with a computer-controlled solenoid valve to control the idle speed. This type of arrangement is often called a ''throttle kicker solenoid.''

IGNITION TIMING CONTROL

Most engine computers control the ignition timing. Conventional vacuum and mechanical advance mechanisms cannot accurately maintain the *exact* ignition timing possible with computer control. The computer controls the timing by controlling the ignition module primary ignition circuit turn-''off'' time. Computer-controlled timing is often called EST (electronic spark timing). The computer uses the various input signals, compares the results with the program (PROM), and changes the ignition instantly if needed to maintain the most ideal spark timing. The spark timing changes according to some of the following factors:

1. *Low engine temperature.* Timing can be advanced.
2. *High engine temperature.* Timing should be retarded.
3. *High altitude.* Timing can be advanced.
4. *Engine load is high* (vacuum low). Timing should be retarded.
5. *Engine load is light* (vacuum high). Timing can be advanced.

The computer can calculate the correct ignition timing considering all of the various input factors. Some computer systems are capable of retarding the ignition timing (up to 30°) if detonation is detected by a knock sensor.

FIGURE 25-18   *Typical idle speed control (ISC) motor. (Courtesy of General Motors Corporation.)*

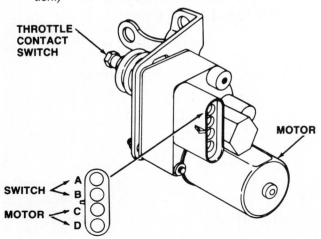

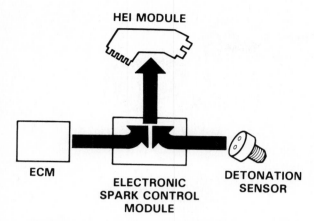

FIGURE 25-19 (Courtesy of General Motors Corporation.)

KNOCK SENSORS

A knock sensor is a piezoelectric sensor that transforms the engine detonation vibrations directly into an electrical signal which is sent to the computer. The knock control, often called ESC (electronic spark control), permits the maximum timing for best gas mileage and performance, yet retards automatically to prevent damaging detonation. Using detonation control, the ignition timing will automatically retard timing if lower-octane fuel is used. See Figure 25-19.

AIR MANAGEMENT

Many computer-equipped engines use an air pump as part of the emission control system. AIR stands for *air injection reaction* and the pump provides the additional air necessary for the oxidizing catalytic converter. The computer controls the airflow from the pump by switching "on" and "off" various solenoid valves. When the engine is cold, the air

FIGURE 25-20 (Courtesy of General Motors Corporation.)

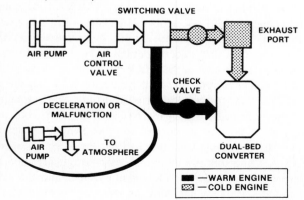

pump output is directed to the exhaust manifold to help provide enough oxygen to convert HC (unburned gasoline) and CO (carbon monoxide) to H_2O (water) and CO_2 (carbon dioxide). When the engine becomes warm, the computer operates the air valves and directs the air pump output to the catalytic converter. See Figure 25-20.

Whenever the vacuum rapidly increases above normal idle vacuum, as during rapid deceleration, the computer diverts the air pump output to the air cleaner assembly to silence the air. Diverting the air to the air cleaner prevents exhaust backfire during deceleration.

OTHER COMPUTER OUTPUTS

The typical automotive computer is capable of processing over 100,000 commands per second. (Some systems are capable of over 1,000,000 commands per second.) Many computers control other functions in addition to those indicated above. Some of these additional functions include:

1. *Electric engine cooling fan.* The computer "knows" the coolant temperature, idle speed, vehicle speed, and turns the electric cooling fan "on" or "off" as required. If the electric fan is off, less current is required from the alternator, and therefore, the engine can achieve better gas mileage.

2. *Torque converter clutch.* The torque converter clutch (lock-up converter) reduces the slippage losses of a normal torque converter. The computer uses the inputs from the throttle position sensor (TPS), engine vacuum (VAC, MAP, BARO), and vehicle speed sensor (VSS) to calculate the correct application time for the clutch.

3. *EGR valve operation.* Exhaust gas recirculation is critical to maintain the proper internal engine fuel-burning rate to maintain lowest NO_x (oxides of nitrogen) emissions and to prevent harmful engine detonation (ping). EGR valves are vacuum operated and the computer can (on some models) control the vacuum going to the EGR valve. The computer-controlled EGR includes a solenoid switch to switch the vacuum "on" and "off." Since this on-and-off signal is based on the pulse width signal from the computer, this type of EGR control is often called *pulse-width modulated* (controlled) EGR.

Some computer systems also control air-conditioning (A/C) clutch operation, electric fuel pump operation, blower motor speed, heater, defroster, and A/C airflow "doors" plus many other nonengine performance-related functions. See Figure 25-21. See Chapter 27 for details on body computers.

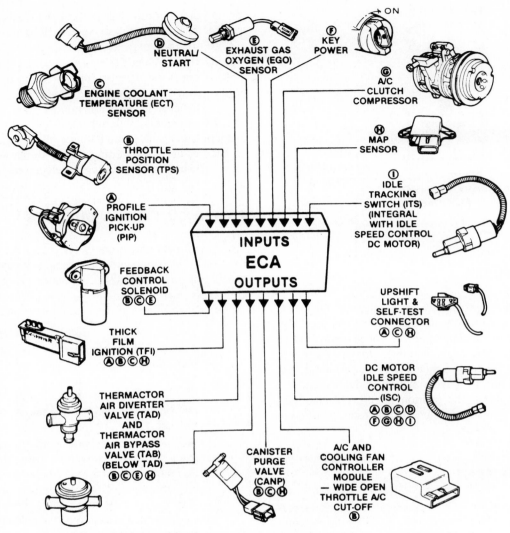

FIGURE 25-21 *Typical computer sensors and controls. (Courtesy of Ford Motor Company.)*

OPEN-AND-CLOSED-LOOP OPERATION

Computer systems do not start controlling all aspects of the fuel and ignition immediately upon startup of a cold engine. The O_2 sensor must be above 600°F (315°C) before it can supply useful exhaust oxygen content information to the computer. Therefore, the PROM supplies values for fuel mixture and ignition timing when the engine is cold. When the engine is operating with these preset values, the engine operation is called *open loop*. It can be easily remembered by the term "open" circuit, meaning that no current flows. In open-loop operation, the system is not closed, meaning that the oxygen sensor does not signal the computer regarding the oxygen content of the exhaust. When the oxygen sensor starts to produce usable signals, the computer is capable of going into *closed-loop* operation.

FUEL INJECTION CONTROLS— OPERATION AND TESTING

Electronic fuel injection uses a small 12-V solenoid to open a valve inside the injector assembly. Whenever the solenoid is energized (12 V applied), the injector allows pressurized fuel [about 10 psi for throttle body injection (TBI) units, 40 psi for port injection units] to be discharged into the low-pressure area in the intake manifold or intake port. The computer controls the "on" time for each fuel injection. The "on" time is called *pulse width*. The computer uses the information from the oxygen (exhaust), coolant, and throttle position sensors (plus other factors) to determine the exact length of time to keep the injectors open. Most port injectors use the basic design originally used in the late 1960s by Bosch. Individual cylinder injectors can be tested simply by touching or listening to them to be

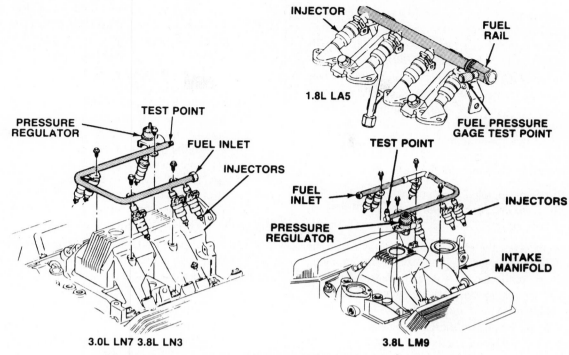

FIGURE 25-22 *Typical port injection systems. (Courtesy of General Motors Corporation.)*

certain that they are working with the engine running. Most Bosch-style injectors should also measure between 1.5 and 16.0 Ω across the wiring terminals. (Check the manufacturer for exact specifications.) Replace any injector not within the resistance limits noted above. All injectors on the same engine should have the same resistance. If an individual injector had a slightly higher resistance, less current flow would occur and the mixture supplied by the injector would be *lean*. An injector with a lower resistance would supply a richer mixture. Unequal or a defective injector(s) can cause a rough idle and missing on acceleration. See Figure 25-22.

> **NOTE:** Some manufacturers wire one-half of the fuel injectors on one fuse and one-half on another fuse. Always check all fuses in the fuse panel if poor performance and rough idle are experienced on a fuel-injected engine.

NORMAL MODES OF OPERATION

Automotive computer systems are programmed to operate under several different starting and driving conditions. See Figure 25-23. These various conditions are called "modes of operation" and include:

> *Starting mode.* The first mode of operation is the starting mode. The computer and PROM provide for starting enrichment based on a signal from the

FIGURE 25-23 *(Courtesy of Ford Motor Company.)*

ENGINE OPERATING MODES

Engine Operating Mode	Air/Fuel Ratio	Engine Temperature	Exhaust Gas Sensor Input	Air/Fuel Temperature
Engine Crank	Fixed 2:1 to 12:1	Cold to cool	None	Cold to cool
Engine Warm-up	Fixed 2:1 to 15:1	Warming	None until engine warm-up	Warming
Open Loop	Fixed 2:1 to 15:1	Cold or warm	May signal, but ignored by processor	Cold or warm
Closed Loop	14.7:1 Depends on exhaust gas sensor input	Warm	Signalling	Warm
Hard Acceleration	Variable rich mixture, depends on driver demands	Warm	Signals, but ignored by processor	Warm
Deceleration	Variable lean mixture	Warm	Signals, but ignored by processor	Warm
Idle	Rich or lean depends on calibration	Warm	Signal, may be ignored (depends on calibration)	Warm

coolant sensor which sends to the computer a true indication of the engine temperature. The air/fuel mixture supplied can vary from 1.5:1 at −40°F (−40°C) to 14.7:1 at 212°F (100°C).

Clear flood. If the engine floods (too much fuel) when attempting to start the engine, depressing the accelerator beyond 80% throttle signals the computer to switch modes. If the RPM is less than 400 and the accelerator pedal is 80 to 100% depressed (as determined by TPS), the air/fuel ratio delivered is 20:1 to help clear a flooded engine.

NOTE: Some port-injected engines supply *no fuel* at all if the engine is cranking and the throttle is over 80% open.

Open loop. After the engine starts, it runs according to a predetermined program set in the computer memory until the O_2 (oxygen) sensor becomes warm enough to start producing usable data for the computer.

Closed loop. Whenever the O_2 sensor starts to provide usable oxygen content information from the exhaust to the computer, and if the coolant sensor indicates above 150°F (66°C), the computer switches to closed-loop operation. In closed loop, ignition timing is optimized for maximum gas mileage and performance. Torque converter clutch and other systems are also engaged to ensure lowest emissions and maximum gas mileage.

Acceleration mode. Whenever operating in closed loop and the accelerator is depressed, the computer provides greater fuel flow for the power necessary for maximum acceleration. The torque converter clutch is also released and other factors are changed for power, not economy.

Deceleration mode. When deceleration occurs, the computer reduces the amount of fuel that would normally be supplied.

COMPUTER PRECAUTIONS

All automotive computers are sensitive to high voltage or high temperatures. There are several service-related activities that should *never* be attempted when working on *any* computer-equipped vehicle. These precautions include:

1. *Never* disconnect *any* 12-V operating unit while the ignition switch is "on" (engine running or not running). Because of the self-induction of any coil, when an electrical unit is disconnected, a "spike" or "transient" of extremely high voltage is created (over 7000 V is possible), which can

severely damage the computer and/or sensors. Following is a partial list of units that should *never* be disconnected while the engine is running or when the ignition switch is "on":
   (a) Either battery cable
   (b) Mixture-control solenoid
   (c) Idle-speed control units (stepper motor)
   (d) Electronic fuel injectors
   (e) Air management solenoids (air pump solenoids)
   (f) Ignition module wiring
   (g) The PROM from the computer
   (h) Any computer wiring
   (i) Blower motor wiring connections
   (j) Air-conditioning clutch wiring

2. *Never* jump start another vehicle or receive a jump start from another car unless the jumper cables are connected or disconnected with the ignition switch "off."

3. Do not mount radio speakers near the computer. The speaker magnets can damage the circuits and components inside the computer.

4. Do not use an electric arc welder on the car body without first disconnecting (unplugging) the computer. Use care when performing body repairs near the computer or sensors.

5. Be certain to ground yourself by touching the car body before removing or installing a PROM. Static electricity can damage computer circuits.

6. Repair windshield leaks as soon as possible to prevent possible damage to kickpanel-mounted computers from moisture.

7. Never use an analog (needle-type) ohmmeter on any computer sensor unless specified in the test procedure. A high-impedance meter is required to prevent the voltage from the test meter from damaging sensors and computer circuits.

8. Do not use a test light while troubleshooting any computer-connected electrical unit. To prevent possible computer or sensor damage, always use a digital high-impedance meter during testing unless otherwise instructed.

9. Be certain to wear a grounding strap whenever working on or near a computer-operated digital dash. A grounding strip that fastens around your wrist and clips to the vehicle body is available commercially for a reasonable cost. The high voltage produced in our bodies as we move into or out of a car, called electrostatic discharge (ESD), is capable of producing voltages as high as 10,000 V.

SUMMARY

1. Computer engine controls are required to provide accurate ignition timing and precise air/fuel mixtures for all driving conditions.

2. All computer engine control units, regardless of name, use various types of memory to function. These include ROM, read-only memory; RAM, random access memory; PROM, programmable read-only memory; EPROM, erasable programmable read-only memory; and EEPROM, electrically erasable programmable read-only memory.

3. Most computer systems measure the operating conditions of the engine and select from memory the correct air/fuel ratio and ignition for maximum power and fuel economy with the lowest possible exhaust emissions. A typical automotive computer system does over 100,000 calculations per second. It bases the calculations on information from various sensors, including:
   (a) Coolant temperature sensor
   (b) Oxygen sensor in the exhaust system
   (c) Throttle position sensor
   (d) Barometric pressure sensor
   (e) Engine vacuum sensor
   (f) Vehicle speed sensor
   (g) Engine knock sensor

4. The automotive computer controls air/fuel ratio and ignition timing to provide the optimum mixture for every driving condition. Some computer systems also control torque converter clutch applications, cooling fan operation, air management systems, and other engine-related systems.

STUDY QUESTIONS

25-1. List and describe five different types of memories commonly used in automotive computers.

25-2. Explain how an oxygen sensor can be tested using a high-impedance voltmeter.

25-3. Explain pressure sensors and why they are important for proper engine operation.

25-4. List four input devices and four output devices used on most automotive computer systems.

25-5. Describe the "modes of operation" of a typical automotive computer system.

25-6. List eight precautions with which all automotive technicians should be familiar in order to avoid damaging automotive computer systems.

MULTIPLE-CHOICE QUESTIONS

25-1. The type of memory that disappears or is lost when the ignition is turned off is called:
   (a) RAM.
   (b) ROM.
   (c) PROM.
   (d) EPROM.
   (e) EEPROM.

25-2. The coolant sensor:
   (a) is a thermistor.
   (b) operates the "hot" light on the dash.
   (c) has a high resistance when hot and a low resistance when cold.
   (d) operates the thermostat.

25-3. Computer *inputs* include:
   (a) BARO, TPS, M/C solenoid.
   (b) MAP, O_2, TPS.
   (c) air management, M/C solenoid, VAC.
   (d) vehicle speed sensor, air management, TPS.

25-4. Computer *outputs* include:
  (a) air management, M/C solenoid, idle speed control.
  (b) TPS, O$_2$, and ignition timing control.
  (c) BARO, MAP, VAC.
  (d) vehicle speed, TPS, M/C solenoid.

25-5. The oxygen sensor:
  (a) supplies a high voltage (1 V) when the exhaust is rich.
  (b) supplies a high voltage (1 V) when the exhaust is lean.
  (c) tells the computer if the engine is burning oil.
  (d) tells the computer the altitude of the engine based on the oxygen content of the exhaust.

25-6. Ignition timing control is affected by:
  (a) the temperature—the lower the temperature, the greater the spark advance.
  (b) the MAP—the lower the engine vacuum, the less the spark advance.
  (c) the knock sensor—the greater the engine knock, the less the spark advance.
  (d) all of the above.

25-7. When the computer is sensing all engine factors and making engine fuel mixture and/or ignition timing changes based on the sensors, it is called:
  (a) stoichiometric.
  (b) open loop.
  (c) closed loop.
  (d) clear flood.

25-8. Technician A says that the throttle position sensor is a variable resistor that sends a high voltage (about 5 V) back to the computer if the throttle is wide open (WOT). Technician B says that the oxygen sensor must be about 600°F before the voltage can be used by the computer. Which technician is correct?
  (a) A only.
  (b) B only.
  (c) both A and B.
  (d) neither A nor B.

25-9. Technician A says that a computer-equipped vehicle should never be jump started. Technician B says that a computer-equipped vehicle should never jump start another vehicle. Which technician is correct?
  (a) A only.
  (b) B only.
  (c) both A and B.
  (d) neither A nor B.

25-10. Computer fuel injection systems:
  (a) use a higher-pressure fuel pump.
  (b) have the computer control the pulse width.
  (c) have the computer operate the fuel injector units.
  (d) all of the above.

26

Computerized Engine Control System Diagnosis and Testing

COMPUTERIZED ENGINE CONTROLS ARE COMPLEX SYSTEMS THAT REQUIRE a systematic approach to diagnosing troubles. Before attempting to correct an engine operation problem that may include the computer system, check *everything* that could cause the engine problem as if the car was not computer controlled. The topics covered in this chapter include:

1. Typical noncomputer problems
2. Fault codes: GM/Ford/Chrysler
3. System performance checks
4. Ignition timing procedures
5. Typical computer problems
6. Scan tool testing

TYPICAL NONCOMPUTER PROBLEMS

If the driver has noticed a ''check engine,'' ''power loss,'' or ''power limited'' light, proceed through the troubleshooting sequences specified by the manufacturer. If the engine running complaint does _not_ trigger a system warning light, proceed with basic engine troubleshooting as if the engine were not computer equipped. Typical noncomputer problems and possible causes include:

Rough Idle (Possible Stall)

1. Idle speed too low
2. Idle mixture incorrect or unequal (vacuum leak)
3. Timing retarded
4. Clogged PCV valve or hoses
5. Defective spark plug wires
6. Worn or cracked spark plug
7. Cracked or defective purge valve for the charcoal canister system
8. Defective power valve in the carburetor (if equipped)
9. EGR valve stuck open

Missing on Acceleration

1. Defective spark plug wire
2. Cracked or defective distributor cap
3. Defective distributor rotor
4. Crossed spark plug wires
5. Shorted or cracked ignition coil
6. Loose condenser (point-type ignition)
7. Loose primary coil lead connections
8. Clogged fuel filter
9. Weak fuel pump
10. Cracked or softened fuel lines

Poor Gas Mileage

1. Retarded ignition timing
2. Clogged exhaust
3. Defective vacuum advance unit on the distributor
4. Vacuum leak around the carburetor, intake manifold, or a defective vacuum hose
5. Defective choke (closed all the time)
6. Defective power valve in the carburetor (if equipped)
7. Clogged air filter
8. Defective thermostatic air cleaner, permitting warm air to enter the carburetor all the time
9. EGR valve stuck open
10. Defective or too low a cooling system thermostat

Pings on Acceleration

1. Ignition timing advanced
2. Too low a grade of fuel
3. Vacuum leak near the carburetor, intake manifold, or a defective vacuum hose
4. EGR valve not opening correctly
5. Clogged heat riser allowing exhaust heat under the carburetor all the time
6. Defective thermostatic air cleaner, allowing warm air to enter the carburetor all the time

THE 3-HOUR FUEL PUMP TEST

Shortly after computer-controlled engines were first produced, a fleet technician spent 3 hours testing an engine. The car performed perfectly unless accelerated rapidly. At high engine speeds (rpms), the engine would cut out, miss, and almost stall. The technician checked the carburetor mixture solenoid, oxygen sensor, TPS, and almost every other computer system component. After 3 hours of troubleshooting, the problem was discovered to be a weak fuel pump. If trouble codes are not present, always check all _noncomputer_ items that can cause the problem before checking the computer-controlled units.

GM COMPUTER DIAGNOSIS AND SERVICE

Many computer systems have on-board diagnosis capability built into the system. By checking the trouble codes, the technician can determine where the problem is located in most cases.

The GM CCC (Computer Command Control) system uses a ''check engine'' or ''check engine soon'' light to notify the driver of possible system failure. If the check engine light is ''on,'' the computer is _not_ working. The engine is running on the backup or ''limp-in'' mode of limited spark advance and rich fuel mixture.

Under the dash (on most GM cars) is an ALCL (Assembly Line Communications Link). The terminals are lettered. The top-right-side terminals are labeled A and B, and when connected (with a wire spade connector, paper clip, or other suitable tool) will start the ''check engine'' light flashing a code when the ignition switch is ''on'' (engine not running). See Figures 26-1 and 26-2. Each code flashes three times, then all the codes repeat in numerical sequence.

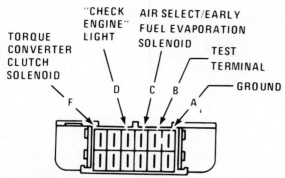

TORQUE CONVERTER CLUTCH SOLENOID — F

"CHECK ENGINE" LIGHT — D

AIR SELECT/EARLY FUEL EVAPORATION SOLENOID — C

TEST TERMINAL — B

GROUND — A

ASSEMBLY LINE COMMUNICATION LINK (ALCL) CONNECTOR

FIGURE 26-1 *(Courtesy of General Motors Corporation.)*

DIAGNOSTIC CODE DISPLAY

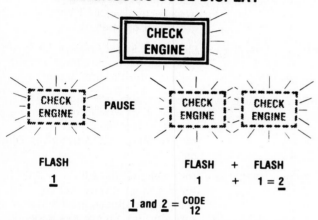

CHECK ENGINE

CHECK ENGINE — PAUSE — CHECK ENGINE — CHECK ENGINE

FLASH
1

FLASH + FLASH
1 + 1 = 2

1 and 2 = CODE 12

FIGURE 26-2 *(Courtesy of General Motors Corporation.)*

NOTE: Trouble codes can vary according to year, make, model, and engine. Always consult the service literature or service manual for the exact vehicle being serviced. See Figures 26-3 and 26-4.

GM SYSTEM PERFORMANCE CHECK

After the trouble codes have been corrected (or no trouble codes are indicated), the operation of the computer can be checked. Checking the operation of the computer is called the system performance check. See Figure 26-5 for detailed procedures.

A carburetor-equipped General Motors vehicle equipped with a computer is checked with a dwell meter. The dwell meter is connected to a separate test lead under the hood, identified by a blue wire with a green open connector. See Figure 26-6. The dwell meter indicates the amount of time that the solenoid is energized. This "on" time is called the "duty cycle." When the MC solenoid is

energized, the meter rods inside the carburetor are forced downward and the carburetor delivers its leanest mixture. When the solenoid is deenergized (no current), a spring(s) raises the metering rods and the carburetor can supply its richest mixture. The dwell meter is usually set on the six-cylinder scale regardless of the number of cylinders. If the solenoid is always energized (metering rods down—lean), the dwell meter would read 60°. If the solenoid were not energized at all (metering rods up—rich), the dwell meter reads 0°. Since the MC solenoid is *always* cycled 10 times a second, it cannot be "on" or "off" more than 90% of the time. Therefore, the highest dwell meter reading possible is 54° (90% of 60°) and the lowest dwell meter reading possible is 6° (10% of 60°).

A high dwell meter reading indicates that a rich mixture is present in the engine and the computer is attempting to compensate by keeping the metering rods down (high dwell). A low dwell meter reading indicates that a lean mixture is present and the computer is attempting to compensate by leaving the metering rods up (no current—low dwell—to the MC solenoid).

A properly running engine (in closed loop) should idle with a dwell close to 30° (25 to 35°) and be varying slightly left and right. With the engine set at *exactly* 3000 rpm, the dwell should indicate 10 to 50° (prefer 35 to 45°). When the engine is still in open loop (not warm enough to go closed loop), the computer is sending a *fixed* dwell to the MC solenoid. The value of the dwell reading is not adjustable or changeable and is based on the temperature of the engine as determined by the coolant sensor. Once the engine reaches normal operating temperature and goes into closed loop, the dwell (fuel mixture control) will vary between 10° and 50° (2 to 4° is the normal dwell meter swing range). The dwell does vary because the computer is reacting to oxygen sensor information and making the necessary changes. The actual dwell swing should be read from the average of the dwell meter readings.

HINT: For the best performance, the larger the engine, the higher the dwell should be at 3000 rpm. For example, a four-cylinder engine will perform satisfactorily at 35° dwell. A larger V-6 runs best around 40° and a V-8 delivers its best performance near 45°. A dwell reading above 30° means that the fuel mixture being delivered to the engine is slightly rich and the computer is driving it leaner. This situation provides improved throttle response over a lean mixture being driven rich.

The dwell of the MC solenoid is changed by adjusting the rich stop and lean stop adjustments inside the carburetor. The dwell at idle can be changed by the adjustment of the idle air bleed and/or idle mixture screws. These adjustments require special gauging tools and should be set to factory-specified limits.

TROUBLE CODE IDENTIFICATION

The "CHECK ENGINE" light will only be "ON" if the malfunction exists under the conditions listed below. It takes up to five seconds minimum for the light to come on when a problem occurs. If the malfunction clears, the light will go out and a trouble code will be set in the ECM. Code 12 does not store in memory. If the light comes "on" intermittently, but no code is stored, go to the "Driver Comments" section. Any codes stored will be erased if no problem reoccurs within 50 engine starts. A specific engine may not use all available codes.

The trouble codes indicate problems as follows:

TROUBLE CODE 12   No distributor reference pulses to the ECM. This code is not stored in memory and will only flash while the fault is present. Normal code with ignition "on," engine not running.

TROUBLE CODE 13   Oxygen Sensor Circuit — The engine must run up to five minutes at part throttle, under road load, before this code will set.

TROUBLE CODE 14   Shorted coolant sensor circuit — The engine must run up to five minutes before this code will set.

TROUBLE CODE 15   Open coolant sensor circuit — The engine must run up to five minutes before this code will set.

TROUBLE CODE 21   Throttle position sensor circuit — The engine must run up to 25 seconds, at specified curb idle speed, before this code will set.

TROUBLE CODE 22 (1984 ONLY)   Throttle Position Sensor (TPS) circuit voltage low (grounded circuit or misadjusted TPS). Engine must run 20 seconds at specified curb idle speed, to set code.

TROUBLE CODE 23   M/C solenoid circuit open or grounded.

TROUBLE CODE 24   Vehicle speed sensor (VSS) circuit — The vehicle must operate up to five minutes, at road speed, before this code will set.

TROUBLE CODE 32   Barometric pressure sensor (BARO) circuit low.

TROUBLE CODE 34   Vacuum sensor or Manifold Absolute Pressure (MAP) circuit — The engine must run up to five minutes, at specified curb idle, before this code will set.

TROUBLE CODE 35   Idle speed control (ISC) switch circuit shorted. (Over 50% throttle for over 2 seconds.)

TROUBLE CODE 41   No distributor reference pulses to the ECM at specified engine vacuum. This code will store in memory.

TROUBLE CODE 42   Electronic spark timing (EST) bypass circuit or EST circuit grounded or open.

TROUBLE CODE 43   Electronic Spark Control (ESC) retard signal for too long a time; causes retard in EST signal.

TROUBLE CODE 44   Lean exhaust indication — The engine must run up to five minutes, in closed loop and at part throttle, before this code will set.

TROUBLE CODE 45   Rich exhaust indication — The engine must run up to five minutes, in closed loop and at part throttle, before this code will set.

TROUBLE CODE 51   Faulty or improperly installed calibration unit (PROM). It takes up to 30 seconds before this code will set.

TROUBLE CODE 53 (1984 ONLY)   Exhaust Gas Recirculation (EGR) valve vacuum sensor has seen improper EGR vacuum.

TROUBLE CODE 54   Shorted M/C solenoid circuit and/or faulty ECM.

TROUBLE CODE 55   Grounded Vref (terminal "21"), high voltage on oxygen sensor circuit or ECM.

FIGURE 26-3   *Trouble codes for General Motors vehicles equipped with a carburetor. (Courtesy of General Motors Corporation.)*

12 NO REFERENCE PULSES TO ECM

13 OXYGEN SENSOR CIRCUIT
FAILED

14 COOLANT READING TOO HIGH

15 COOLANT READING TOO LOW

21 THROTTLE POSITION SENSOR
READING TOO HIGH

22 THROTTLE POSITION SENSOR
READING TOO LOW

24 VEHICLE SPEED SENSOR FAILED

31 WASTEGATE ELECTRICAL
SIGNAL OPEN OR GROUNDED

32 EGR ELECTRICAL OR VACUUM
CIRCUIT MALFUNCTION

33 MASS FLOW SENSOR READING
TOO HIGH

34 MASS FLOW SENSOR READING
TOO LOW OR NO SIGNAL FROM
SENSOR

41 CAM SENSOR FAILED

42 ERROR IN DISTRIBUTOR OR C^3
SYSTEM

43 ELECTRONIC SPARK CONTROL
FAILURE

44 OXYGEN SENSOR LEAN TOO
LONG

45 OXYGEN SENSOR RICH TOO
LONG

51 CALIBRATION PROM ERROR

52 CALPAK MISSING

55 INTERNAL ECM ERROR
(A/D CONVERTER)

FIGURE 26-4 *Trouble codes for fuel-injected General Motors engines. (Courtesy of General Motors Corporation.)*

FIGURE 26-5 *The system performance check should be used to determine any problem in the computer system. (Courtesy of General Motors Corporation.)*

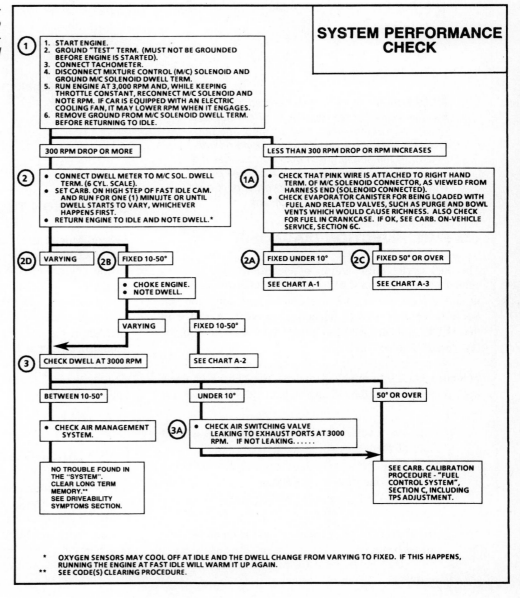

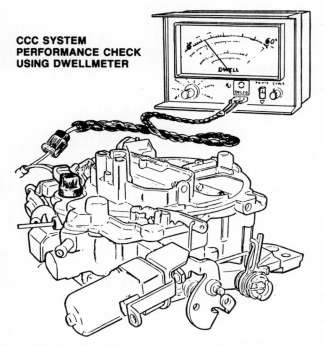

CCC SYSTEM PERFORMANCE CHECK USING DWELLMETER

FIGURE 26-6 *A dwell meter should be set on the six-cylinder scale regardless of the number of cylinders of the engine being tested. The positive (+) lead of the dwell meter should be connected to the single green connector under the hood. The negative (–) lead of the dwell meter should be connected to a good ground. (Courtesy of General Motors Corporation.)*

GM FUEL INJECTION DIAGNOSIS

The duty cycle (pulse width) of electronic fuel injection cannot be read with a dwell meter. Whether the engine is in closed loop or open loop, a rich or lean condition can easily be determined by connecting terminals A and B in the ALCL and starting the engine. Check the operation of the ''check engine'' light.

With the engine running and the diagnostic terminal grounded (ALCL terminals A and B), the ''check engine'' light is ''off'' when the exhaust is lean and ''on'' when it is rich.

> *Open loop.* ''Check engine'' light flashes at a rapid rate of two times per second.
>
> *Closed loop.* ''Check engine'' light flashes slower at a rate of one time per second.
>
> *Lean exhaust.* ''Check engine'' light is out all or most of the time.
>
> *Rich exhaust.* ''Check engine'' light is on all or most of the time.

HOW CAN A DISCHARGED BATTERY CAUSE FLOODING OF A FUEL-INJECTED ENGINE?

Engines equipped with electronic fuel injection commonly use electric fuel pumps. If battery voltage drops below a pre-determined value, the fuel injectors are pulsed on for a slightly longer time to compensate for the lower fuel pump pressure supplied to the injectors. As the battery voltage decreases, the pulse time of the injectors increases. Therefore, if an engine fails to start, the starter amperage drain lowers the battery voltage, and could cause excessive fuel delivery from the injectors and foul the spark plugs. The battery voltage correction factor normally works correctly and does not cause engine flooding. However, a defective or discharged battery can often cause flooding, fouled spark plugs, and a no-start situation, even though the battery is still capable of cranking the engine.

GM IGNITION TIMING

General Motors uses more than eight different methods and test procedures for testing the ignition timing, depending on the engine, engine equipment, and body style. Most methods involve disconnecting the computer from the ignition distributor. This usually involves disconnecting the four-way conductor at the junction a few inches from the distributor. Other engines require that the computer be in diagnostics (A and B terminals together at ALCL) or use the average timing methods. Consult under the hood emission decal for the *exact* procedure before attempting to set ignition timing.

FORD DIAGNOSIS AND SERVICE

An MCU (microprocessor control unit) is used on selected Ford cars and trucks, while the EEC (electronic engine control) system of engine control is used on all other models. Both the MCU and the EEC systems have built-in diagnostic codes. See Figures 26-7, 26-8, and 26-9.

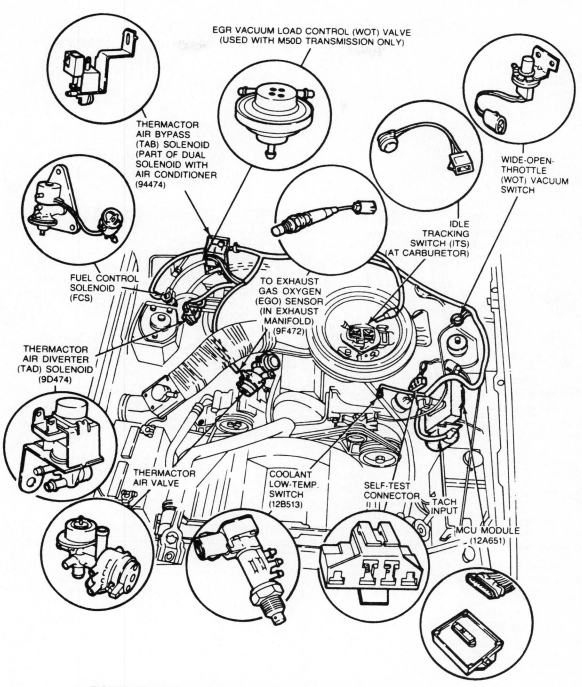

EGR VACUUM LOAD CONTROL (WOT) VALVE
(USED WITH M50D TRANSMISSION ONLY)

THERMACTOR
AIR BYPASS
(TAB) SOLENOID
(PART OF DUAL
SOLENOID WITH
AIR CONDITIONER)
(94474)

WIDE-OPEN-
THROTTLE
(WOT) VACUUM
SWITCH

IDLE
TRACKING
SWITCH (ITS)
(AT CARBURETOR)

FUEL CONTROL
SOLENOID
(FCS)

TO EXHAUST
GAS OXYGEN
(EGO) SENSOR
(IN EXHAUST
MANIFOLD)
(9F472)

THERMACTOR
AIR DIVERTER
(TAD) SOLENOID
(9D474)

THERMACTOR
AIR VALVE

COOLANT
LOW-TEMP.
SWITCH
(12B513)

SELF-TEST
CONNECTOR

TACH
INPUT

MCU MODULE
(12A651)

FIGURE 26-7 *Typical 2.3-liter four-cylinder MCU component locations.
(Courtesy of Ford Motor Company.)*

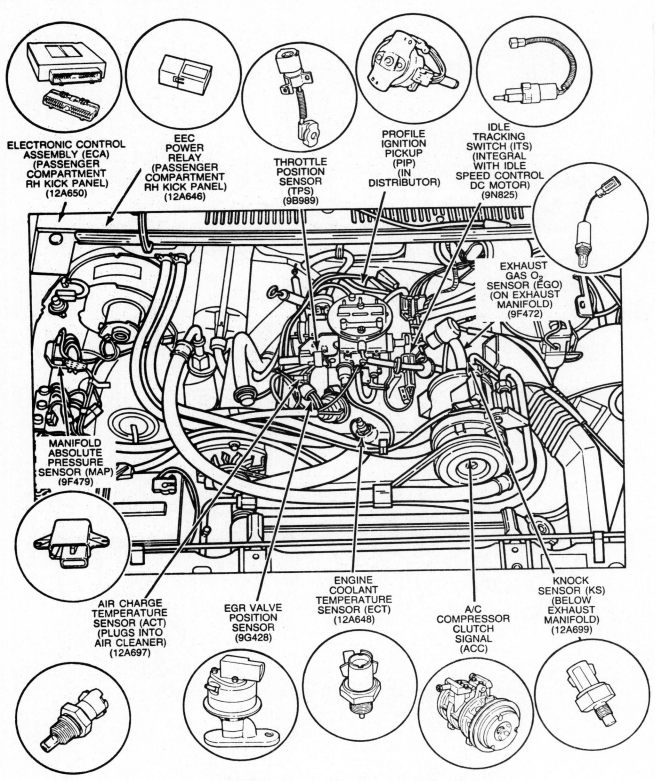

FIGURE 26-8 *Typical EEC IV component locations.* *(Courtesy of Ford Motor Company.)*

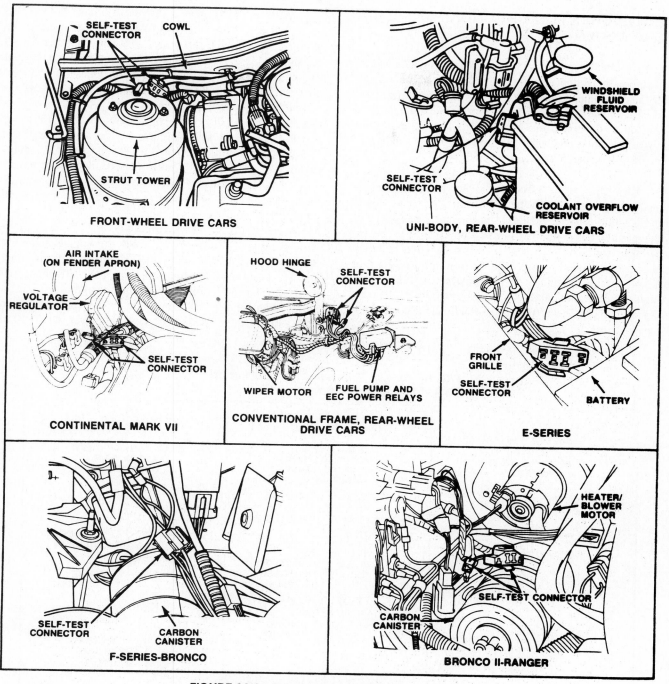

FIGURE 26-9 *Typical self-test connector locations.*
(Courtesy of Ford Motor Company.)

PROCEDURES FOR OBTAINING FORD CODES

The best tool to use during troubleshooting is a *STAR* (self-test automatic readout) tester. If a STAR tester is not available, a needle (analog)-type voltmeter can be used for MCU and EEE IV systems. Connect a jumper lead and an analog voltmeter as illustrated in Figure 26-10. The test connector is usually located under the hood on the driver's side (except EEC III). See Figures 26-10 and 26-11.

Key On/Engine Off Test. With the ignition key "on" (engine off), watch the voltmeter pulses that should appear within 5 to 30 seconds. (Ignore any initial surge of voltage when the ignition is turned "on.").

The computer will send a two digit code that will cause the voltmeter to pulse or move from left to right. For example, if the voltmeter needle pulses two times, then pauses for 2 seconds, and then pulses three times, the code is 23. See Figure 26-11. There is normally a 4-second pause between codes.

After all of the codes have been reported, the computer will pause for about 6 to 9 seconds, then cause the voltmeter needle to pulse once, and pause another 6 to 9 seconds. This is the normal separation between current trouble codes and continuous memory codes (intermittent problems). Code 11 is the normal pass code, which means that no fault has been stored in memory. Therefore, normal operation of the diagnostic procedure using a voltmeter should indicate the following if no codes are set: 1 pulse (2-second pause), 1 pulse (6- to 9-second pause), 1 pulse (6- to 9-second pause), 1 pulse (2-second pause), and finally, 1 pulse. The pulses separated by a 2-second internal mean code 11, which is the code used between current and intermittent trouble codes.

Engine Running Test. Start the engine and raise the speed to 2500 to 3000 rpm within 20 seconds after starting. Hold a steady high engine speed until the initial pulses appear ("2" for a four-cylinder engine, "3" for a six-cylinder, and "4" for an eight-cylinder). Continue to hold a high engine speed until the code pulses begin (10 to

FIGURE 26-10 *Test equipment hookup required to determine Ford trouble codes.*

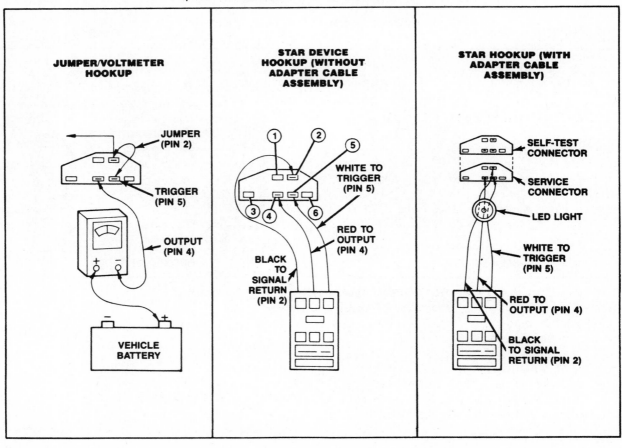

SUMMARY

1. Diagnosis and testing of computer-equipped engines should begin with visual and other checks of the basic systems. If trouble codes are not present, check all non-computer items that could cause the engine problem before checking the computer-controlled units.

2. Most computer-controlled engine systems have built-in self-diagnosis capabilities. Trouble codes may be either current problems (''hard'' codes) or intermediate codes which are not presently current (''soft'' or history codes). Each automotive manufacturer specifies exact procedures for determining and interrupting trouble codes.

3. The correct operation of the computer can often be checked by measuring output voltages, dwell readings, or other indications for ignition and/or fuel mixture control.

4. Most sensors, such as coolant, oxygen, MAP, TPS, and others, can be tested individually. Computer-controlled output devices such as mixture control solenoids, air management solenoids, timing control, and other computer control units can also be tested individually.

5. Exact test procedures as specified by the manufacturer should be followed whenever working with any computer system.

STUDY QUESTIONS

26-1. If no trouble codes are present, explain why noncomputer items should be checked before attempting to find the problem in a computer-controlled unit.

26-2. Describe the procedure for accessing GM trouble codes, Ford trouble codes, and Chrysler codes.

26-3. Explain how the ignition timing can be checked on computer-controlled engines.

26-4. Explain how using the incorrect silicone sealer on a valve cover gasket can cause oxygen sensor damage.

MULTIPLE-CHOICE QUESTIONS

26-1. Technician A says that if no trouble codes are present and the engine runs poorly, the coolant sensor should be checked first. Technician B says to check all noncomputer units first. Which technician is correct?
   (a) A only.
   (b) B only.
   (c) both A and B.
   (d) neither A nor B.

26-2. Trouble codes are determined by:
   (a) grounding the A and B terminals together in the ALCL on GM cars.
   (b) unplugging the diagnostic connector link on GM cars.
   (c) applying 20 in. Hg to the MAP sensor and reading flashes of ''power loss'' light on Chrysler cars.
   (d) turning the ignition switch ''on'' and ''off'' twice within 5 seconds on Ford cars.

26-3. All trouble codes:
   (a) tell the technician exactly what part requires replacement.
   (b) tell the technician what circuit may have problems.
   (c) tell the technician exactly which component or wire is defective.
   (d) none of the above.

26-4. In the GM system performance check:
   (a) a high dwell reading indicates a rich mixture being driven leaner.
   (b) a high dwell reading indicates a lean mixture being driven richer.
   (c) a fixed dwell means that the computer is in a closed loop.
   (d) both (b) and (c) are correct.

26-5. Ignition timing checks on computer-equipped cars:
   (a) are the same as noncomputer engines.
   (b) are not adjustable.
   (c) require special equipment such as magnetic offset timing units.
   (d) none of the above.

26-6. Oxygen sensors:
   (a) are ruined if leaded fuel is used.
   (b) are ruined if some silicone sealers are used on some parts of the engine.
   (c) must be replaced every 5000 miles for proper operation.
   (d) both (a) and (b).

26-7. Coolant sensors:
   (a) warn the driver when coolant temperatures are too high.
   (b) provide an output from the computer.
   (c) can be tested using an ohmmeter.
   (d) should be replaced every year for best performance.

26-8. Throttle position sensors:
   (a) must be adjusted correctly.
   (b) if not correct, can cause engine ping.
   (c) if not correct, can cause lack of power.
   (d) all of the above.

26-9. Scan tools:
   (a) will always find the computer problem.
   (b) "scan" the various input and output readings.
   (c) are required tools to adjust or service any computer-equipped car.
   (d) can be used to check previous gas mileage history on all computer engines.

26-10. Technician A says that if there are no trouble codes, the computer system and engine must be functioning correctly. Technician B says that many engine problems, such as ignition module problems, often are not part of the computer engine control diagnostics. Which technician is correct?
   (a) A only.
   (b) B only.
   (c) both A and B.
   (d) neither A nor B.

27

Body Computers

BODY COMPUTERS ARE COMPUTERS OR MICROPROCESSORS THAT CONtrol various functions of a vehicle and do not usually control the engine operation. This chapter includes the basic operation of several automotive computer systems which are commonly referred to as body computers. The topics covered in this chapter include:

1. Body computer inputs
2. Body computer outputs
3. Heating, ventilation, and air-conditioning controls
4. Body computer power supply
5. Service diagnostics

INTRODUCTION TO BODY COMPUTERS

Body computer is a generic term commonly used to describe microprocessor-controlled functions other than engine operation. Body computers are usually used to control air-conditioning and related functions. A body computer uses information about the engine and the vehicle from the engine computer. Some of the inputs to the body computer from the engine computer include (see Figure 27-1):

1. Engine speed (rpm)
2. Engine coolant temperature
3. Vehicle speed

Other inputs that are used by the body computer to control air-conditioning and other functions include (see Figure 27-2):

1. Battery voltage
2. Oil pressure
3. Fuel level
4. Power steering switch positions
5. Air-conditioning clutch status
6. Air management door positions for the ventilation system

FIGURE 27-1 *IPC indicates "instrument panel cluster"; BCM, "body control module"; ECM, "electronic control module" (engine computer); ECC, "electronic comfort control" (heating and air conditioning control panel); and CRT, "cathode ray tube display" (if equipped). The sensor and switch inputs represent any sensor or switch monitored by the body computer. (Courtesy of General Motors Corporation.)*

SOURCES OF INPUT

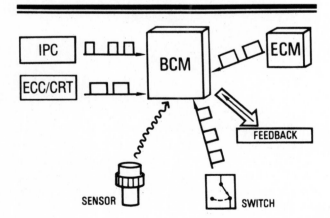

BCM/ECM DATA TRANSFER

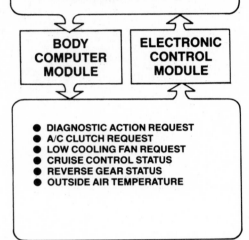

FIGURE 27-2 *(Courtesy of General Motors Corporation.)*

7. Outside air temperature
8. Inside air temperature

The body computer also communicates and controls other systems of the car, including the heating, ventilation, and air-conditioning (HVAC) system; instrument panel display, including dash instruments; and tell-tale warning lights. See Figure 27-3.

FIGURE 27-3 *(Courtesy of General Motors Corporation.)*

TYPICAL BCM OUTPUT DESTINATIONS

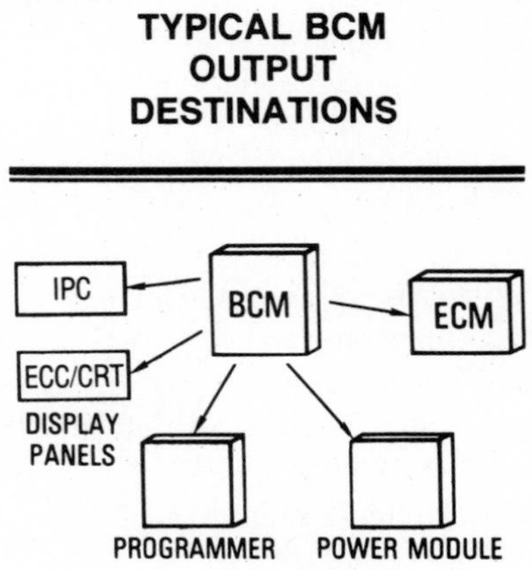

COMPUTER CLOCKS

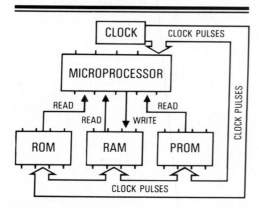

FIGURE 27-4 *(Courtesy of General Motors Corporation.)*

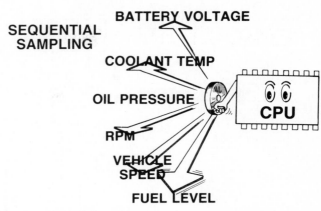

FIGURE 27-5 *(Courtesy of General Motors Corporation.)*

BODY COMPUTER OPERATION

Since the body computer is receiving information from a variety of sources, a system of data processing is used to handle all of the information. Many sensors produce a changing (analog) signal, and since the body computer can use only "on" and "off" signals, a circuit is used to convert analog signals to digital signals. This circuit is called an analog-to-digital (A-to-D) converter. Special crystals inside the body converter produce a cadence or "clock" pulse of known length. These clock cadence counts permit all information to be sequenced correctly for proper processing of the data from the A-to-D converter. The engine computer contains similar crystal clock circuits, and this permits the transfer of data between computers because the clock pulses are timed together. See Figure 27-4. A typical automotive computer system can communicate between computers at the rate of over 8000 bits of information per second. See Figure 27-5.

Since the body computer must look at information from a variety of sources, a system of information sampling is used called *multiplexing*. For example, the body computer or central processing unit (CPU) checks the coolant temperature, engine rpm, and many other inputs. Some of these inputs do not change rapidly, such as coolant temperature, while inputs such as engine rpm must be checked constantly. Therefore, the body computer checks on the inputs of all its inputs on a regular sampling sequence and acts on these inputs based on the program in the computer PROM. See Figure 27-6. Body computers also use EEPROMs to store odometer readings for the electronic instrument display unit as well as the necessary RAM for calculations of input data.

POWER STEERING INPUT

A typical body computer receives a voltage signal from a power steering switch if the pressure in the high-pressure side exceeds 300 psi (207 kPa). This high pressure indicates that the power steering is starting to drain engine power. This signal is processed by the body computer and signals the engine computer to increase the idle speed to compensate for the additional engine load.

AIR-CONDITIONING CLUTCH CONTROL

The body computer also controls the air-conditioning compressor clutch operation. The compressor clutch is left on continuously at idle, for example, to prevent engine speed variations that could result if the air-conditioning clutch were permitted to cycle. Using inputs from the air-conditioning system pressures and temperatures inside the car, the body computer selects the most efficient operation of the compressor clutch. See Figure 27-7.

ELECTRICAL SYSTEM CONTROL

Alternator field current is also controlled by many computer systems to provide the best balance between efficient battery voltage control and smooth engine operation. For example, whenever an electrical load is added to the electrical system, such as when a blower motor is turned on at idle, the body computer will gradually cause the alternator to increase its charging rate, thereby reducing the effects on engine speed. The body computer is also used to control cruise (speed) control (if equipped).

TYPICAL DATA PROCESSING

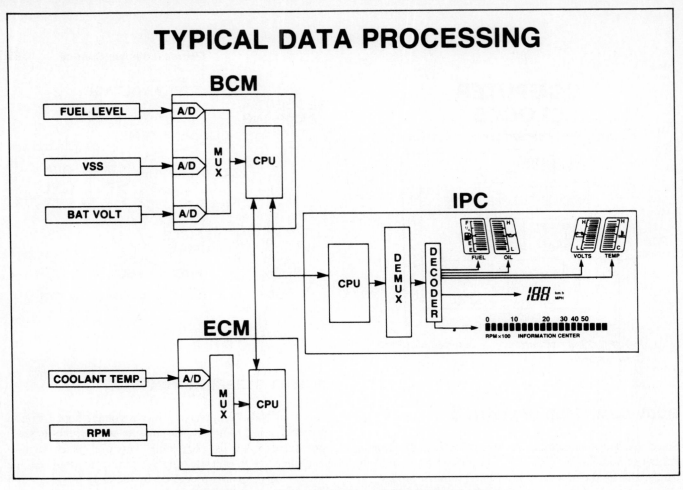

FIGURE 27-6 (Courtesy of General Motors Corporation.)

COMPRESSOR CLUTCH CONTROL

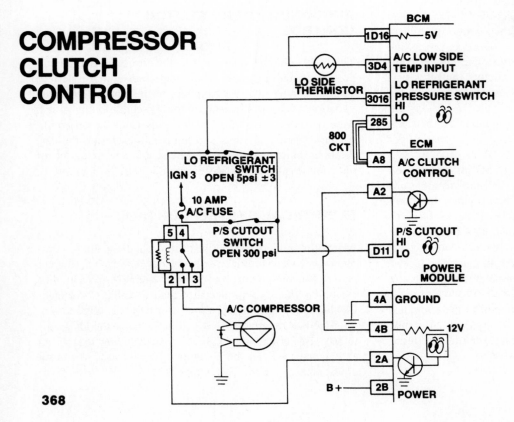

FIGURE 27-7 (Courtesy of General Motors Corporation.)

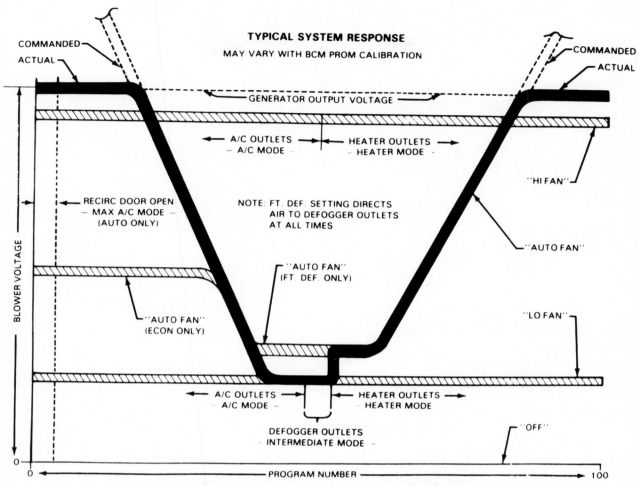

TYPICAL SYSTEM RESPONSE

MAY VARY WITH BCM PROM CALIBRATION

COMMANDED
ACTUAL

COMMANDED
ACTUAL

GENERATOR OUTPUT VOLTAGE

← A/C OUTLETS → ← HEATER OUTLETS →
– A/C MODE –   – HEATER MODE –

''HI FAN''

RECIRC DOOR OPEN
– MAX A/C MODE –
(AUTO ONLY)

NOTE: FT. DEF. SETTING DIRECTS
AIR TO DEFOGGER OUTLETS
AT ALL TIMES

''AUTO FAN''

BLOWER VOLTAGE

''AUTO FAN''
(FT. DEF. ONLY)

''AUTO FAN''
(ECON ONLY)

''LO FAN''

← A/C OUTLETS → ← HEATER OUTLETS →
– A/C MODE –   – HEATER MODE –

DEFOGGER OUTLETS
– INTERMEDIATE MODE –

''OFF''

0

0 ← PROGRAM NUMBER → 100

FIGURE 27-8   *(Courtesy of General Motors Corporation.)*

VENTILATION SYSTEM CONTROL

Most body computers control the speed of the blower motor by varying the voltage sent to the motor. The ventilation, heating, and air-conditioning system is controlled by the body computer not only by controlling the blower motor voltage, but also by controlling the air-mixing (blend) doors and heater valve operation. A special electronic unit contains vacuum valves and a stepper motor to control the vacuum-controlled ventilation doors and fresh-air-blend door. A special circuit built into the controller sends a signal back to the body computer of the position of the blend door. This information is used by the computer to determine the location, and therefore the changes in the position of the blend door, that may be necessary for proper air management temperatures. See Figure 27-8.

WHAT IS A ''WAKE-UP'' SIGNAL?

A ''wake-up'' signal is a signal sent to the body computer power supply from the door handle or door jam switch. As soon as the door is opened, the computer assumes that an engine start, climate control, and dash display will be happening in a short time. The signal from the door opening starts (wakes up) the power supply and starts to warm up the various circuits necessary to receive and process information such as the digital dash display information. Therefore, a defective door switch could prevent the operation of the dash display. See Figure 27-9.

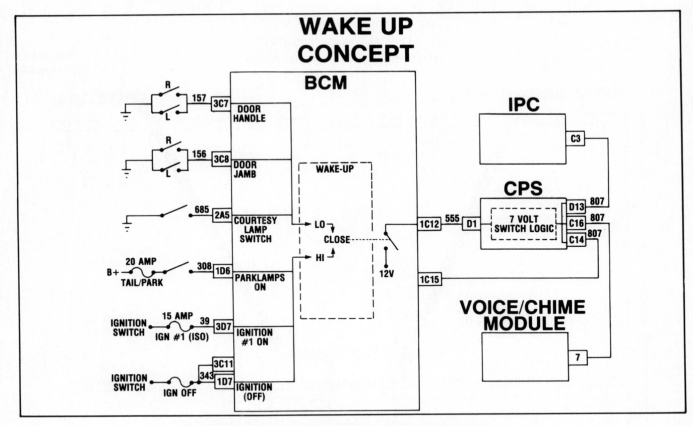

FIGURE 27-9 *(Courtesy of General Motors Corporation.)*

The body computer can also use the actual position of the door information as a basis for setting a trouble message, and can relay this information to a technician if the correct sequence of dash buttons is pushed. The body computer also controls the electronic dash and message center displays based on inputs from all its sensors.

BODY COMPUTER PRECAUTIONS

Most cars equipped with body computers use a separate power supply to provide electrical power for all of the various circuits and sensors. This power supply must be as free of electrical interference as possible. A typical power supply furnishes 5 to 7 V to the engine and body computer as well as to other units and sensors needed for proper operation. The power and ground connections to this power supply are usually separate from the main battery source connections.

These separate electrical connections are called "quiet" or "noise free" and are "isolated" power and "isolated" grounds. No other electrical accessories should be attached to isolated power and ground connections. The purpose of keeping all other accessories away from the

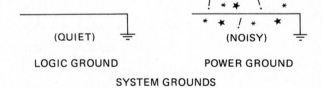

FIGURE 27-10 *(Courtesy of General Motors Corporation.)*

computer power supply is to be certain that voltage spikes created in other circuits will not affect the body computer operation. Some body computer–equipped cars use two separate battery cables for both the power "hot"-side and the ground-side connections. See Figure 27-10.

BODY COMPUTER DIAGNOSTICS

Most computer-equipped vehicles have a built-in diagnostic system for trouble codes. Most cars equipped with body computers not only can display trouble codes, but are also capable of displaying over 100 trouble codes, input data, output data, and switch positions. This information is displayed either as a code number or actually displayed in numbers and letters on a digital dash. See Figures 27-11 and 27-12.

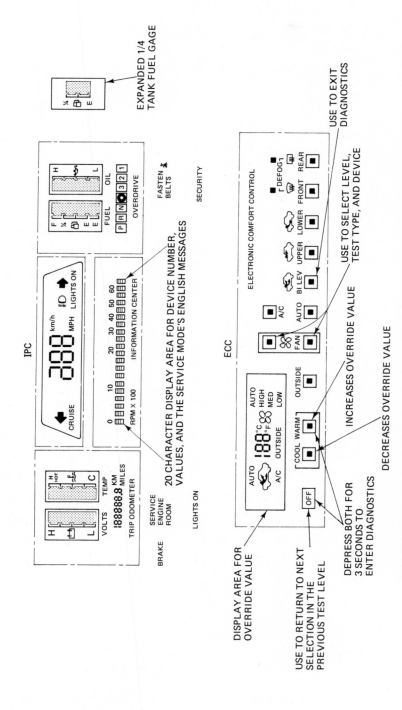

FIGURE 27-11 *Typical dash display of a car equipped with a body computer. Note how the controls can be used to enter diagnostics. (Courtesy of General Motors Corporation.)*

SERVICE DIAGNOSTICS
SYSTEM OVERVIEW

SERVICE DIAGNOSTICS
BASIC OPERATION

- ● ENTER DIAGNOSTICS BY SIMULTANEOUSLY PRESSING THE "OFF" AND "WARM" BUTTONS ON THE ECC FOR 3 SECONDS.

- ● DIAGNOSTIC CODES BEGIN WITH ANY ECM CODES FOLLOWED BY BCM CODES.

- ● TO PROCEED WITH DIAGNOSTICS PRESS THE INDICATED ECC BUTTONS.

- ● HI AND LO REFER TO THE FAN UP AND FAN DOWN BUTTONS.

- ● EXIT DIAGNOSTICS BY PRESSING "BI–LEV"

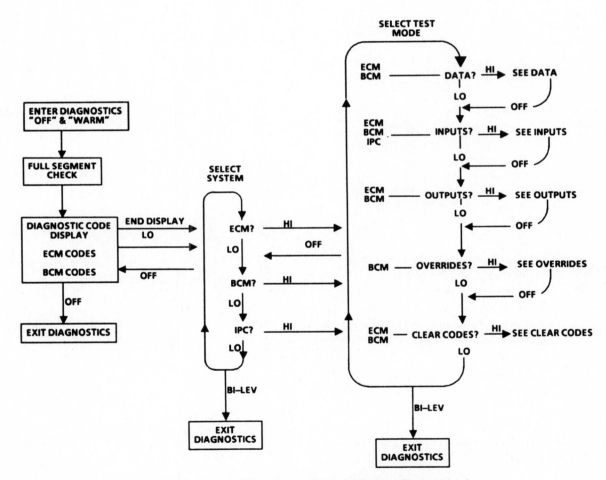

FIGURE 27-12   *(Courtesy of General Motors Corporation.)*

Even though an assembly line communication link (ALCL) is furnished under the dash of most cars with body computers, more information is available through the body computer's own diagnostic program than could ever be accessed by a scan tool. The diagnostic information is accessible on many cars equipped with a body computer by pushing various heating and air-conditioning controls in a certain sequence. For example, most General Motor's cars equipped by a body computer are accessed by pushing "off" and "warmer" at the same time. Consult the service literature for diagnostic codes and procedures.

BINARY NUMBERS

Most computers use the binary number system in which 8 bits or the signals "on" or "off" are used to represent numbers and letters of the alphabet. See Figure 27-13. The numbers that can be represented range from 0 to 255. Often, the status of a computer-controlled unit is displayed as a number ranging from 0 to 255. The number 128 represents the midpoint between 0 and 255, and generally represents normal operation. If the numbers displayed are greatly different from 128, consult the service literature for the car being serviced to determine the possible causes and corrections that may be required.

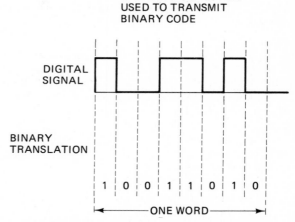

FIGURE 27-13 *(Courtesy of General Motors Corporation.)*

SUMMARY

1. Body computers are used to control non-engine-controlled functions such as heating, ventilation, and air conditioning, and digital dash operation.

2. A body computer uses information from its own sensors, plus information from the engine control computer, to control its output devices.

3. The body computer also usually contains an EEPROM used for storing miles traveled and for displaying the odometer reading.

4. Most body computers use separate power and ground connections that should not be used by any other accessory, to make certain that interference does not cause damage to the body computer.

5. Some computers use a "wake-up" signal from a door switch to be certain that all displays are functioning as soon as the driver turns on the ignition.

6. Body computer diagnostics are usually accessed by pushing heating and air-conditioning control buttons in a certain sequence.

7. The information displayed for some systems may be given in binary numbers which range from 0 to 255, with 128 being the midpoint.

STUDY QUESTIONS

27-1. Name 10 typical body computer inputs.

27-2. Explain why computers use crystal-controlled clock circuits.

27-3. Describe how diagnostic information can be brought up on the dash display.

27-4. Explain binary counts.

MULTIPLE-CHOICE QUESTIONS

27-1. Technician A says that the engine control computer and the body computer are not connected. Technician B says that accessories must never be connected to a computer power supply or ground connection. Which technician is correct?
- **(a)** A only.
- **(b)** B only.
- **(c)** both A and B.
- **(d)** neither A nor B.

27-2. The heating, ventilation, and air-conditioning system controls operated by the body computer include:
- **(a)** the blower motor speed.
- **(b)** the hot water valve.
- **(c)** the blend door position.
- **(d)** all of the above.

27-3. A ''wake-up'' signal is sent by:
- **(a)** the ignition switch.
- **(b)** the door switch.
- **(c)** the seat switch.
- **(d)** the accelerator pedal switch.

27-4. Body computer diagnostics can usually be accessed by pushing _____ and _____ at the same time.
- **(a)** on, off.
- **(b)** off, warmer.
- **(c)** on, high.
- **(d)** bilevel, warmer.

27-5. A binary number of 128 means:
- **(a)** the timing is set at 28° BTDC.
- **(b)** the engine coolant temperature is 28°C.
- **(c)** the engine coolant temperature is 128°C.
- **(d)** none of the above.

27-6. The process of sampling information from more than one source is called:
- **(a)** sampling.
- **(b)** multiplexing.
- **(c)** using EEPROMS.
- **(d)** clock circuits.

27-7. Most body computers:
- **(a)** communicate with engine control computers.
- **(b)** use replaceable PROMS.
- **(c)** have built-in diagnostics.
- **(d)** all of the above.

27-8. Clock circuits are built into automatic computers to:
- **(a)** display time of day.
- **(b)** tell the driver when to change oil.
- **(c)** wake up the computer at the correct time.
- **(d)** none of the above.

Glossary

AIR Air injection reaction emission control system.

Air management system The system of solenoids and valves that control the output of the air pump to the catalytic converter, air cleaner housing, or exhaust manifold.

ALCL Assembly line communications link.

ALDL Assembly line diagnostic link.

Alnico A permanent-magnet alloy of *al*uminum, *ni*ckel, and *co*balt.

Alternator An electric generator that produces alternating current.

AM Amplitude modulation.

Ammeter An electrical test instrument used to measure amperes (unit of the amount of current flow). An ammeter is connected in series with the circuit being tested.

Ampere The unit of the amount of current flow. Named for André Ampère (1775–1836).

Ampere turns The unit of measurement for electrical magnetic field strength.

Analog A type of dash instrument that indicates values by use of the movement of a needle or similar device. An analog signal is continuous and variable.

Anode The positive electrode; the electrode toward which electrons flow.

ANSI American National Standards Institute.

Antenna trimmer A method used to calibrate the antenna for an AM radio.

Antimony A metal added to non-maintenance-free or hybrid battery grids to add strength.

Armature The rotating unit inside a dc generator or starter, consisting of a series of coils of insulating wire wound around a laminated iron core.

ATC After top center.

ATDC After top dead center.

Atom An atom is the smallest unit of matter that still retains its separate unique characteristic.

AWG American wire gauge system.

Backlight The rear window of a car.

Bakelite A brand name of the Union Carbide Co. for phenol-formaldehyde resin plastic.

Ballast resistor A variable resistor used to control the primary ignition current through the ignition coil. At lower engine speed the temperature of the ballast resistor is high and its resistance is high. When engine rpm is high, the ballast resistance is low, permitting maximum current through the ignition coil.

BARO sensor A sensor used to measure barometric pressure.

Base The name for the section of a transistor that controls the current flow through the transistor.

Battery A chemical device that produces a voltage created by two dissimilar metals submerged in an electrolyte.

Bendix drive An inertia-type starter engagement mechanism not used on cars since the early 1960s.

Bias In electrical terms, bias is the voltage applied to a device or component to establish the reference point for operation.

Blower-motor An electric motor and squirrel cage type of fan moving air inside the car for heating, cooling, and defrosting.

Brushes A copper or carbon conductor used to transfer electrical current from or to a revolving electrical part such as that used in an electric motor or generator.

BTDC Before top dead center.

CAFE Corporate average fuel economy.

Calcium A metallic chemical element added to the grids of a maintenance-free battery to add strength.

Candlepower The amount of light produced by a bulb is measured in candlepower.

Capacitance Electrical capacitance is a term used to measure or describe how much charge can be stored in a capacitor (condenser) for a given voltage potential difference. Capacitance is measured in farads or smaller increments of farads, such as microfarads.

Capacitor A condenser: an electrical unit that can pass alternating current, yet block direct current. Used in electrical circuits to control fluctuations in voltage.

Carbon pile An electrical test instrument used to provide an electrical load for testing batteries and the charging circuit.

Catalytic converter An emission control device located in the exhaust system that changes HC and CO into harmless H_2O and CO_2. If a three-way catalyst, NO_x is also separated into harmless separate hydrogen (N) and oxygen (O).

Cathode The negative electrode.

CCC Computer command control is the name of General Motors' computer engine control system.

Cell A group of negative and positive plates to form a cell capable of producing 2.1 V. Each cell contains one more negative plate than positive plate.

CEMF Counter electromotive force.

Centrifugal advance A spark advance mechanism that uses centrifugal force (outward force increases with rotational speed) to increase timing advance in proportion to engine speed.

Charging circuit Electrical components and connections necessary to keep a battery fully charged. Components include the alternator, voltage regulator, battery, and interconnecting wires.

Chassis ground In electrical terms, a ground is the desirable return circuit path. Ground can also be undesirable and provide a shortcut path for a defective electrical circuit.

Circuit A circuit is the path that electrons travel from a power source, through a resistance, and back to the power source.

Circuit breaker A mechanical unit that opens an electrical circuit in the event of excessive flow.

CO Carbon monoxide.

Cold cranking amperes (CCA) The rating of a battery's ability to provide battery voltage during cold-weather operation. CCA is the number of amperes that a battery can supply at 0 °F (-18 °C) for 30 seconds and still maintain a voltage of 1.2 V per cell (7.2 V for a 12-V battery).

Collector The name of one section of a transistor.

Commutator The name for the copper segments of the armature of a starter or dc generator. The revolving segments of the commutator collects the current from or distributes it to the brushes.

Composite headlights A type of headlights that uses a separate replaceable bulb.

Compound-wound A type of electric motor where some field coils are wired in series and some field coils are wired in parallel with the armature.

Conductor A material that conducts electricity and heat. A metal that contains fewer than four electrons in its atom's outer shell.

Conventional theory The theory that electricity flows from positive ($+$) to negative ($-$).

Coulomb A coulomb is 6.28×10^{18} (6.28 billion billion) electrons.

Courtesy light General term used to describe all interior lights.

Cranking circuit Electrical components and connections required to crank the engine to start. Includes starter motor, starter solenoid/relay, battery, neutral safety switch, ignition control switch, and connecting wires and cables.

CRT Cathode ray tube.

Cunife A magnetic alloy made from copper (Cu), nickel (Ni), and iron (Fe).

Current Electron flow through an electrical circuit; measured in amperes.

Current limiter One section of a voltage regulator for a dc generator charging system. The current limiter section opens the field current circuit whenever generator amperage output exceeds safe limits to protect the generator from overheating damage.

Cutout relay One section of a voltage regulator for a dc generator charging system. The cutout relay section prevents battery current from flowing through the dc generator toward ground whenever generator voltage is lower than battery voltage.

Darlington pair Two transistors electrically connected to form an amplifier. This permits a very small current flow to control a large current flow. Named for Sidney Darlington, a physicist at Bell Laboratories from 1929 to 1971.

Deep cycling The full discharge and then the full recharge of a battery.

Delta wound A type of stator winding where all three coils are connected in a triangle shape. Named for the triangle-shaped Greek capital letter.

Digital A method of display that uses numbers instead of a needle or similar device.

Dimmer switch An electrical switch used to direct the current to either bright or dim headlight filaments.

Diode An electrical one-way check valve made from combining a P-material and an N-material.

Diode trio A group of three diodes grouped together with one output used to put out the charge indicator lamp and provide current for the field from the stator windings on many alternators.

Direct current Electric current that flows in one direction.

Distributor Electromechanical unit used to help create and distribute the high voltage necessary for spark ignition.

Doping The adding of impurities to pure silicon or germanium to form either P or N semiconductor materials.

DPDT switch Double-pole, double-throw switch.

Dwell The number of degrees of distributor cam rotation that the points are closed.

Earth ground The most grounded ground. A ground is commonly used as a return current path for an electrical circuit.

EEPROM Electronically erasable programmable read-only memory.

EFI Electronic fuel injection.

EGR Exhaust gas recirculation. An emission control device to reduce NO_x (oxides of nitrogen).

Electricity Electricity is the movement of free electrons from one atom to another.

Electrolyte Any substance which, in solution, is separated into ions and is made capable of conducting an electric current. The acid solution of a lead-acid battery.

Electromagnetic gauges A type of dash instrument gauge that uses small electromagnetic coils for the needle movement of the gauge.

Electromagnetic induction Electromagnetic induction was discovered in 1831 by Michael Faraday and is the generation of a current in a conductor that is moved through a magnetic field.

Electromagnetism A magnetic field created by current flow through a conductor.

Electromotive force The force (pressure) that can move electrons through a conductor.

Electron A negative-charged particle: 1/1800 the mass of a proton.

Electron theory The theory that electricity flows from negative ($-$) to positive ($+$).

Electronic ignition General term used to describe any of various types of ignition systems that use electronic instead of mechanical components, such as contact points.

Element Any substance that cannot be separated into different substances.

EMF Electromotive force.

Emitter The name of one section of a transistor. The arrow used on a symbol for a transistor is on the emitter and the arrow points toward the negative section of the transistor.

EPROM Erasable programmable read-only memory.

ESC Electronic spark control means that the computer system is equipped with a knock sensor that can retard spark advance if necessary to eliminate spark knock.

EST Electronic spark timing; the computer controls spark timing advance.

Farad A unit of capacitance named for Michael Faraday (1791–1867), an English physicist. A farad is the capacity to store 1 coulomb of electrons at 1 volt of potential difference.

Feedback The reverse flow of electrical current through a circuit or electrical unit that should not normally be operating. This feedback current (reverse-bias current flow) is most often caused by a poor ground connection for the same normally operating circuit.

Fiber optics The transmission of light through special plastic that keeps the light rays parallel even if the plastic is tied in a knot.

Field coils Coils or wire wound around metal pole shoes to form the electromagnetic field inside an electric motor.

Filament The light-producing wire inside a light bulb.

FM Frequency modulation.

Forward bias Current flow in normal direction.

Full fielding The method of supplying full battery voltage to the magnetic field of a generator as part of the troubleshooting procedure for the charging system.

Fuse An electrical safety unit constructed of a fine tin conductor that will melt and open the electrical circuit if excessive current flows through the fuse.

Fusible link A type of fuse that will melt and open the protected circuit in the event of a short circuit, which could cause excessive current flow through the fusible link. Most fusible links are actually wires four gauge sizes smaller than the wire of the circuits being protects.

Gassing The release of hydrogen and oxygen gas from the plates of a battery during charging or discharging.

Gauge Wire sizes as assigned by the American wire gauge system; the smaller the gauge number, the larger the wire.

Gauss A unit of magnetic induction or magnetic intensity named for Karl Friedrich Gauss (1777–1855), a German mathematician.

Generator A device that converts mechanical energy into electrical energy.

Grid The lead-alloy framework (support) for the active materials of an automotive battery.

Ground The lowest possible voltage potential in a circuit. In electrical terms, a ground is the desirable return circuit path. Ground can also be undesirable and provide a shortcut path for a defective electrical circuit.

Growler Electrical tester designed to test starter and dc generator armatures.

Hall-effect sensor A type of electromagnetic sensor used in electronic ignition and other systems. Named for Edward H. Hall, who discovered the Hall effect in 1879.

Hash An unclear or a messy section of a scope pattern.

Hazard flasher Emergency warning flashers; lights at all four corners of the vehicle flash on and off.

HC Hydrocarbons (unburned fuel); when combined with NO_x and sunlight, they form smog.

Heat sink Usually, a metallic finned unit used to keep electronic components cool.

HEI General Motors' name for their electronic ignition: High Energy Ignition.

Hold-in winding One of two electromagnetic windings inside a solenoid; used to hold the movable core into the solenoid.

Hole theory A theory which states that as an electron flows from negative $(-)$ to positive $(+)$, it leaves behind a hole. According to the hole theory, the hole would move from positive $(+)$ to negative $(-)$.

Horsepower A unit of power; 33,000 foot-pounds per minute. One horsepower equals 746 W.

Hybrid Something (such as a battery) made from more than one different elements.

Hydrometer An instrument used to measure the specific gravity of a liquid. A battery hydrometer is calibrated to read the expected specific gravity of battery electrolyte.

IAC Idle air control.

Ignition circuit Electrical components and connections that produce and distribute high-voltage electricity to ignite the air/fuel mixture inside the engine.

Ignition coil An electrical device consists of two separate coils of wire: a primary and a secondary winding. The purpose of an ignition is to produce a high-voltage (20,000 to 40,000 V), low-amperage (about 80 mA) current necessary for spark ignition.

Ignition timing The exact point of ignition in relation to piston position.

ILC Idle load control.

Inductive reactance An opposing current created in a conductor whenever there is a charging current flow in a conductor.

Insulator A material that does not readily conduct electricity and heat. A nonmetal material that contains more than four electrons in its atom's outer shell.

Ion An atom with an excess or deficiency of electrons forming either a negative or a positive charged particle.

ISC Idle speed control.

IVR Instrument voltage regulator. An IVR is used to maintain constant voltage to thermoelectric gauges to maintain accuracy.

Joule A unit of electrical energy. One joule equals 1 watt $\times$ 1 second (1 V $\times$ 1 A $\times$ 1 s).

Jumper cables Heavy-gauge (4 to 00) electrical cables with large clamps, used to connect a vehicle that has a discharged battery to a vehicle that has a good battery.

Kicker A throttle kicker is used on some computer engine control systems to increase engine speed (rpm) during certain operating conditions, such as when the air-conditioning system is on.

Kilo Means 1000; abbreviated k or K.

Knock sensor A sensor that can detect engine spark knock.

LCD Liquid-crystal display.

Lead peroxide The positive plate of an automotive-style battery; the chemical symbol is PbO_2.

Lead sulfate Both battery plates become lead sulfate when the battery is discharged. The chemical symbol for lead sulfate is $PbSO_4$.

LED Light-emitting diode.

Lumbar The lower section of the back.

Magnequench A magnetic alloy made from neodymium, iron, and boron.

Magnetic timing A method of measuring ignition that uses a magnetic pickup tool to sense the location of a magnet on the harmonic balancer.

MAP Manifold absolute pressure.

Manifold vacuum Low pressure (vacuum) measured at the intake manifold of a running engine (normally between 17 and 21 in. Hg at idle).

M/C solenoid Mixture control solenoid.

Meniscus The puckering or curvature of a liquid in a tube. A battery is properly filled with water when the electrolyte first becomes puckered.

Module A group of electronic components functioning as a component of a larger system.

Mutual induction The generation of an electric current due to a changing magnetic field of an adjacent coil.

Neutrons A neutral-charged particle; one of the basic particles of the nucleus of an atom.

NO$_x$ Oxides of nitrogen; when combined with HC and sunlight, form smog.

N-type material Silicon or germanium doped with phosphorus, arsenic, or antimony.

Nucleus The central part of an atom which has a positive $(+)$ charge and contains almost all of the mass of the atom.

NVRAM Nonvolatile random access memory.

Ohm The unit of electrical resistance. Named for Georg Simon Ohm (1787–1854).

Ohmmeter An electrical test instrument used to measure ohms (unit of electrical resistance). An ohmmeter uses an internal battery for power and must never be used when current is flowing through a circuit or component.

Ohm's law An electrical law that states: "it requires 1 volt to push 1 ampere through 1 ohm of resistance."

Omega The last letter of the Greek alphabet; a symbol for ohm, the unit for electrical resistance.

Open circuit An open circuit is any circuit that is not complete and in which no current flows.

Oscilloscope A visual display of electrical waves on a fluorescent screen or cathode ray tube.

Partitions Separations between the cells of a battery. Partitions are made of the same material as that of the outside case of the battery.

Pasting The process of applying active battery materials onto the grid framework of each plate.

PCV Positive crankcase ventilation.

Permalloy A permanent-magnet alloy of nickel and iron.

Permeability The measure of a material's ability to conduct magnetic lines of force.

Photoelectric principle The production of electricity created by light striking certain sensitive materials, such as selenium or cesium.

Piezoelectric principle The principle by which certain crystals become electrically charged when pressure is applied.

Pinion gear A small gear on the end of the starter drive which rotates the engine flywheel ring gear for starting.

PM motor A permanent-magnet electric motor.

Polarity The condition of being positive or negative in relation to a magnetic pole.

Porous lead Lead with many small holes to make a surface porous for use in battery negative plates; the chemical symbol for lead is Pb.

Ported vacuum Low pressure (vacuum) measured above the throttle plates. As the throttle plates open, the vacuum increases and becomes of the same value as the manifold vacuum.

Power side The wires leading from the power source (battery) to the resistance (load) of a circuit.

PROM Programmable read-only memory.

Proton A positive-charged particle; one of the basic particles of the nucleus of an atom.

P-type material Silicon or germanium doped with boron or indium.

Pull-in windings One of two electromagnetic windings inside a solenoid used to move a movable core.

Pulse generators An electromagnetic unit that generates a voltage signal used to trigger the ignition control module that controls (turns on and off) the primary ignition current of an electronic ignition system.

Pulse width The amount of ''on'' time of an electronic fuel injector.

Radial grid A lead-alloy framework for the active materials of a battery that has radial support spokes to add strength and to improve battery efficiency.

Radio choke A small coil of wire installed in the power lead leading to a pulsing unit such as an IVR to prevent radio interference.

RAM Random access memory.

Rectifier An electronic device that converts alternating current into direct current.

Rectifier bridge A group of six diodes, three positive (+) and three negative (−) commonly used in alternators.

Relay An electromagnetic switch that uses a movable arm.

Reluctance The resistance to the movement of magnetic lines of force.

Reserve capacity The number of minutes a battery can produce 25 A and still maintain a battery voltage of 1.75 V per cell (10.5 V for a 12-V battery).

Residual magnetism Magnetism remaining after the magnetizing force is removed.

Resistance The opposition to current flow.

Reverse bias Current flow in the opposite direction from normal.

Rheostat An adjustable variable resistor.

Rise time The time, measured in microseconds, for the output of a coil to rise from 10% to 90% of its maximum output.

ROM Read-only memory.

RPM Revolutions per minute.

RTV Room-temperature vulcanization.

Saturation The point of maximum magnetic field strength of a coil.

Sediment chamber A space below the cell plates of some batteries to permit the accumulation of sediment deposits flaking from the battery plates. Use of a sediment chamber keeps the sediment from shorting the battery plates.

Self-induction The generation of an electric current in the wires of a coil created when the current is first connected or disconnected.

Semiconductor A material that is neither a conductor nor an insulator; has exactly four electrons in the atom's outer shell.

Separators In a battery, nonconducting porous, thin materials used to separate positive and negative plates.

Series wound In a starter motor, the field coils and the armature are wired in series. All of the current flows through the field coils, through the hot brushes, through the armature, then to the ground through the ground brushes.

Servo unit A vacuum-operated unit that attaches to the throttle linkage to move the throttle on a cruise control system.

Shelf life The length of time that something can remain on a storage shelf and not be reduced in performance level from that of a newly manufactured product.

Short circuit A circuit in which current flows, but bypasses some or all of the resistance in the circuit. A connection that results in a ''copper-to-copper'' connection.

Short to ground A short circuit in which the current bypasses some or all of the resistance of the circuit and flows to ground. Since ground is usually steel in automotive electricity, a short to ground (grounded) is a ''copper-to-steel'' connection.

Shunt A device used to divert or bypass part of the current from the main circuit.

Smog The term used to describe a combination of *sm*oke and *fo*g. Formed by NO_x and HC with sunlight.

Solenoid An electromagnetic switch that uses a movable core.

Specific gravity The ratio of the weight of a given volume of a liquid divided by the weight of an equal volume of water.

Sponge lead Lead with many small holes used to make a surface porous or spongelike for use in battery negative plates; the chemical symbol for lead is Pb.

Starter drive A term used to describe the starter motor drive pinion gear with overrunning clutch.

State of charge The degree or the amount that a battery is charged. A fully charged battery would be 100% charged.

Stator A name for three interconnected windings inside an alternator. A rotating rotor provides a moving magnetic field and induces a current in the windings of the stator.

Stepper motor A motor that moves a specified amount of rotation.

Stoichiometric An air/fuel ratio of exactly 14.7:1.

Stroboscopic A very bright pulsing light triggered from the firing of one spark plug. Used to check and adjust ignition timing.

Tach Abbreviation for tachometer, instrument, or gauge used to measure rpm (revolutions per minute).

TBI Throttle body injection.

TDC Top dead center.

Tell-tale light Dash warning light.

TFI Thick-film integration; the name of the type of Ford electronic ignition system.

Thermistor A resistor that changes resistance with temperature. A positive-coefficient thermistor has increased resistance with an increase in temperature. A negative-coefficient thermistor has increased resistance with a decrease in temperature.

Thermoelectric principle The production of current flow created by heating the connection of two dissimilar metals.

Thermoelectric meters A type of dash instrument that uses heat created by current flow through the gauge to deflect the indicator needle.

Torque A twisting force which may or may not result in motion.

Trade number The number stamped on an automotive light bulb. All bulbs of the same trade number have the same candlepower and wattage, regardless of the manufacturer of the bulb.

Transducer An electrical and mechanical speed sensing and control unit used on cruise control systems.

Transistor A semiconductor device that can operate as an amplifier or an electrical switch.

VAC Vacuum sensor.

Vacuum Pressure below atmospheric, measured in units of inches of mercury (in. Hg).

Vacuum advance A spark advance unit that advances the ignition timing in relation to engine vacuum.

Vacuum kicker A computer-controlled throttle device used to increase idle rpm during certain operating conditions, such as when the air-conditioning system is operating.

Volt The unit of electrical pressure; named for Alessandro Volta (1745–1827).

Voltage regulator An electronic or mechanical unit that controls the output voltage of an electrical generator or alternator by controlling the field current of the generator.

Voltmeter An electrical test instrument used to measure volts (unit of electrical pressure). A voltmeter is connected in parallel with the unit or circuit being tested.

VTF Vacuum-tube fluorescent.

Watt An electrical unit of power; 1 watt equals current (amperes) × voltage (1/746 hp). Named after James Watt, a Scottish inventor.

WOT Wide-open throttle.

Wye wound A type of stator winding in which all three coils are connected to a common center connection. Called a wye because the connections look like the letter Y.

Zener diode A specially constructed (heavily doped) diode designed to operate with a reverse-bias current after a certain voltage has been reached. Named for Clarence Melvin Zener.

Answers to Multiple-Choice Questions

Chapter 1: 1-1. b, 1-2. b, 1-3. d, 1-4. a, 1-5. b, 1-6. b, 1-7. a, 1-8. b

Chapter 2: 2-1. b, 2-2. d, 2-3. b, 2-4. c, 2-5. a, 2-6. d, 2-7. b, 2-8. d

Chapter 3: 3-1. b, 3-2. b, 3-3. a, 3-4. b, 3-5. a, 3-6. c, 3-7. c, 3-8. a, 3-9. a, 3-10. d

Chapter 4: 4-1. d, 4-2. a, 4-3. d, 4-4. b, 4-5. d, 4-6. b, 4-7. a, 4-8. b

Chapter 5: 5-1. c, 5-2. b, 5-3. d, 5-4. c, 5-5. d, 5-6. d, 5-7. a, 5-8. a

Chapter 6: 6-1. a, 6-2. d, 6-3. a, 6-4. c, 6-5. b, 6-6. b, 6-7. d, 6-8. b

Chapter 7: 7-1. a, 7-2. b, 7-3. c, 7-4. d, 7-5. e, 7-6. a, 7-7. d, 7-8. d

Chapter 8: 8-1. c, 8-2. c, 8-3. c, 8-4. b, 8-5. a, 8-6. a, 8-7. d, 8-8. d, 8-9. c, 8-10. a

Chapter 9: 9-1. c, 9-2. d, 9-3. c, 9-4. c, 9-5. a, 9-6. a, 9-7. a, 9-8. b, 9-9. d, 9-10. a

Chapter 10: 10-1. d, 10-2. b, 10-3. d, 10-4. c, 10-5. d, 10-6. a, 10-7. d, 10-8. d

Chapter 11: 11-1. c, 11-2. b, 11-3. c, 11-4. a, 11-5. d, 11-6. b, 11-7. a, 11-8. b, 11-9. c, 11-10. d

Chapter 12: 12-1. a, 12-2. b, 12-3. c, 12-4. c, 12-5. d, 12-6. d, 12-7. b, 12-8. d, 12-9. b, 12-10. a

Chapter 13: 13-1. c, 13-2. d, 13-3. c, 13-4. c, 13-5. d, 13-6. d, 13-7. b, 13-8. e, 13-9. b, 13-10. c

Chapter 14: 14-1. d, 14-2. a, 14-3. d, 14-4. b, 14-5. d, 14-6. b, 14-7. d, 14-8. a, 14-9. e, 14-10. c

Chapter 15: 15-1. a, 15-2. d, 15-3. c, 15-4. c, 15-5. c, 15-6. a, 15-7. c, 15-8. d, 15-9. a, 15-10. b

Chapter 16: 16-1. b, 16-2. c, 16-3. b, 16-4. a, 16-5. c, 16-6. d, 16-7. c, 16-8. d, 16-9. b, 16-10. b

Chapter 17: 17-1. d, 17-2. c, 17-3. c, 17-4. d, 17-5. b, 17-6. c, 17-7. a, 17-8. b, 17-9. d, 17-10. d

Chapter 18: 18-1. d, 18-2. b, 18-3. d, 18-4. c, 18-5. b, 18-6. b, 18-7. c, 18-8. d, 18-9. a, 18-10. b

Chapter 19: 19-1. b, 19-2. a, 19-3. b, 19-4. b, 19-5. a, 19-6. b, 19-7. b, 19-8. d, 19-9. b, 19-10. d

Chapter 20: 20-1. b, 20-2. b, 20-3. d, 20-4. d, 20-5. b, 20-6. b, 20-7. a, 20-8. a, 20-9. b, 20-10. c

Chapter 21: 21-1. d, 21-2. d, 21-3. b, 21-4. d, 21-5. d, 21-6. b, 21-7. d, 21-8. c

Chapter 22: 22-1. d, 22-2. a, 22-3. c, 22-4. d, 22-5. c, 22-6. d, 22-7. b, 22-8. d

Chapter 23: 23-1. c, 23-2. c, 23-3. b, 23-4. c, 23-5. d, 23-6. a, 23-7. c, 23-8. d, 23-9. d, 23-10. c

Chapter 24: 24-1. a, 24-2. b, 24-3. d, 24-4. b, 24-5. c, 24-6. b, 24-7. b, 24-8. a, 24-9. c, 24-10. b

Chapter 25: 25-1. a, 25-2. a, 25-3. b, 25-4. a, 25-5. a, 25-6. d, 25-7. c, 25-8. c, 25-9, d, 25-10. d

Chapter 26: 26-1. b, 26-2. a, 26-3. b, 26-4. a, 26-5. d, 26-6. d, 26-7. c, 26-8. d, 26-9. b, 26-10. b

Chapter 27: 27-1. c, 27-2. d, 27-3. b, 27-4. b, 27-5. d, 27-6. b, 27-7. d, 27-8. d

Index

Table
of
Sidebars